AF389365

Laser-Induced Periodic Surface Nano- and Microstructures for Tribological Applications

Laser-Induced Periodic Surface Nano- and Microstructures for Tribological Applications

Editors

Jörn Bonse
Dirk Spaltmann

MDPI • Basel • Beijing • Wuhan • Barcelona • Belgrade • Manchester • Tokyo • Cluj • Tianjin

Editors
Jörn Bonse
Bundesanstalt für Materialforschung und
-prüfung (BAM)
Germany

Dirk Spaltmann
Bundesanstalt für Materialforschung und
-prüfung (BAM)
Germany

Editorial Office
MDPI
St. Alban-Anlage 66
4052 Basel, Switzerland

This is a reprint of articles from the Special Issue published online in the open access journal *Lubricants* (ISSN 2075-4442) (available at: https://www.mdpi.com/journal/lubricants/special_issues/laser_periodic).

For citation purposes, cite each article independently as indicated on the article page online and as indicated below:

LastName, A.A.; LastName, B.B.; LastName, C.C. Article Title. *Journal Name* **Year**, *Article Number*, Page Range.

ISBN 978-3-03943-523-4 (Hbk)
ISBN 978-3-03943-524-1 (PDF)

Cover image courtesy of Jörn Bonse and Dirk Spaltmann.

© 2020 by the authors. Articles in this book are Open Access and distributed under the Creative Commons Attribution (CC BY) license, which allows users to download, copy and build upon published articles, as long as the author and publisher are properly credited, which ensures maximum dissemination and a wider impact of our publications.

The book as a whole is distributed by MDPI under the terms and conditions of the Creative Commons license CC BY-NC-ND.

Contents

About the Editors . . . . **vii**

Jörn Bonse and Dirk Spaltmann
Editorial: Special Issue "Laser-Induced Periodic Surface Nano- and Microstructures for
Tribological Applications"
Reprinted from: *Lubricants* **2020**, *8*, 34, doi:10.3390/lubricants8030034 . . . . **1**

Philipp G. Grützmacher, Francisco J. Profito and Andreas Rosenkranz
Multi-Scale Surface Texturing in Tribology—Current Knowledge and Future Perspectives
Reprinted from: *Lubricants* **2019**, *7*, 95, doi:10.3390/lubricants7110095 . . . . **5**

Jörg Schille, Lutz Schneider, Stefan Mauersberger, Sylvia Szokup, Sören Höhn,
Johannes Pötschke, Friedemann Reiß, Erhard Leidich and Udo Löschner
High-Rate Laser Surface Texturing for Advanced Tribological Functionality
Reprinted from: *Lubricants* **2020**, *8*, 33, doi:10.3390/lubricants8030033 . . . . **47**

Stefan Rung, Kevin Bokan, Frederick Kleinwort, Simon Schwarz, Peter Simon,
Jan-Hendrik Klein-Wiele, Cemal Esen and Ralf Hellmann
Possibilities of Dry and Lubricated Friction Modification Enabled by Different Ultrashort
Laser-Based Surface Structuring Methods
Reprinted from: *Lubricants* **2019**, *7*, 43, doi:10.3390/lubricants7050043 . . . . **67**

Tobias Stark, Thomas Kiedrowski, Holger Marschall and Andrés Fabián Lasagni
Avoiding Starvation in Tribocontact Through Active Lubricant Transport in Laser Textured
Surfaces
Reprinted from: *Lubricants* **2019**, *7*, 54, doi:10.3390/lubricants7060054 . . . . **81**

Dariush Bijani, Elena L. Deladi, Matthijn B. de Rooij and Dirk J. Schipper
The Influence of Surface Texturing on the Frictional Behaviour in Starved Lubricated
Parallel Sliding Contacts
Reprinted from: *Lubricants* **2019**, *7*, 68, doi:10.3390/lubricants7080068 . . . . **99**

Joel Voyer, Johann Zehetner, Stefan Klien, Florian Ausserer and Igor Velkavrh
Production and Tribological Characterization of Tailored Laser-Induced Surface
3D Microtextures
Reprinted from: *Lubricants* **2019**, *7*, 67, doi:10.3390/lubricants7080067 . . . . **125**

Sanne H. van der Poel, Marek Mezera, Gert-willem R. B. E. Römer, Erik G. de Vries
and Dave T. A. Matthews
Fabricating Laser-Induced Periodic Surface Structures on Medical Grade
Cobalt–Chrome–Molybdenum: Tribological, Wetting and Leaching Properties
Reprinted from: *Lubricants* **2019**, *7*, 70, doi:10.3390/lubricants7080070 . . . . **145**

Jon Joseba Ayerdi, Nadine Slachciak, Iñigo Llavori, Alaitz Zabala, Andrea Aginagalde,
Jörn Bonse and Dirk Spaltmann
On the Role of a ZDDP in the Tribological Performance of Femtosecond Laser-Induced Periodic
Surface Structures on Titanium Alloy against Different Counterbody Materials
Reprinted from: *Lubricants* **2019**, *7*, 79, doi:10.3390/lubricants7090079 . . . . **159**

Iaroslav Gnilitskyi, Alberto Rota, Enrico Gualtieri, Sergio Valeri and Leonardo Orazi
Tribological Properties of High-Speed Uniform Femtosecond Laser Patterning on Stainless Steel
Reprinted from: *Lubricants* **2019**, *7*, 83, doi:10.3390/lubricants7100083 . . . . **173**

About the Editors

Jörn Bonse (Dr.) is a tenured scientist at the German Federal Institute for Materials Research and Testing (BAM) in Berlin, Germany. His research interests include the fundamentals and applications of laser–matter interaction, especially with respect to ultrashort laser pulses, laser-induced periodic nanostructures, time-resolved optical techniques, laser processes in photovoltaics, surface functionalization, and laser safety. He received a doctoral degree in Physics from the Technical University of Berlin (Germany) in 2001 and a diploma degree in Physics from the University of Hannover (Germany) in 1996. Dr. Bonse has occupied various research positions at institutions such as the Max Born Institute for Nonlinear Optics and Short Pulse Spectroscopy (MBI) in Berlin, the Spanish Research Council (CSIC) in Madrid (Spain), and the Laser Zentrum Hannover (LZH) in Hannover. He was appointed as a senior laser application specialist at Newport's Spectra-Physics Lasers Division in Stahnsdorf, Germany. Dr. Bonse is a member of the German Physical Society (DPG) and of the European Materials Research Society (E-MRS). In 1999, he was a recipient of an award for applied research, presented by the federal German state Thuringia, for the development of high-power fiber lasers. He received an OSA Outstanding Reviewer Award from the Optical Society of America in 2013. Between 2014 and 2017 he served as an Associate Editor for the journal Optics Express of the Optical Society of America. He has authored more than 150 refereed journal publications and 2 patents related to his research activities.

Dirk Spaltmann (Dr.) is a senior scientist at the German Federal Institute for Materials Research and Testing (BAM) in Berlin, Germany. His research interests include the tribological performance and wear resistance of (DLC-) coatings as well as new materials under slip-rolling, reciprocating–sliding, and extreme-pressure conditions, giving special consideration to tribometric, standardization, and data management aspects. He received a doctoral degree (Ph.D.) in Physics from the University of Wales, College of Cardiff (UWCC, Great Britain) in 1993 and a diploma degree in Physics from the RWTH Aachen University (Germany) in 1989. After a two-year post-doctoral position at UWCC, Dr. Spaltmann took up work at a start-up company near Berlin (Germany) as head of research and development as well as production. In 1999, he joined the Macro-Tribology and Wear Protection group of BAM (Germany). Dr. Spaltmann is a member of the German Physical Society (DPG), the German Tribology Association (GFT), and the German Hydrogen and Fuel-Cell Association (DWV). He is a recognized reviewer for several well-known journals, has authored more than 70 refereed journal publications, and holds 9 patents related to his career activities.

 *lubricants*

Editorial

Editorial: Special Issue "Laser-Induced Periodic Surface Nano- and Microstructures for Tribological Applications"

Jörn Bonse * and Dirk Spaltmann

Bundesanstalt für Materialforschung und -prüfung (BAM), Unter den Eichen 87, 12205 Berlin, Germany;
dirk.spaltmann@bam.de
* Correspondence: joern.bonse@bam.de

Received: 9 March 2020; Accepted: 16 March 2020; Published: 22 March 2020

Keywords: laser-induced periodic surface structures (LIPSS); tribology; applications; wear; friction; lubricant; nanostructure; microstructure

---

Laser material processing is an innovative technology that generates surface functionalities on the basis of optical, mechanical, or chemical properties [1]. In the form of laser surface texturing (LST), it has attracted a remarkable amount of research to tailor surface properties towards various tribological applications [2]. Of this single-step, laser-based technology, the main advantages are the contactless machining, featuring a high flexibility, efficiency, and speed, along with the excellent quality of the processed products. LST can be applied precisely, localized to sub-micrometric areas, but, via laser beam scanning, it is also feasible for structuring large surface areas the size of square-meters.

This Special Issue, "Laser-Induced Periodic Surface Nano- and Microstructures for Tribological Applications" (https://www.mdpi.com/journal/lubricants/special_issues/laser_periodic) [3–11] focuses on the latest developments concerning the tribological performance of laser-generated periodic surface nano-/microstructures and their applications. This includes the laser-based processing of different surface patterns, such as "self-organized" laser-induced periodic surface structures (LIPSS, ripples) [12], grooves, micro-spikes, hierarchical hybrid nano-/microstructures, microfeatures generated by direct laser interference patterning (DLIP) [13], and even dimples or other topographic geometries shaped by direct laser modification or ablation. The applications of these periodically nano-/micropatterned surfaces are aimed at improving the lubricated or non-lubricated tribological performance of surfaces in conformal [4,7,11] and even non-conformal contacts [5,6,9,10] through a reduction of wear, a variation of the coefficient of friction (CoF) [4,5,8,11], altered load carrying capacity, and more. Such enhanced tribological performances result in energy savings, improved reliability, increased lifetimes, as well as durability, leading in turn to extended maintenance intervals and, thus, reduced down-time. The latter is especially beneficial for bearings, gears, engines, seals, cutting tools, and other tribological components. Fundamental aspects involve relevant physical [7] and chemical effects [4,10] accompanying the laser-generated nano-/microscale topographies. Such aspects are, e.g., alterations of the material structures [10], hardness, superficial oxidation [4,10], the role of additives contained in lubricants [10], surface wettability and directed liquid transport [6,7], micro-hydrodynamic effects [7], and more.

Two feature papers [3,4] were invited for the adequate framing of this Special Issue. Seven more papers were submitted as regular contributions [5–11]. Both academic and industrial researchers were attracted, providing a bridge between research in the fields of tribology and laser material processing. This allowed us to foster the current knowledge and present new ideas for future applications and new technologies.

Grützmacher et al. [3] provided a review article on multi-scale surface texturing, discussing the current knowledge and future perspectives in tribology. The authors underline that numerical methods and experiments should be suitably combined to further push the development and design of multi-scale surfaces to enable lower friction and wear over a broader range of tribological conditions.

Schille et al. [4] presented a new world-record in high-rate laser surface texturing for enabling tribological functionalities. The authors demonstrated large-area surface texturing at 3.8 m^2/min. maximum area processing rate. They achieved this by employing ultrafast lasers with MHz pulse repetition rates in combination with polygon scanner technologies. This allowed for unprecedented 950 m/s beam moving speeds. This extremely high processing rate could be achieved by splitting the laser beam for parallel and ultrafast processing at 560 m/s scan speed. In practice, when considering the 40% facet utilization rate of the installed polygon scanner setup, the effective processing rate was still 1.5 m^2/min. for large-area surface texturing. This is at least twice the surface processing rates for metals reported elsewhere.

Rung et al. [5] compared the tribological performance of dry and oil-lubricated periodic surface structures generated on steel surfaces either via self-organization (LIPSS) or by DLIP. Although structures with similar periodicity (0.9–1.5 µm) were produced, their tribological behaviour in a linear reciprocating ball-on-disc configuration was different [attributed to the different modulation depths of either ~0.2 µm (LIPSS) or ~1.5 µm (DLIP)]. For both, the dry and lubricated test conditions, a reduction of the coefficient of friction could always be reached by at least one type of the laser-generated periodic surface structures.

Stark et al. [6] studied the effects of periodic surface textures for the avoidance of starvation in the lubricated tribo-contact. In numerical simulations and experiments, the beneficial effect of a directed lubricant (oil) transport was demonstrated through periodic DLIP-generated micro-groove patterns.

Bijani et al. [7] developed a theoretical model to study the effects of surface texturing on the frictional behaviour of parallel sliding contacts under starved lubrication conditions. Applying a deterministic asperity contact model, which considered the effects of different scales of surface features (texture and roughness), it was shown that surface texturing may improve lubricant film formation under the conditions of starvation. Moreover, under certain conditions, the surface texturing can reduce the CoF.

Voyer et al. [8] presented an industrial approach which determined the manufacturing feasibility of a specific 3D surface micro-texture on hardened steel through ultrashort pulsed laser processing. These surface structures were then qualified regarding their tribological properties under different lubrication conditions with either oil or a solid "anti-friction" coating.

Van der Poel et al. [9] investigated the ps-laser based fabrication of different types of LIPSS on medical grade cobalt-chrome-molybdenum alloy for hip prosthesis applications. Emphasis was laid on their associated tribological, surface wetting, and leaching properties. Based on their experimental results, the authors concluded that the laser textured surfaces on CoCrMo alloy are not suitable for bearing surfaces in a metal-on-plastic contact. However, they gave recommendations for future improvements.

Ayerdi et al. [10] confirmed a prior hypothesis on the role of the lubricant ZDDP (zinc-dialkyl-dithiophosphate), present in many commercial engine oils. A ZDDP-additivated engine oil was "re-build" from its basic constituents. Its beneficial behaviour was verified in linear, reciprocating tribological tests of LIPSS-covered titanium alloy surfaces against balls made of different materials.

Gnilitskyi et al. [11] measured the Stribeck curve of X5CrNi18010 stainless steel against 100Cr6 steel in engine oil lubricated pin-on-flat tribological tests. The rotating disks were covered by fs-laser processed LIPSS. Over the entire range investigated (Stribeck numbers between 5×10^{-7} and 8×10^{-6} m^{-1}), the CoF of the LIPSS-patterned surface was substantially lower than of the non-patterned one, with the reduction ranging from 10% up to 25% (at $5–9 \times 10^{-7}$ m^{-1}). Simultaneously, wear reductions of 65% were obtained.

Finally, the Guest Editors would like to express their sincere gratitude to all authors and reviewers of this Special Issue for their intense efforts and to the editorial staff of Lubricants for their professional support and guidance.

Conflicts of Interest: The authors declare no conflict of interest.

References

1. Bäuerle, D. *Laser Processing and Chemistry*, 4th ed.; Springer: Berlin, Germany, 2011.
2. Etsion, I. State of the art in laser surface texturing. *J. Tribol.* **2005**, *127*, 248–253. [CrossRef]
3. Grützmacher, P.G.; Profito, F.J.; Rosenkranz, A. Multi-Scale Surface Texturing in Tribology—Current Knowledge and Future Perspectives. *Lubricants* **2019**, *7*, 95. [CrossRef]
4. Schille, J.; Schneider, L.; Mauersberger, S.; Szokup, S.; Höhn, S.; Pötschke, J.; Reiß, F.; Leidich, E.; Löschner, U. High-rate Laser Surface Texturing for Advanced Tribological Functionality. *Lubricants* **2020**, *8*, 33. [CrossRef]
5. Rung, S.; Bokan, K.; Kleinwort, F.; Schwarz, S.; Simon, P.; Klein-Wiele, J.-H.; Esen, C.; Hellmann, R. Possibilities of Dry and Lubricated Friction Modification Enabled by Different Ultrashort Laser-Based Surface Structuring Methods. *Lubricants* **2019**, *7*, 43. [CrossRef]
6. Stark, T.; Kiedrowski, T.; Marschall, H.; Lasagni, A.F. Avoiding Starvation in Tribocontact Through Active Lubricant Transport in Laser Textured Surfaces. *Lubricants* **2019**, *7*, 54. [CrossRef]
7. Bijani, D.; Deladi, E.L.; de Rooij, M.B.; Schipper, D.J. The Influence of Surface Texturing on the Frictional Behaviour in Starved Lubricated Parallel Sliding Contacts. *Lubricants* **2019**, *7*, 68. [CrossRef]
8. Voyer, J.; Zehetner, J.; Klien, S.; Ausserer, F.; Velkavrh, I. Production and Tribological Characterization of Tailored Laser-Induced Surface 3D Microtextures. *Lubricants* **2019**, *7*, 67. [CrossRef]
9. Van der Poel, S.H.; Mezera, M.; Römer, G.-W.R.B.E.; de Vries, E.G.; Matthews, D.T.A. Fabricating Laser-Induced Periodic Surface Structures on Medical Grade Cobalt–Chrome–Molybdenum: Tribological, Wetting and Leaching Properties. *Lubricants* **2019**, *7*, 70. [CrossRef]
10. Ayerdi, J.J.; Slachciak, N.; Llavori, I.; Zabala, A.; Aginagalde, A.; Bonse, J.; Spaltmann, D. On the Role of a ZDDP in the Tribological Performance of Femtosecond Laser-Induced Periodic Surface Structures on Titanium Alloy against Different Counterbody Materials. *Lubricants* **2019**, *7*, 79. [CrossRef]
11. Gnilitskyi, I.; Rota, A.; Gualtieri, E.; Valeri, S.; Orazi, L. Tribological Properties of High-Speed Uniform Femtosecond Laser Patterning on Stainless Steel. *Lubricants* **2019**, *7*, 83. [CrossRef]
12. Bonse, J.; Höhm, S.; Kirner, S.V.; Rosenfeld, A.; Krüger, J. Laser-induced periodic surface structures—A scientific evergreen. *IEEE J. Sel. Top. Quantum Electron.* **2017**, *23*, 9000615. [CrossRef]
13. Mücklich, F.; Lasagni, A.; Daniel, C. Laser interference metallurgy—using interference as a tool for micro/nano structuring. *Int. J. Mater. Res.* **2006**, *97*, 1337–1344. [CrossRef]

© 2020 by the authors. Licensee MDPI, Basel, Switzerland. This article is an open access article distributed under the terms and conditions of the Creative Commons Attribution (CC BY) license (http://creativecommons.org/licenses/by/4.0/).

 *lubricants* 

Review

Multi-Scale Surface Texturing in Tribology—Current Knowledge and Future Perspectives

Philipp G. Grützmacher [1,2], **Francisco J. Profito** [3] **and Andreas Rosenkranz** [4,*]

[1] Institute for Engineering Design and Product Development, Tribology Research Group, TU Wien, 1040 Vienna, Austria; philipp.gruetzmacher@tuwien.ac.at

[2] Chair of Functional Materials, Department of Material Science and Engineering, Campus D33, 66123 Saarbrücken, Germany

[3] Department of Mechanical Engineering, Polytechnic School of the University of São Paulo, São Paulo 17033360, Brazil; fprofito@usp.br

[4] Department of Chemical Engineering, Biotechnology and Materials, University of Chile, Santiago 8370415, Chile

* Correspondence: arosenkranz@ing.uchile.cl; Tel.: +56-995-463-769

Received: 30 September 2019; Accepted: 25 October 2019; Published: 28 October 2019

Abstract: Surface texturing has been frequently used for tribological purposes in the last three decades due to its great potential to reduce friction and wear. Although biological systems advocate the use of hierarchical, multi-scale surface textures, most of the published experimental and numerical works have mainly addressed effects induced by single-scale surface textures. Therefore, it can be assumed that the potential of multi-scale surface texturing to further optimize friction and wear is underexplored. The aim of this review article is to shed some light on the current knowledge in the field of multi-scale surface textures applied to tribological systems from an experimental and numerical point of view. Initially, fabrication techniques with their respective advantages and disadvantages regarding the ability to create multi-scale surface textures are summarized. Afterwards, the existing state-of-the-art regarding experimental work performed to explore the potential, as well as the underlying effects of multi-scale textures under dry and lubricated conditions, is presented. Subsequently, numerical approaches to predict the behavior of multi-scale surface texturing under lubricated conditions are elucidated. Finally, the existing knowledge and hypotheses about the underlying driven mechanisms responsible for the improved tribological performance of multi-scale textures are summarized, and future trends in this research direction are emphasized.

Keywords: multi-scale surface texturing; hierarchical surfaces; numerical approaches; friction reduction; wear reduction

---

1. Introduction

As early as in ancient Egypt, tribology, which includes friction, wear, and lubrication, has been put to use to reduce frictional losses and mitigate interfacial damage between rubbing surfaces. During the construction of the pyramids, the Egyptians had already realized that the efforts involved in transporting heavy stones could be significantly reduced when putting rolling elements underneath the stones. Most probably, this can be considered as the first evidence that rolling friction is lower than pure sliding friction. Over the centuries, prominent researchers including Da Vinci, Newton, Amontons, Coulomb among many others have studied frictional phenomena under different aspects [1,2]. During their investigations, they realized that friction is governed by the interplay of various mechanochemical phenomena taking place simultaneously on the contact interface. To shed more light on the involved processes, they initially tried to describe the frictional behavior of dry sliding contacts from a phenomenological point of view by establishing a mathematical relationship between the applied normal load and the resulting friction force. The definition of the coefficient of kinetic friction (COF) goes back to Amontons and Coulomb, who independently figured out that the resulting friction force is proportional to the normal load for dry sliding contacts [3]. The constant relating both forces is the kinetic COF. This simple, linear relationship often oversimplifies the complexities found in actual tribological interfaces, where time-dependent, non-equilibrium thermodynamic processes contribute to the resulting COF [4]. The Amontons-Coulomb's model of dry friction assumes that the kinetic COF does not depend on the nominal contact area and on the magnitude of the sliding velocity. These observations have been refined by Bowden and Tabor, who realized that the COF, in fact, does not depend on the nominal contact area but rather on the real contact area between the asperity peaks, thus connecting the resulting friction force with the real contact area and shear strength of the interface [5].

The real contact area strongly depends on the roughness of the interacting surfaces [6,7]. Since the surfaces are effectively in contact only at the tips of distinct asperities, all individual contact spots have to be summed up to determine the entire contact area. Having a closer look at the surface roughness, it becomes obvious that multiple scales are involved. Dependent on the respective instrument and magnification used to measure the surface topography, different geometric features become visible ranging from nanometer- to millimeter-scale [8]. This immediately demonstrates that the frictional properties of a contact pair depend on multi-physics phenomena occurring in multiple length scales, and therefore the accurate determination of the friction force relies on the solution of a multi-scale problem. Apart from the geometric and fractal aspects of surface roughness, mechanochemical effects on different scales also affect the resulting frictional behavior. Thinking about the atomic and molecular levels, the bonding strength directly correlates with the friction force. Furthermore, the atomic/molecular arrangement and order phenomena influence friction and wear. Considering order and crystal structures, chemical imperfections and lattice defects play a significant role in the subsequent materials and hence frictional properties. This impressively underlines that frictional behavior is also influenced by chemical contributions occurring on different scales. From an engineering point of view, friction and wear are also affected on micrometer- and millimeter-scale when tolerances of manufacturing processes are considered. This short paragraph ultimately demonstrates that tribology is highly inter-disciplinary and definitely encompasses the research and solution of multi-physics and multi-scale problems.

A detailed look into nature shows that it often makes use of multi-scale/hierarchical surfaces to optimize physical properties [8–11]. A well-studied phenomenon is the drag reduction observed for dolphins and sharks induced by their skin containing features of at least two different scales [12,13]. In addition, the ability to tune adhesion properties by hierarchical surface textures as observed for beetles, flies, geckos among others, is another prominent example of the important role of interfacial geometry [14–16]. Moreover, the lotus leaf, for instance, combines a specific surface roughness on different scales with modified surface chemistry to enable self-cleaning properties and superhydrophobicity [17,18].

Inspired by these strategies provided by nature and the direct impact of the surface topography on tribological performances, specific surface textures have been utilized for more than three decades to optimize friction and wear. The pioneering systematic work in this field has been carried out by Etsion and co-workers showing the potential benefits of surface textures in different lubricated machine components (seals and piston rings among others) [19,20]. Afterwards, this topic has experienced tremendous interest and attention in the tribological community. In the last years, several review papers have summarized the state-of-the-art regarding the influence of surface texturing on friction and wear in laboratory experiments and machine components [21–24]. Interestingly, the majority of the published research works make use of purely single-scale textures. The positive effects of single-scale textures depend to a large extent on the type of contact (conformal or nonconformal) and lubrication regime. Under full-film hydrodynamic lubrication, surface textures may function as micro-hydrodynamic bearings which boost the fluid pressure, thus increasing the overall load-carrying capacity [25–28]. This is particularly significant for contacts with parallel and flat surfaces, such as those encountered in mechanical seals and parallel thrust bearings. Furthermore, the subambient pressure zones formed in textures close to the contact inlet due to the fluid cavitation phenomenon responsible for the aforementioned micro-hydrodynamic bearing effect may also contribute to drawing additional lubricant into the contact (inlet-suction effect) [27,29,30]. However, it needs to be pointed out that an inappropriate choice of texture parameters (i.e., texture depth, width, and density) may lead to detrimental effects due to the excessive increase of the cavitation zones, which may reduce the local film thickness and the load carrying capacity [31]. Under mixed lubrication, surface textures typically fulfill multiple functions. Besides contributing to improved hydrodynamic pressure they also reduce the real contact area and they are able to store lubricant thus acting as a secondary oil supply. Moreover, textures can trap wear particles thus reducing abrasive wear. Under boundary lubrication, surface texturing reduces the real area of contact and can induce the formation of pressure-induced boundary layers, thus lowering friction and wear [32,33]. The proper combination of the aforementioned aspects leads to a significant reduction of friction and wear, and/or to shift the transition between different lubrication regimes [34–36]. Considering dry friction, the reduction of the real contact area as well as the storage of wear particles can be named as the main effects contributing toward improved tribological properties [37–39].

Although nature provides numerous examples of the successful use of multi-scale textures for friction reduction, the transfer of these ideas to engineering applications has been scarcely realized. Therefore, it can be expected that multi-scale surface textures bear the tremendous potential to further optimize friction and wear properties in machine components. In this context, this review article aims at summarizing the existing articles in the field of multi-scale surface textures to improve friction and wear. First, potentially suited fabrication techniques will be reviewed, as well as discussing the advantages and shortcomings of these techniques. Afterwards, the article reviews the research conducted in biologically inspired multi-scale surface textures and their effect on friction and wear. The next section will summarize the numerical methods used to model multi-scale surfaces to predict potential friction and wear reductions. Finally, this article intends to outline the mechanisms responsible for the improved friction and wear behavior promoted by multi-scale textures as well as provide future research directions for improved texture designs.

2. Fabrication Strategies for Multi-Scale Surface Textures

Historically, numerous methods have been utilized to create surface textures for tribological applications. In this context, surface textures can be defined as geometric features following a deterministic pattern, which is intended to be engineered to induce certain surface functionalities [40]. Moreover, textures can have a preferential direction or be arranged in a random fashion [40]. The fabrication methods can be roughly divided into mechanical, chemical, physical, and thermal methods [41]. Examples for each group can be found in Table 1. A detailed overview of the individual families of surface texturing methods has been given in the form of tree structures by Costa and Hutchings [41]. The benefits and limitations of each texturing technique have to be considered regarding productivity, efficiency, geometric flexibility, accuracy, material flexibility, among others. The general geometry and accuracy of the produced textures significantly affect the tribological behavior of the generated surfaces [22,23]. In this regard, the pitch, depth, edge angle, and line accuracy of the texture features are all factors influencing the tribological performance [22]. Accuracy is especially important for multi-scale textures since textures have to be combined suitably and both patterns should be left intact during the texturing process. For patterned surfaces in general and particularly for multi-scale surfaces, the real area of contact and hence the resulting friction force is significantly influenced by the arrangement of the textural features [40]. Efficiency is a critical factor when thinking about large-scale production. This is especially true in case of patterning many components in industrial chains, which directly asks for a rather cheap and efficient texturing method [41]. Of a vast variety of texturing methods, laser texturing is the most advanced technique, which can be traced back to its high flexibility in terms of materials and geometries, high speed, good accuracy, and excellent control over the textures' geometry [20,22,41–43]. Furthermore, laser texturing belongs to the methods which remove material from the surface and therefore creates more durable and shear resistant textures compared to other methods that add material to the surface (i.e., protrusions) [22].

Table 1. General classification and examples for each type of texturing method [41].

Type	Methods
Mechanical	Microcutting, Microcoining
Chemical	Chemical etching, chemical vapor deposition (CVD)
Physical	Physical vapor deposition (PVD), focused ion beam (FIB) texturing
Thermal	Laser interference texturing, electrical beam texturing

Having the generation of multi-scale surface textures in mind, they can be fabricated in a multi-step [44–49] or single-step [50–52] process. Naturally, single-step processes are less time-consuming, easier to integrate into production lines, and therefore more efficient. Moreover, multi-step processes, which consist of different steps but using the same equipment to produce textures on different scales, are more practical and efficient than combining two or more different texturing techniques to produce multi-scale surfaces. It has to be pronounced that every surface inherently shows texture features on several scales. During laser surface texturing, for example, ablated particles can be redeposited on the surface, creating a random nano-roughness [53]. However, this review shall be limited to such techniques, which deliberately engineer multi-scale surfaces.

In the literature, several review papers can be found summarizing methods to create surface textures for tribological applications [40–42]. Hence, here only techniques with a special connection to multi-scale surface texturing will be presented.

2.1. Multi-Step Processes

Generally speaking, almost every two texturing techniques can be combined to create multi-scale surface textures. Thereby, upon combining different texturing techniques, three factors have to be considered in particular: (i) the complexity of the individual processes and, hence, the overall effort for the texturing approach, (ii) the sequence of the texturing steps [54], and (iii) the respective size limitations of each method. Regarding tribologically effective surface textures, suitable texturing techniques are those which generate surface textures on significantly different scales. According to Hsu et al., wide and shallow textures improve the tribological behavior under hydrodynamic lubrication, whereas narrow and deep textures can reduce friction under mixed or boundary lubrication [55]. In their multi-scale approach, they used a triboindenter to fabricate the textures individually by nano-scratching. As shown in Figure 1, they manufactured a polished reference sample, single-scale textures, a mixture of different dimples as well as different overlapping dimples. Such mechanical texturing methods for which a specific pattern can be programmed and then machined (CNC machining) offer the advantage of very flexible texture arrangements. Nevertheless, such processes are rather slow, since every texture feature has to be created individually, which results in strong limitations regarding their industrial applicability [22].

Similarly, laser surface texturing can be used to create multi-scale textures by ablating predefined surface areas, whereby each feature is created individually. Therefore, the laser beam is directed to the desired surface areas either by moving the sample via a translation stage or by scanning the beam over the sample surface with a galvanometric scanner [42,56–58]. In between the spots where the textures are to be created, the laser beam can be blocked by a shutter system. The general texture shape and their morphology can be influenced by the laser parameters like laser fluence, wavelength, and pulse duration as well as the focusing system [59–63]. The technique for which the laser beam is merely focused and scanned over the sample surface is called direct laser writing (DLW). Using laser surface texturing and specifically DLW, Segu et al. created multi-scale textures on steel samples for tribological purposes [64–66]. Examples of these textures are depicted in Figure 2. In this regard, it needs to be highlighted that again the advantage of this technique is the flexibility in terms of geometrical shapes, but it lacks speed, especially when creating very small texture features.

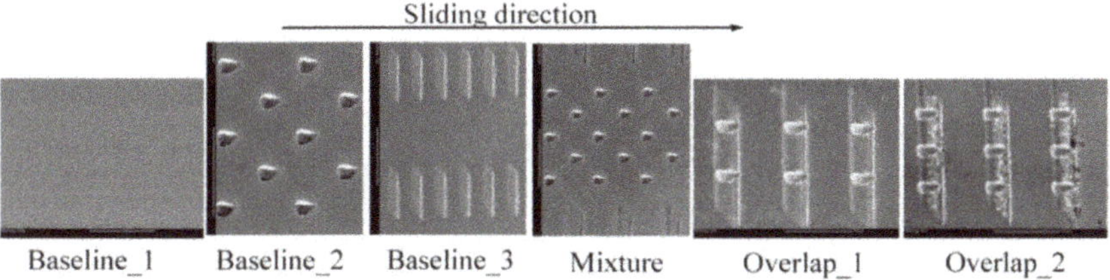

Figure 1. Multi-scale surface textures fabricated by nano-scratching using a triboindenter. In this work, single-scale, mixed, and overlapping textures have been produced. Adapted from [55].

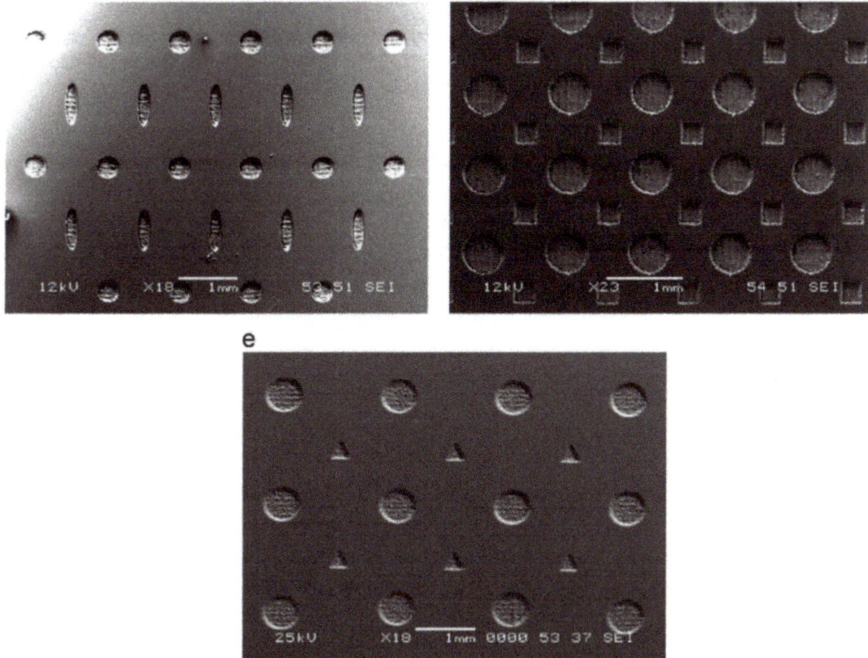

Figure 2. Multi-scale surface textures created by laser surface texturing using a pulsed Q-switched Nd:YAG laser with a power of 24 W and a pulse duration of 200 ns. The smaller textures (squares, triangles) have a side length of 250 μm while the bigger circles have a diameter of 500 μm. Adapted from [66].

Additionally, chemical methods like etching processes can be used to create multi-scale textures [67,68]. Wang et al. used a combination of lithography and reactive ion etching to create multi-scale textures consisting of bigger circular dimples with a diameter of 350 μm and smaller square dimples having a side length of 40 μm on silicon carbide to improve its tribological properties. The advantage of these chemical methods is that the surface topography is modified while the chemical and mechanical properties of the surface remain fairly constant [42]. Furthermore, material removal can be efficiently controlled, the design of the surface textures is flexible, and irregular shapes and complex geometries can be textured. However, the method is constituted by rather complex procedures like lithographic methods and deep textures are rather difficult to obtain due to electrolyte diffusion and ohmic polarization. Moreover, it is expensive to create high-resolution textures, and less versatile regarding possible materials than laser surface texturing [22,42].

Furthermore, different texturing methods can be combined to create multi-scale textures [47–49,54,69–74]. Resendiz et al. used end milling with a single crystal diamond cutter and micro shot-blasting to produce multi-scale surface textures on aluminum [47]. The bigger circular dimples fabricated by machining had a diameter of 150 μm and a depth of 30 μm. The smaller textures created by shot blasting with 10 μm aluminum oxide particles are completely covering the underlying primary machined textures and can be found inside the dimples and on the non-textured portions between the dimples. Their depth is not given but the roughness increases by 88% to 0.33 μm compared to the flat surface. With a similar method using laser surface texturing and micro shot-blasting with 25 μm alumina particles, Kim et al. manufactured multi-scale surfaces on sapphire wafers [73]. Thereby, the micro-patterns created by the laser process had a depth of 100–300 μm and a diameter of 40–160 μm. In contrast, the shot blasted textures are with a surface roughness (Rq) of 450 nm much smaller, even though their exact dimensions are not given. Gachot et al. and later Grützmacher et al. combined micro-coining with laser surface texturing, specifically direct laser interference patterning (DLIP), to create multi-scale surfaces [48,49,54,70,72]. DLIP is a technique, which uses overlapping laser beams to modulate the laser intensity spatially by interference. By applying this technique, multiple texture features can be created on the surface in a single laser shot [62]. Micro-coining is a fast process, which allows for the generation of high-quality textures in the range of 20–200 μm at low cost, while DLIP is a fast and versatile technique especially suited to create smaller textures with feature sizes between 200 nm and 30 μm [54]. It is shown that the process sequence is important for this combination. Thereby, micro-coining should precede the laser process because the inverted process sequence may lead to the partial destruction of the laser textures especially in highly deformed areas, such as the flanks of the micro-coined textures [54]. If performed correctly, however, this approach can lead to pronounced multi-scale textures with a homogenous distribution of both texture types over the surface, as can be seen in Figure 3.

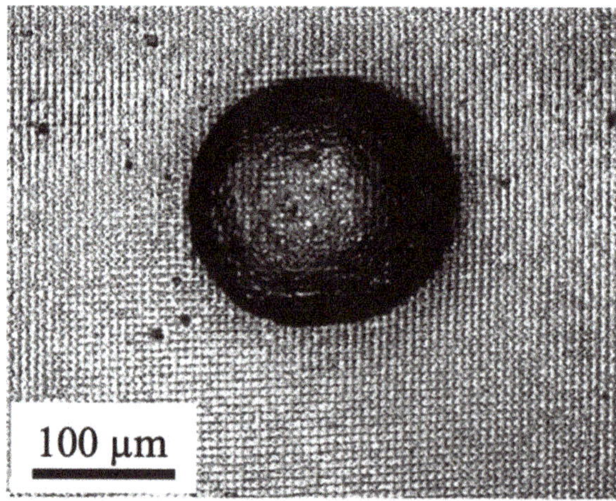

Figure 3. Multi-scale surface textures created by a combination of micro-coining and direct laser interference patterning. The bigger micro-coined textures have a diameter of 180 μm and a depth of 43 μm, whereas the cross-like laser pattern has a width of 6 μm and a depth of 0.6 μm. Adapted from [70].

Additionally, multi-scale textures can be manufactured by replicating natural surfaces inhibiting a multi-scale surface topography like the lotus leaf [44–46,75]. Shafiei and Alpas mimicked the natural surface texture of lotus leaf and boa's skin using a cellulose acetate film. In a subsequent step, a nanocrystalline nickel layer is deposited onto this film to obtain the positive surface texture of the natural sample [45,46]. To add surface textures on another scale to the lotus leaf replicas, an additional electrodeposition step was used, which led to the deposition of spherical nickel droplets on the tips of the micro-textures as shown in Figure 4 [46].

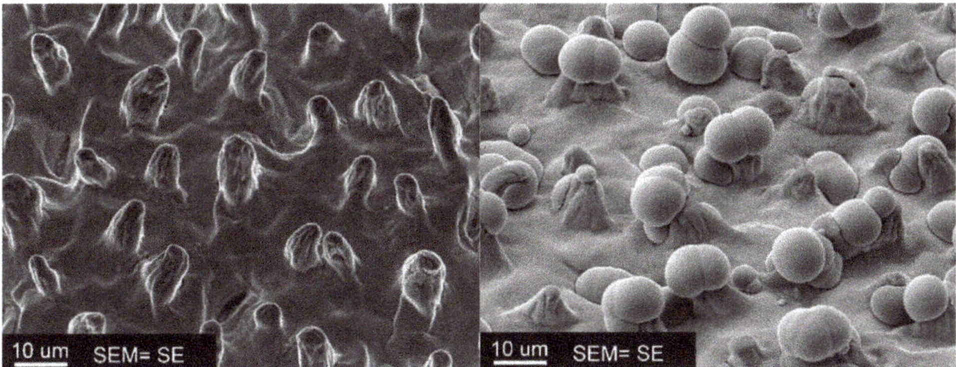

Figure 4. Multi-scale surface textures on nickel created by replicating the natural surface of the lotus leaf and a subsequent electrodeposition step. Adapted from [46].

Finally, an innovative method to create multi-scale surface textures is to combine two different laser processes, namely DLW and DLIP [76,77]. Thereby, the bigger surface textures are created by DLW and the smaller textures by DLIP. This eliminates the disadvantage of low productivity for DLW when creating small textures and offers the advantage of creating both texture types with the same tool even though a two-step fabrication process is used. Thinking about the industrial application of such textures, this means that only one tool needs to be integrated into the production process leading to smaller costs and higher production speeds. Cardoso et al. used this technique to create multi-scale surface textures. They used a solid state laser having a pulse duration of 30 ns for the DLW process fabricating cross-like textures having a periodicity (distance from channel to channel) of 15–35 µm, a depth of 3–5 µm and a width of 15 µm. Subsequently, the smaller textures were fabricated with a DLIP system on top of the bigger DLW textures using an Nd:YVO4 laser with a pulse duration of 10 ps. The smaller textures showed a periodicity of 2.6 µm and a depth of roughly 1 µm [76]. Another laser process, which has been proposed to create multi-scale textures is the combination of DLW with the controlled generation of laser induced periodic surface structures (LIPSSs) [78]. Zhang et al. used this approach to fabricate multi-scale textures. The primary textures fabricated by DLW have a width of 40 µm and a depth of 50 µm, whereas the LIPSSs on top of these textures have a depth of merely 150 nm and a period of 550 nm [78].

It is worth mentioning that several other techniques like self-assembly of nanostructures by thermal evaporation [17,79], nano-imprinting [80] or special etching techniques [67] can be utilized. However, these techniques are quite complicated and not well suited for industrial applications.

2.2. Single-Step Processes

Single-step processes to create multi-scale patterns are scarcely spread. However, some can be found in the literature [59,81–83]. However, often these multi-scale patterns occur rather randomly. Qiu et al. used an electrochemical growth approach to fabricate multi-scale cobalt textures [82]. In their work, they demonstrated cobalt crystals with several levels of hierarchy having a flower-like morphology.

However, a single-step method to fabricate multi-scale samples in a controlled fashion is again the laser process. By making use of the formation of LIPSSs during laser processing with suitable laser parameters, bigger laser textures can be created by DLW, while simultaneously smaller LIPPS are formed [81]. Ahmmed and Kietzig used a femtosecond Ti:sapphire laser system with a wavelength of 800 nm and a pulse duration <85 fs to fabricate multi-scale textures on copper [81]. The laser inscribed primary pattern demonstrated a width of 14–80 µm and a depth of 60–130 µm whereas the LIPSSs were on a scale of several hundred nanometers. Thereby, the primary textures can be controlled by the laser parameters (i.e., laser power, pulse duration, scanning speed, etc.), whereas the secondary LIPSSs textures can be controlled by the laser wavelength, the incidence angle, and the polarization of the laser beam [57].

3. Effect of Multi-Scale Textures on Friction and Wear—Experimental Studies

A significant amount of research work has been conducted to replicate the multi-scale surface topographies of the Lotus leaf, the Rice leaf, shark skin, snakeskin, among others and test these surface with regard to their potential to optimize friction and wear under dry and lubricated conditions. Shafiei and Alpas fabricated bio-inspired multi-scale surface textures by mimicking biological surfaces such as the Lotus leaf and the snakeskin using replica film followed by the electrodeposition nanocrystalline nickel. For the replicated, multi-scale Lotus leaf sample, a 30% reduction of the maximum COF under dry sliding was observed compared to a flat surface. The authors attributed the observed effect to a reduction of the real contact area [45]. Following the same approach, Shafiei and Alpas fabricated multi-scale surfaces by replicating the Lotus leaf and combining it with a chemical surface treatment (PFPE) to achieve superhydrophobicity and low friction. Using this combined treatment, a maximum friction reduction of 60% under dry friction was shown, which was explained by the reduction of the real area of contact [46]. Similarly, Wang et al. fabricated bio-inspired superhydrophobic surface textures in nickel (Lotus- and Rice-leaf) by combining a replicating technique with nickel electroplating. Though this approach, multi-scale surface textures with protruding and depressing textures were realized. The final fabrication step was a chemical modification of the resulting surface textures, applying PFPE as a lubricant. Afterwards, they studied the tribological behavior of the fabricated surface textures with and without final PFPE surface treatment. The best performance with significantly reduced COFs over the entire sliding time was found for the multi-scale surface textures with subsequent PFPE treatment irrespective of the type of textures (protruding or depressing). Additionally, they verified that the wear resistance was improved when using a final chemical PFPE treatment. The observed results were traced back to the good lubrication abilities of PFPE films as well as the possibility to reduce the real area of contact and to trap wear particles in the surface textures [44]. Using a combination of nanocasting, electroplating, and physical vapor deposition, Wang et al. fabricated diamond-like carbon films with Lotus leaf-like textures thus aiming at generating hard and flexible coatings with superhydrophobicity and good tribological properties. The main drawback of this innovative idea is the number of steps necessary to fabricate these surfaces. The fabricated textures showed a pronounced reduction in the COF over time with PFPE lubrication (without any further oil), which was traced back to the possibility to store wear particles [84]. However, it remains questionable if this is really the only contributing aspect leading to the improved friction and wear performance. The research group of El Mansori has put considerable attention to the analysis and imitation of python skin [43,85,86]. In their tribological investigations under dry conditions, they impressively demonstrated that pythons naturally have tribologically optimized surface features, which would be worth to mimic by laser surface texturing. The main conclusion of these works is that reptile skin typically follows an aperiodic and asymmetric pattern, which is in contrast to the deterministic idea of surface texturing such as arranging dimples or other shapes in a regular matrix. It can be figured that by copying more ideas from nature, improved texture designs with superior friction and wear performance can be achieved [43,87]. The research group led by Greiner conducted research towards a similar direction thus copying the surface texture of snakes and lizards to optimize the friction response under dry and lubricated conditions. They verified that these bio-inspired surface textures reduce the COF by about 40% under dry conditions, while a 3-fold friction reduction was observed in lubricated systems [88,89]. Moreover, Greiner et al. observed a pronounced size effect when fabricating and testing multi-scale surface textures with variable diameter. Surface textures with the biggest diameter showed the lowest frictional results. The explanation of the obtained results is not straight-forward and requests more in-depth analysis of the underlying phenomena, which will be subject of ongoing research work [89].

Wang et al. studied the tribological performance of textured SiC contacts under water lubrication. Initially, they verified a 2.5-fold increase in the critical load (for the transition from full-film to mixed lubrication) for single-scale textures with the lowest area density, a low depth, and an intermediate diameter [90]. In a follow-up paper, they used reactive ion etching to fabricate multi-scale textures having small and large dimples to optimize the texture effect of SiC–SiC pairings by increasing both the hydrodynamic pressure (big dimples) and the lubricant supply (small dimples). The multi-scale texture showed the best tribological performance under water lubrication with a 3.3-fold increase in the resulting load carrying capacity (Figure 5). The authors attributed the positive effects induced by surface textures to an additional hydrodynamic pressure and lubricant reservoir effect. Additionally, surface textures helped to improve the running-in process, thus leading to a smoother surface with lower roughness values. In the case of the multi-scale surface, it was speculated that the finer textures improved the water supply, thus offering more water in the tribological contact zone, which is beneficial to induce tribochemical reactions between water and SiC [91]. A summary of the research conducted by Wang et al. on textured SiC surfaces can be found in [92].

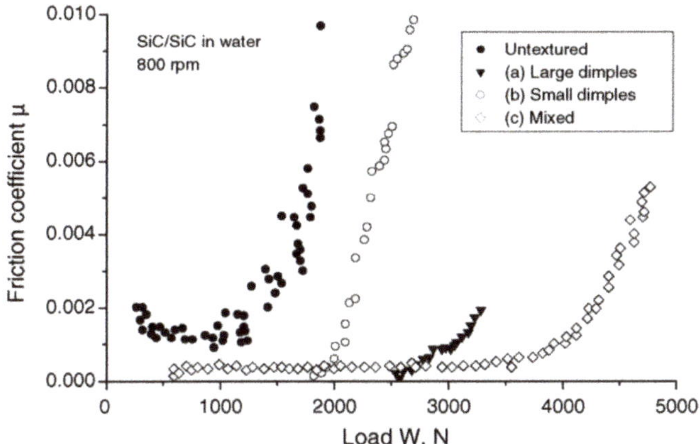

Figure 5. Coefficient of kinetic friction (COF) versus load curves in order to determine the critical load for an untextured reference as well as single-scale (large and small dimples) and multi-scale surface textures (mixed). Adapted from [91].

Segu et al. studied the effect of multi-shape textures combining circular, elliptical and triangular textures using a pin-on-disk set-up. The texture combination was fabricated by laser surface texturing using a nanosecond ND-YAG laser with a pulse duration of 200 ns. For all combinations, the structural depth was kept constant at 6.5 microns. Moreover, two area densities, namely 12 and 20%, were selected. The authors recorded Stribeck-like curves and compared the frictional behavior of the multi-shape surfaces with two reference surfaces (grinded and polished). All texture combinations showed beneficial frictional properties with a faster transition (i.e., at lower sliding speed) to low frictional values indicating hydrodynamic lubrication. In addition, they also proved the possibility to positively affect the friction and wear performance under dry sliding conditions using multi-shape textures. The textures showed a reduced averaged COF, which was traced back to the possibility to store wear debris in the textures thus removing it from the tribological contact. Furthermore, they also demonstrated beneficial frictional properties under lubricated conditions for longer sliding times. The beneficial effects of these multi-shape surfaces under lubricated conditions were explained by the increased possibility to build-up additional hydrodynamic pressure [66]. The contribution presented by Segu et al. gives interesting insights into the frictional behavior of multi-shape surfaces under dry and lubricated conditions but lacks on the presentation of carefully selected reference measurements. The data of the grinded and polished references are not sufficient to really justify the frictional efficiency of the multi-shape surfaces since the tribological results of single-shape textures (purely circular, elliptical, and triangular) have not been presented for comparison. Following the promising results of the combination of circular and elliptical multi-shape textures, Segu et al. fabricated multi-shape textures consisting of circular and elliptical textures having a depth between 3.5 and 7.5 microns as well as a density between 5 and 20%. With regard to the structural depth, they verified the best frictional behavior with the lowest COF for an intermediate depth of 5.5 microns. They explained this observation with the interplay between the oil film thickness, the structural depth of the textures and the potential pressure build-up. Although demonstrating interesting results, this study again lacks the presentation of suitable reference data as well as a deep elucidation of the obtained results [65].

The research group of Hsu conducted also important research related to the effects and mechanisms of surface textures in tribological contacts. An initial study aimed at investigating different texture geometries under low load and high-speed conditions as well as under high load and low speed conditions. Related to the first conditions, elliptical textures oriented perpendicular to the sliding direction led to the best results with the greatest friction reduction under boundary and mixed lubrication. Different contributions such as reserve lubricant flow, cavitation, the storage of lubricant inside the textures as well as a squeeze effect must be taken into consideration and may act simultaneously. Under high load and low speed conditions, all surface textures induced detrimental effects with increased friction, which has been mainly traced back to undesired edge effects [93]. Following this systematic study on single-scale surface textures with the main conclusions that large/shallow textures reduce friction under full-film lubrication and small/deep textures are effective under mixed and boundary lubrication, Hsu et al. extended their texture design to multi-scale surface textures. The general idea was to combine small but deep textures with large but shallow textures to be efficient under different lubricated conditions. Hence, Hsu et al. created a mixture of textures on different scales but also followed design rules found in nature, which favor an overlapping of textures on different scales, as shown in Figure 1 in Section 2. For experiments performed under a low contact pressure, all textured samples irrespective of single- or multi-scale were efficient to reduce friction. Particularly, the multi-scale textures with overlapping features showed a maximum friction reduction of up to 80% (Figure 6). Even under higher contact pressure, this multi-scale texture reduced friction by 70%, which is a significant advance in the design of surface textures for high contact pressure applications. Moreover, the multi-scale samples showed negligible wear features, which was traced to the transition from mixed to full-film lubrication even under higher contact pressures. Generally speaking, this study impressively demonstrated that by combining surface textures on different scales (each one optimized for a different lubrication regime) can bring an overall improvement of the frictional performance with a significant friction reduction across all lubrication regimes [55].

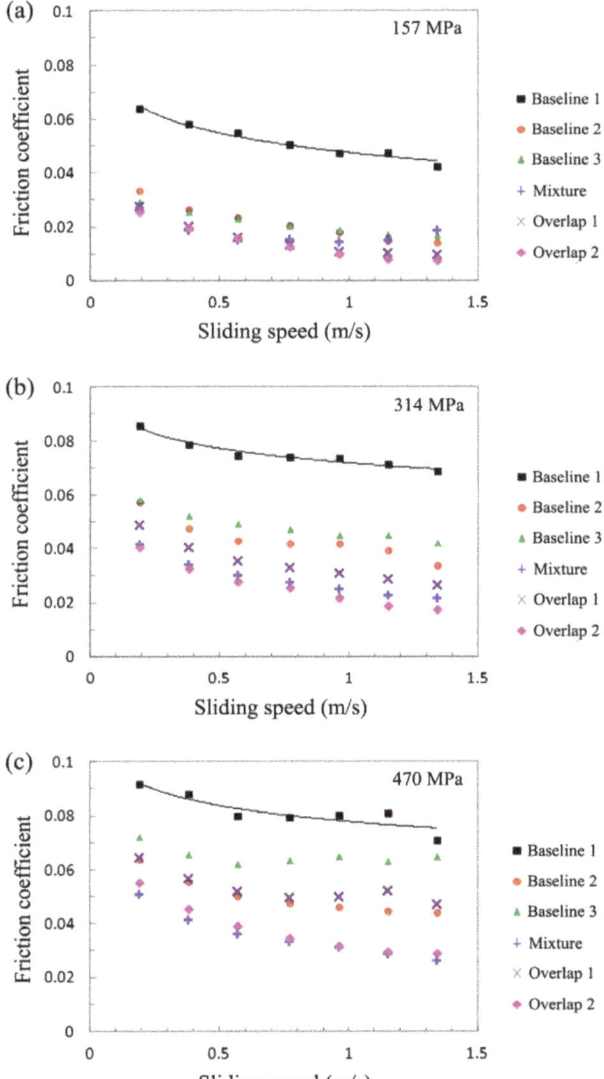

Figure 6. COF versus sliding speed for different reference cases (polished and single-scale textures) and different designs for multi-scale surface textures (mixture and overlapping textures) depending on the acting contact pressure (**a**) 157 MPa, (**b**) 314 MPa and (**c**) 470 MPa. The black baseline represents the polished reference surface, whereas baseline 2 and 3 stand for the single-scale surface textures. Adapted from [55].

Inspired by a significant friction reduction induced by micro-coined dimples as verified by [94], Grützmacher et al. overlapped micro-coined surfaces with finer textures fabricated by direct laser interference patterning (DLIP). For their study, they selected hemispherical micro-coined surfaces having a depth of 50 and 95 microns. Single-scale micro-coined samples with a depth of 50 microns showed beneficial frictional results, while deeper micro-coined samples demonstrated detrimental results regarding friction and wear. With this selection of the micro-coined geometries, Grützmacher et al.

aimed at addressing the effect of additional finer cross-like laser textures thus answering the question whether this kind of texture pattern may improve the frictional behavior of single-scale samples and either compensate the negative effects of deep dimples or further improve beneficial samples. Interestingly, the overlapped laser textures downgraded the frictional behavior of the initially beneficial micro-coined sample with a depth of 50 microns. In contrast, the additional laser textures helped to significantly improve the frictional behavior of the sample with a depth of 95 microns. Another interesting aspect has been realized during the analysis of the resulting wear scars. For both multi-scale samples, the wear scars show a rather irregular behavior with deflections from its original, circular trajectory. For this to happen, the tribological counter-body (ball) needs to interact with the underlying surface textures, which reflects the potential pressure build-up induced by the textures. Grützmacher et al. interpreted the obtained results in the following way. The improved friction behavior of the multi-scale texture was traced back to potentially reduced cavitation. For deep structures, it is well known that cavitation is more likely to occur. Combining deeper, coarser textures with fine cross-like textures may, therefore, help to reduce cavitation thus improving the distribution of lubricant in the contact area. This may induce a larger oil film thickness, thus increasing the resulting load-bearing capacity and reducing friction [48]. Following this approach, Grützmacher et al. investigated the effect of single- and multi-scale surface textures applied on the shaft of journal bearings by recording Stribeck-like curves. In order to manufacture these textures, DLIP and roller-coining have been utilized. Though DLIP, finer cross-like textures with a periodicity of 6 microns and a depth of about 1 micron have been realized, while roller-coining aimed at fabricating coarser textures with depths of up to 45 microns. Compared to the polished reference shaft, all textured single- and multi-scale surfaces led to a significant improvement of the frictional performance. As can be seen in Figure 7, under mixed lubrication, a reduction of friction by a factor of about 2–3 was observed, whereas, under hydrodynamic lubrication, a 4.6 fold decrease of the resulting COF was observed for the multi-scale texture combing the finer cross-like laser textures with the deeper micro-coined textures. The improved friction behavior of the aforementioned multi-scale texture was traced back to potentially reduced cavitation. Combining deeper, coarser textures with fine cross-like textures may help to reduce cavitation thus increasing the load bearing capacity as well as reducing the COF under hydrodynamic lubrication [72]. A comprehensive overview of the research efforts of Grützmacher et al. related to the effect of multi-scale surface textures in tribological contacts can be found in [71].

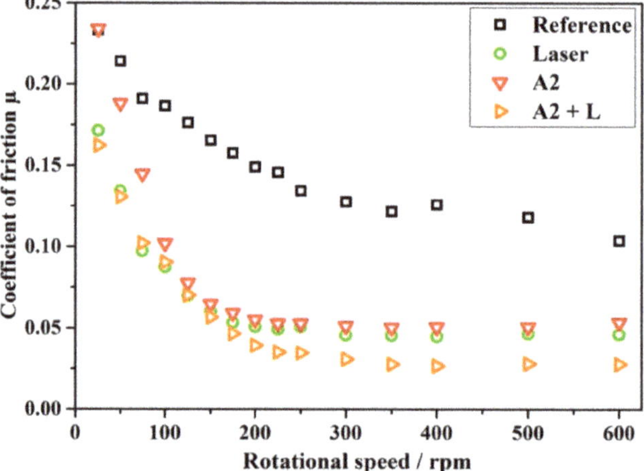

Figure 7. COF versus rotational speed for an unpolished reference, a purely laser-textured sample (laser), a solely micro-coined sample (A2) and the respective multi-scale samples combining direct laser interference patterning (DLIP) and micro-coining. Adapted from [72].

Inspired by the positive effects of multi-scale textures related to the running-in behavior and the transition between mixed and full-film hydrodynamic lubrication, Rosenkranz et al. studied the friction and wear performance of these textures under mixed lubrication. Following the beneficial effects observed for single-scale textures fabricated by DLIP [34] and micro-coining [94], they combined cross-like DLIP textures (depth about 0.6 microns and periodicity about 6 microns) with hemispherical micro-coined textures inhibiting two different depths and periodicities. Using an additive-free PAO oil, Rosenkranz et al. investigated the temporal evolution of the COF over time with a special emphasis on the time when the COF suddenly increases, which has been defined as the maximum oil film lifetime. The largest effect in terms of extending the oil film lifetime has been found for the sample with the deepest and widest micro-coined textures combined with the cross-like laser texture (Figure 8). The obtained results were attributed to an improvement of the lubricant's distribution and the additional pressure build-up due to the finer laser texture, while the coarser micro-coined texture tended to store produced wear particles thus removing them from the tribological contact zone [70].

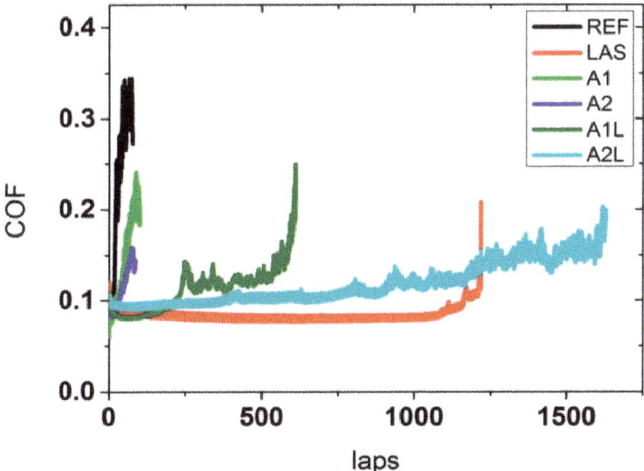

Figure 8. COF versus laps in order to determine the maximum oil film lifetime of the polished reference, the cross-like laser textures (LAS), the micro-coined surfaces (A series) and both corresponding multi-scale textures (A1L and A2L). Adapted from [70].

As already outlined in Section 2, Resendiz et al. have combined inclined end milling and micro shot blasting to create multi-scale surface textures in aluminum samples. Using end milling, circular-shaped dimples with a diameter of about 75 microns and a depth of 30 microns were created, while shot blasting with aluminum oxide particles (diameter of 10 microns) superimposed a finer roughness of the coarser milled textures. Using experimental and numerical approaches, the authors tried to evaluate the respective effect of each texturing method as well as the combination of both under lubricated conditions. In Stribeck-like curves, surface textures fabricated by the combination of end milling and shot blasting showed the best frictional behavior with a significant friction reduction compared to the untreated reference surface (Figure 9). The observed experimental findings were addressed by simulations thus verifying a cavitation effect inducing an additional pressure build-up around the textured surfaces. For the textures fabricated by a combination of both techniques, the greatest film thickness was found. Additionally, they proved that the tribological performance was notably improved by the storage of wear debris, which underlines that two effects are responsible for the superior friction and wear behavior of the multi-scale surface textures [47].

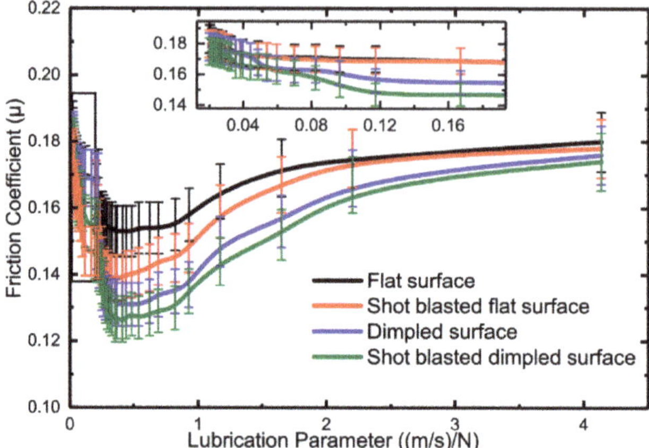

Figure 9. COF versus lubrication parameter for a flat surface, a shot-blasted sample, a dimpled sample, and a shot-blasted dimpled (i.e., multi-scale) surface. Adapted from [47].

Zhang et al. used pulsed nano-second and femto-second lasers to fabricate micro- and nano-scale surface textures as well as a combination of both ending up in a multi-scale texture. After having fabricated the respective textures, all surfaces were covered by a TiAlN coating. While the general idea is promising, the tribological characterization of the fabricated samples is not sufficient to draw any significant conclusion. Additionally, the combination of texturing and coatings is further complicated by the use of MoS2 as a solid lubricant. As already outlined in Section 2, the combination of lasers having a different pulse duration is promising in the context of creating multi-scale textures, but simpler tribological experiments need to be conducted in order to explore the full potential of this approach [78].

4. Effect of Multi-Scale Textures on Friction and Wear—Numerical Approaches

The design of optimum surface textures devised to improve the tribological performance of contacting interfaces can be considerably improved by using numerical simulation tools. Using numerical approaches to simulate the frictional behavior of multi-scale textures, the interplay of several physical factors (i.e., thermal effects, complex rheology, asperity contacts, surface wettability and the intricate lubricant flow as well as percolation) acting simultaneously on different lengths scales in the tribological contact zone can be investigated to facilitate the design of effective multi-scale textures. Complementary, numerical results can be useful to shed light on the underlying mechanisms responsible for the observed friction and wear reduction (e.g., micro-hydrodynamic bearing, inlet-suction, oil reservoir, and debris trapping effects), the improved sealing performance (e.g., lubricant channeling and percolation effects) and the finer contact temperature control induced by multi-scale textures. Therefore, optimum texture parameters (e.g., texture shape, size, depth, density) could be determined for specific applications and working conditions.

In this regard, three main modeling approaches, namely (i) computational fluid dynamics (CFD), (ii) Reynolds-type equation formulations based upon deterministic and averaging/homogenization methods, and (iii) numerical multi-scale techniques, are frequently used to simulate the tribological behavior of textured surfaces in different lubrication regimes. It is important to notice that not all modeling approaches presented in the following sections have already been applied specifically to the simulation of multi-scale textures. Nevertheless, most of the discussed methods and techniques can be extended to accomplish advanced analysis concerning the lubrication performance of contact interfaces with multi-scale textures. Thus, this review also intends to pave the way for future developments of more sophisticated approaches to model multi-scale surface textures.

4.1. CFD Modeling

The first modeling approach makes use of CFD simulations based upon the full solution of the Navier–Stokes (N–S) equations. The main advantage of CFD is the possibility of considering advanced mathematical models and complex flow phenomena, such as inertia and thermal effects, turbulence, cavitation, fluid compressibility and rheology, wall slip, fluid-structure interaction, among others. The major drawback of CFD simulations resides in the huge, and often prohibitive, computational efforts necessary to simulate problems involving textured surfaces due to the inherent fine meshes needed to properly discretize local geometric features. Particularly considering multi-scale textures, several works investigated the influence of hierarchical structures of bio-inspired shark-skin surfaces on friction drag reduction under turbulent conditions. Choi et al. explored the behavior of the turbulent micro flow field on bio-inspired micro-grooved surfaces using direct numerical simulation (DNS) [95,96]. The distribution of the micro flow field over the real shark-skin surface and its effect on the drag reduction were also analyzed through CFD simulations by Zhang et al. [97] and Luo et al. [98,99]. Figure 10 shows the morphology of scales of shark-skin surfaces studied by Luo et al. [98,99], including a schematic diagram illustrating one of the factors responsible for the drag reduction (Figure 10c) and numerical results obtained from CFD simulations (Figure 10d). The authors concluded that the drag-reduction mechanism of shark-skin is a combination of four factors: (i) a decrease of the wall viscous friction due to reduced turbulence next to the walls induced by the micro-groove tips, which stick out the viscous sub-layer, (ii) a decrease of the turbulence intensity near the wall due to the back-flowing phenomenon associated with the micro-droplets with opposite direction to the main flow (see Figure 10c), (iii) a super-hydrophobic effect produced by the boundary slipping phenomenon on the fluid-solid interface, which significantly decreases the velocity gradient and the local viscous resistance on the surface, and (iv) the presence of a nano-chain of mucus covering the wall, which increases the thickness of the viscous sublayer thus producing the aforementioned slipping phenomenon.

More recently, Martin and Bhushan conducted large-scale CFD simulations to optimize shark-inspired riblet geometries and dimensions for low drags. In that work, it was highlighted that the underlying mechanism responsible for the drag reduction was also associated with vortices lifted away from the surface and hence formed over the riblets (see Figure 11a) under turbulent flow conditions thus decreasing the overall shear stress. Furthermore, it has been identified that the optimum size of riblet design features for low-drag and anti-fouling surfaces can range from nano- to micro-scale depending on the size of the physical components for specific applications [100]. A numerical and experimental investigation of different marine drag reduction technologies based upon shark-skin inspired riblet surfaces was carried out by Fu et al. Illustrative examples of CFD analyses conducted by the authors for herringbone riblets are shown in Figure 11b. They demonstrated that triangular-shaped riblets presented a better trade-off between manufacturing and drag reduction [101].

CFD simulations were also applied by Belhadjamor et al. to study the effect of texturing on the anti-fingerprint and self-cleaning performance. It was verified that multi-scale textures are capable to decrease the finger contact area and promote hydrophobicity thus reducing the surface affinity to skin oil [102]. An example of finite element simulation and contact angle and wettability analysis of a hierarchical textured surface for anti-fingerprint and self-cleaning applications is illustrated in Figure 12. Regarding tribological applications, Brajdic-Mitidieri et al. used a CFD model with cavitation to analyze the lubricant flow behavior, load support and friction of linear, convergent pad bearings having a closed pocket. Depending on the bearing's convergence ratio and the pocket's location, the authors identified two different mechanisms responsible for friction reduction: (i) at moderate to high convergence ratios, the reduction of the shear stress is more pronounced than the pressure build-up within the pocket when the textures are suitably positioned in the high pressure region of the bearing, thus reducing the COF, and (ii) at low convergence ratios, the boost in hydrodynamic pressure within the pocket due to its convergent geometry generates higher load support (and lower friction) compared to the non-textured case [103]. A CFD-based thermo-hydrodynamic study was carried out by Vakilian et al. to explore the characteristics of Rayleigh step bearings under different steady conditions [104].

Many other works have been published in the literature involving the use of CFD analysis with different model complexities to investigate the lubrication performance of single scale textured thrust and journal bearings. The reader is referred to [21] for a thorough review of recent works based on CFD analysis to investigate the lubrication performance of single scale textured bearings. Furthermore, despite the studies of [103,104] were not directly associated with multi-scale textures, the methodologies adopted can be used as a reference for more advanced CFD analysis involving multi-scale textures.

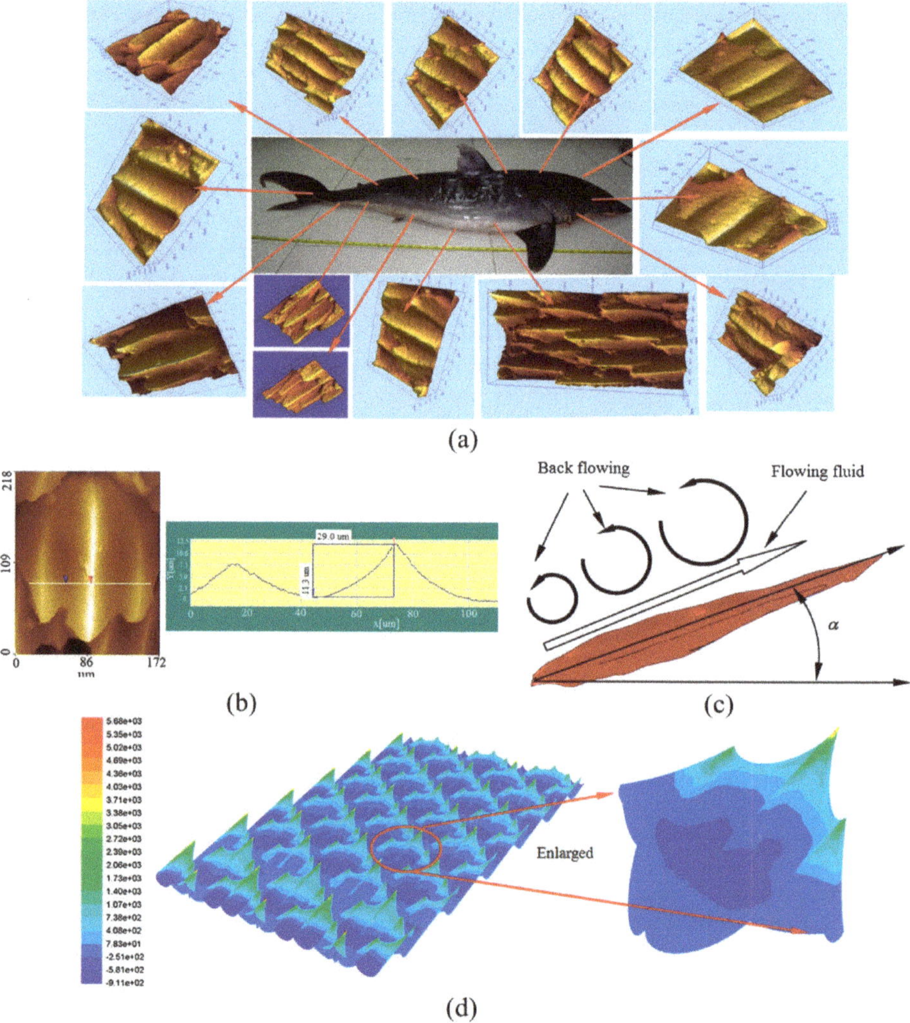

Figure 10. Hierarchical structures of bio-inspired shark-skin surfaces for friction drag reduction. (**a**) Morphology of scales on different locations of the sharkskin. (**b**) 3D image of a biological single shark-skin scale and the corresponding cross-section profile. (**c**) Schematic diagram of the back-lowing phenomenon responsible for the drag reduction due to the attenuation of the turbulence intensity. (**d**) Computational fluid dynamics (CFD) simulation of the resulting shear stress over a real shark-skin surface as well as detailed distribution over a single scale surface. Adapted from [99].

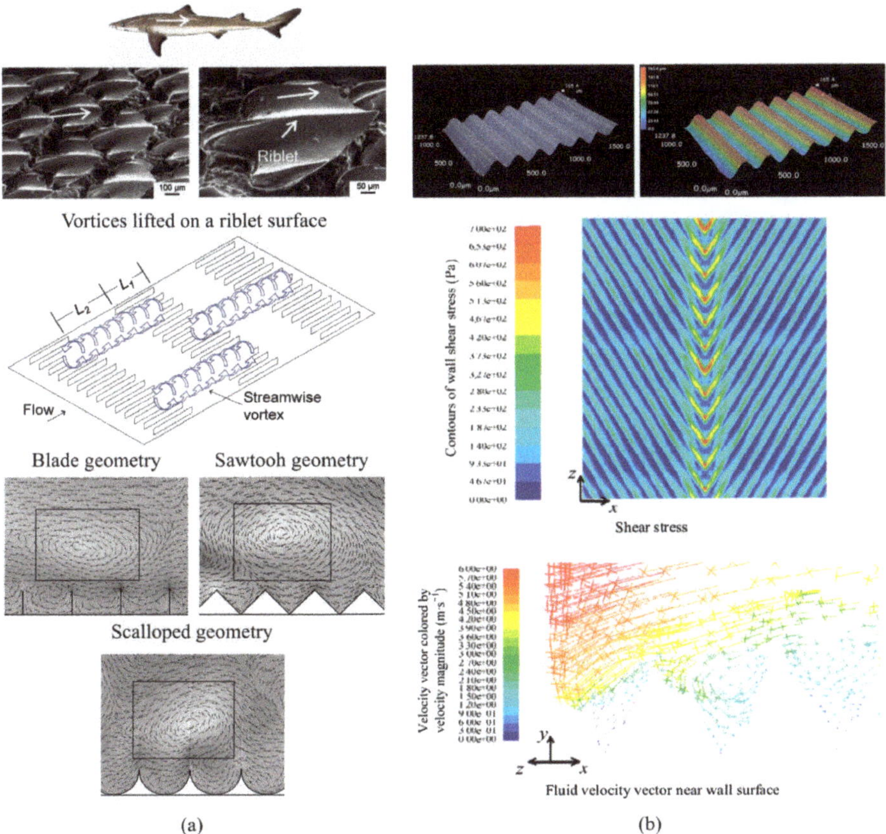

Figure 11. Shark-inspired riblet geometries for low drag applications. (**a**) Top: micrographs of samples acquired by scanning electron microscopy. Middle: schematic of the streamwise vortices lifted mechanism responsible for drag reduction on riblet surfaces. Bottom: velocity fields obtained by CFD simulations showing the vortices lifted on blade, sawtooth and scalloped riblet geometries. (**b**) Top: 3D surface topography measurement of micro-riblets applied to marine drag reduction technologies. Shear stress distribution (middle) and velocity field (bottom) of herringbone riblets obtained by CFD simulations. Adapted from [100,101].

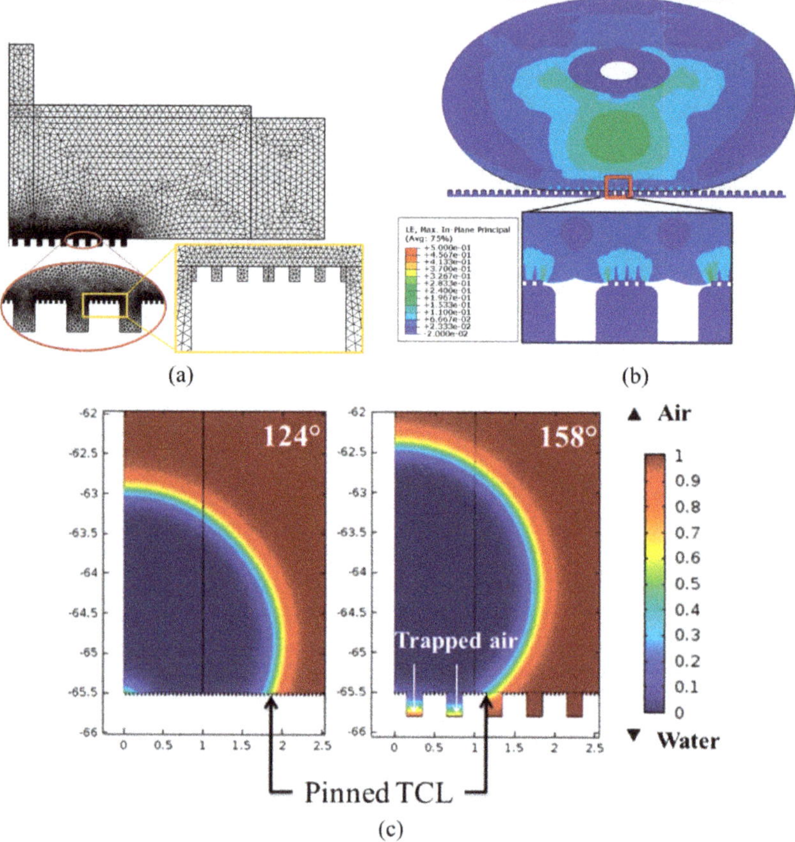

Figure 12. Hierarchical textured surface for anti-fingerprint and self-cleaning applications. (**a**) Finite element mesh. (**b**) Deformed shape and strain distribution. (**c**) Contact angle and wetting state for hydrophobicity analysis. Adapted from [102].

4.2. Reynolds-Type Equation Modeling

The second modeling approach is based upon the solution of the Reynolds equation derived from the thin fluid film lubrication theory. The advantage of this approach is the lower computational cost required to simulate textured contacts compared to full CFD methods. Furthermore, many modified versions of the Reynolds equation exist, in which specific physical aspects such as thermal effects, cavitation, turbulence, lubricant rheology, as well as the influence of surface roughness on the lubricant flow (important for mixed lubrication analysis) have been incorporated. Particularly, some methods commonly used to account for the influence of surface roughness on lubrication can be extended to predict the lubrication performance of multi-scale surface textures. In this sense, a brief explanation of the most important models used for mixed lubrication analysis shall be presented here. Special emphasis has been put on works in which such models have been applied to investigate the lubrication behavior of multi-scale surface textures. In this regard, two classes of methods based upon the Reynolds equation approach, namely deterministic and multi-scale, need to be distinguished. Each class is defined according to the way the components of the surface topography are considered for the mathematical representation of the fluid film gap geometry.

4.2.1. Full Deterministic Methods

In deterministic methods, full-scale representation of the surface topography, including both micro- and macro-scale features of the surface texture, alongside roughness, are considered in the definition of the lubricant film thickness and solution of the Reynolds equation. In the last few decades, quickly advancing computational power and the development of improved numerical techniques (e.g., multi-grid methods and parallel computing) permitted the effective digitalization of engineering surfaces, as well as the efficient solution of the coupled contact mechanics and lubrication problems. Therefore, increasing attention has been devoted to deterministic simulations. Early deterministic solutions of full-film EHL under limited rough contact were proposed for artificial topographies with simple irregularities, such as dimples and sinusoidal waviness [105–109]. Actual 2D roughness profiles were then employed by [110,111] and later 3D measured topographies were considered by [112,113]. More recently, Hu and Zhu [114–125] proposed a fully coupled mixed-TEHL model assuming a continuous lubricant film in the non-contacting regions as well as asperity contact wherever the local fluid film is sufficiently thin. In this model, the film thickness is computed from the deformed average gap, while the lubricant flow and asperity interactions are accounted for in a unified solution framework. Using this formulation, different types of multi-scale surface textures can be used as input data to deterministically simulate the entire transitions from full-film and mixed-TEHL as well as boundary lubrication under more severe conditions. However, since no averaging technique is considered in this formulation, it can only be applied to relatively small regions such as point contacts. The Hu and Zhu deterministic model was applied by [121,126] to propose a virtual surface texturing simulation tool being able to provide comparative information and directions for innovative texture design and optimization, including the relationship between textured surfaces and mixed lubrication characteristics of non-conformal contacts. Figure 13 illustrates examples of different groove textures evaluated with the deterministic mixed-TEHL model proposed by Hu and Zhu. The full-scale mixed-TEHL model developed by Hu and Zhu was also coupled to different multi-scale surface texture decomposition models to investigate the influence of surface roughness [127] and groove texture patterns placed on cylinder surfaces of internal combustion engines [128,129] on the COF.

Li and Chen proposed a deterministic mixed-lubrication model applied to the simulation of the piston ring cylinder liner contact of internal combustion engines. The model is based upon the calculation of the oil transport and the hydrodynamic pressure generation for the contact between a parallel and flat rigid plane sliding against a rough surface [131–135]. Similarly, Profito et al. presented a deterministic mixed-lubrication model based upon the simultaneous solution of the asperity contact and fluid flow problems at the roughness scale considering inter-asperity mass-conservative cavitation. The influence of the cylinder liner wear on the lubrication performance of a Twin Land Oil Control Ring (TLOCR) was analyzed using measured surface topographies of a honed cylinder liner prior to and after 100 h engine tests (see Figure 14a). The results showed that under mixed lubrication, the worn liner surface yielded to an increase of the average hydrodynamic load capacity and a decrease of the asperity contact pressures compared to the unworn liner surface. As illustrated in Figure 14b,c, this was traced back to the smoothing of the plateau regions caused by the wear-out of the highest asperities and the general decrease in the summits curvature, which contributed to facilitate the pressure-driven lubricant flow throughout the inter-grooves zones thus intensifying the role of the honing grooves in the hydrodynamic pressure generation [130]. Afterwards, Tomanik et al. applied the deterministic mixed-lubrication model proposed by Profito et al. to investigate the effect of waviness and roughness of two measured mirror-like coated bore topographies on the hydrodynamic and asperity pressure distributions. The simulation results revealed that most of the fluid pressure was generated by the honing grooves rather than by the localized pores on the coated bore surfaces [136]. Moreover, Biboulet et al. and Noutary et al. proposed multi-grid techniques to solve deterministic hydrodynamic lubrication models with non-mass-conservative cavitation for the piston ring cylinder liner contact with measured textured surfaces. It was shown that the groove depth and density are important factors determining the load carrying capacity, whereas the groove shape has only a minor influence [137–139].

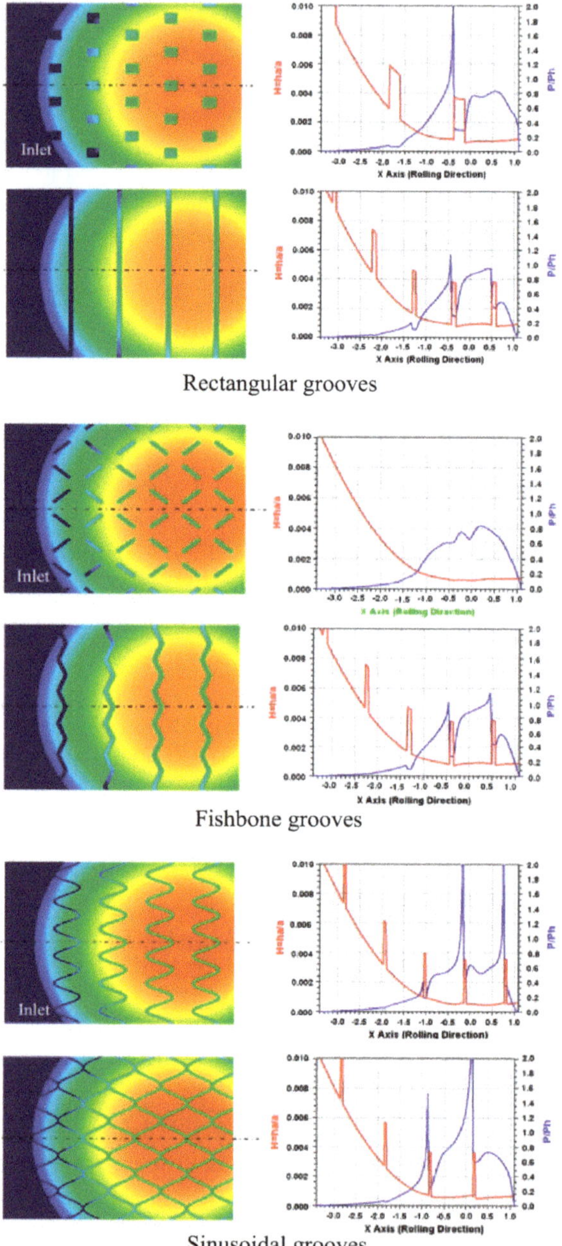

Figure 13. Deterministic simulations obtained with the fully coupled mixed-TEHL model proposed by Hu and Zhu. Sample cases with textured surfaces: rectangular (left), fishbone (middle) and sinusoidal (right) grooves with corresponding film thickness countors, centerline pressures, and normalized film thicknesses. Adapted from [118].

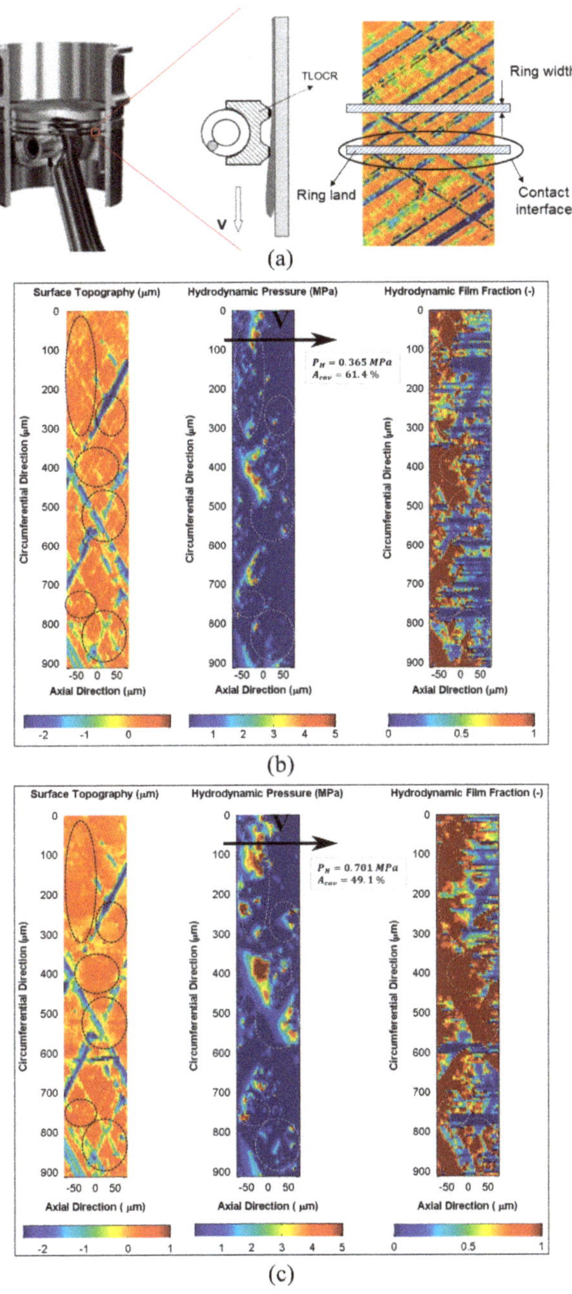

Figure 14. Influence of cylinder liner wear on the lubrication performance of a Twin Land Oil Control Ring (TLOCR) investigated by deterministic simulations. (**a**) Schematic of the TLOCR system and cylinder liner topography. (**b**) Field results for an instantaneous position of the TLOCR land on the liner prior to the engine test. (**c**) Field results for the same instantaneous position of the TLOCR land on the liner after 100 h engine test. Dashed circles highlight regions with most significant changes after wear. Adapted from [130].

A deterministic mixed lubrication model was also proposed by Minet et al. for mechanical seals applications. The model is based on the simultaneous solution of the Reynolds equation with mass-conservative cavitation and asperity contact considered through the Hertzian contact model. The results reproduced numerically the hydrodynamic load carrying capacity between nominally parallel surfaces promoted by the surface roughness, as well as the transition from mixed to full hydrodynamic lubrication regimes in face seals [140].

4.2.2. Analytical Multi-Scale Methods

Despite the better accuracy of fully deterministic methods, the computational efforts required in these cases are often prohibitive in practical applications due to the fine meshes needed to properly capture the local features. This is especially true for multi-scale surface textures, for which very fine meshes would be necessary to discretize the lubrication domain to entirely represent the geometric details of all scales. Therefore, different analytical multi-scale methods, especially averaging and homogenization methods, have been proposed to avoid dense discretization grids. In these methods, the overall influence of the surface texture features, along with the roughness in mixed lubrication analysis, are represented in terms of averaging parameters (flow factors and homogenization factors) introduced in the governing equations (e.g., averaged Reynolds equation) defined over the entire macro-scale lubrication domain. Thus, only the overall macroscopic geometry of the contacting surfaces is effectively considered in this analysis. The overall calculation process of the flow factors used in the averaged Reynolds equation is schematically illustrated in Figure 15. It is important to emphasize that most of the analytical multi-scale methods discussed in the following paragraphs have been initially proposed to model solely the effect of roughness on lubrication. Nevertheless, they can be extended to deal with the different length scales of multi-scale textures. For a deeper understanding of the principles and fundamentals of multi-scale modeling in science and engineering, the interested reader is referred to the comprehensive textbooks [141–144]. Particularly with respect to tribology, the reader is refereed to [145] for an extensive review on modeling and simulation of various physical, chemical and mechanical phenomena across different scales.

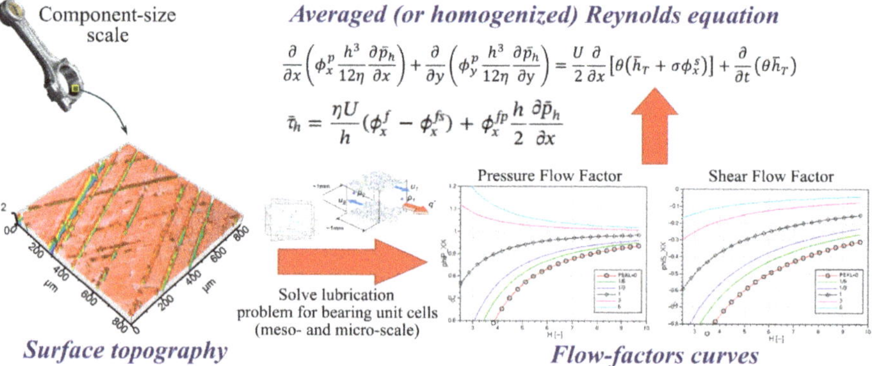

$$\frac{\partial}{\partial x}\left(\phi_x^p \frac{h^3}{12\eta}\frac{\partial \bar{p}_h}{\partial x}\right) + \frac{\partial}{\partial y}\left(\phi_y^p \frac{h^3}{12\eta}\frac{\partial \bar{p}_h}{\partial y}\right) = \frac{U}{2}\frac{\partial}{\partial x}[\theta(\bar{h}_T + \sigma\phi_x^s)] + \frac{\partial}{\partial t}(\theta\bar{h}_T)$$

$$\bar{\tau}_h = \frac{\eta U}{h}(\phi_x^f - \phi_x^{fs}) + \phi_x^{fp}\frac{h}{2}\frac{\partial \bar{p}_h}{\partial x}$$

Figure 15. Schematic illustrating the calculation process of the flow factors (or homogenization factors) used in the averaged (or homogenized) Reynolds equation which, underlay most analytical multi-scale methods. The flow factors can be calculated either from measured surfaces of actual engineering components or from virtually generated topographies. The flow-factor curve on the left shows the variation of the pressure flow-factor as a function of the dimensionless average interfacial separation. The pressure flow-factor has significant impact on the calculation of the hydrodynamic pressure under mixed lubrication conditions. Similarly, the flow-factor curve on the right represents the variation of the shear flow-factor, which influences the shear stress (and thus friction force) calculations.

Averaging Flow Methods

The modeling of the effect of surface roughness patterns on lubrication was first reported by Tzeng and Saibel [146] for one-dimensional transversal roughness and followed by Christensen [147–149] and Chow and Cheng [150] for two-dimensional transverse and longitudinal topographies. All of these pioneering works have been established within the framework of the stochastic process theory, which is based upon the concept of viewing the film thickness as a stochastic process that results in a Reynolds-type equation for the mean or expected fluid pressure.

Patir and Cheng [151,152] proposed an average flow model for general roughness patterns by incorporating "flow factors" coefficients directly in a modified Reynolds equation that is solved in the smooth global domain. The flow factors are determined independently by solving the local deterministic flow problem for a specified rough surface. Unlike the Christensen's methodology that particularly weights the film thickness oscillations according to an expectancy operator defined for a given roughness height distribution, the Patir and Cheng model is derived by "locally averaging the lubricant flows at the microscopic scale" for a representative rough bearing cell. This provides specific flow factors coefficients allowing for the consideration of the roughness (or other length scale components) induced flow perturbation effects directly on the global lubricated domain. The major drawback of this approach is a consequence of its heuristic derivation, which poses limitations on dealing with eventual crossflow produced in the case of surface roughness anisotropy. Such lack of generality was first highlighted by Elrod [153] and subsequently by different authors, including Tripp [154], who proposed a more complete tensor form of the averaged Reynolds equation. In the latter, the effects of roughness anisotropy on lubrication are accounted for in the off-diagonal terms of the diffusion and convective flow tensors, in particular, when the off-diagonal terms are negligible, the Tripp model is essentially identical to the model proposed by Patir and Cheng.

The Patir and Cheng average flow model has proven to be effective to predict the mixed lubrication performance for a wide range of applications, such as thrust and journal bearings, piston and piston-ring cylinder bore systems, mechanical seals, rolling element bearings, gears, and cam-tappet contacts, etc. Regardless of the tribological application, an important aspect for the efficacious use of any average flow model refers to the proper calculation of the flow factor coefficients for a specified topography. Furthermore, the consideration of rough contact mechanics and micro-cavitation effects also significantly affect the accuracy of the flow factor coefficients. In many successful cases, deterministic simulations of representative bearing unit cells at different length scales have been carried out to estimate the flow factors, as reported in several publications involving surface textures [155,156], honing grooves [157,158] and general roughness patterns [159–162].

A comparison between a deterministic hydrodynamic model and the stochastic solution based upon Patir and Cheng approach was undertaken by Dobrica et al. [163] for a partial journal bearing operating under mixed-EHL. The results underlined that the stochastic model correctly predicted the tendencies produced by the different roughness patterns in the fluid pressure distribution and average minimum film thickness, but underestimated the friction torques on both shaft and pad.

A multi-scale method based on the averaging flow concept was proposed by de Kraker et al. [155,156] for surface textures under mixed lubrication including micro-cavitation. In their approach, the local (micro) flow effects for a single micro-scale texture unit cell were evaluated through CFD simulations, and the results were then averaged to flow factors to be used with an averaged Reynolds equation on the macro-scale bearing level. The flow factors are dependent on the ratio between film thickness and texture dimensions, surface velocities and pressure gradient over a texture cell. Additionally, the method presented has no restrictions to the texture dimensions and shape, so that it could be well extended to model multi-scale surface textures.

Homogenization Methods

The modeling of the fluid flow problem in mixed lubrication has been addressed by Bayada [164–168], Jai and Bou-Saïd [169–171] and Buscaglia [172–174] within the framework of the homogenization theory for spatially periodic roughness. This approach is based upon the derivation of a homogenized Reynolds equation, defined at the macroscopic global scale, which captures the overall effects of the surface roughness on the lubricant flow. Besides posing the average flow model on a more rigorous mathematical base that overcomes the pure heuristic induction of the Patir and Cheng model, one additional feature of these techniques is the proper definition of the local (or auxiliary) problem, which has to be solved over a periodic unit cell to compute the average flow tensors [175]. Thus, similarly to the average flow models, the effects occurring at different length scales are incorporated in the homogenized (averaged) Reynolds equation from the solution of well-posed local problems.

A series of works have been published by Almqvist which contributed to the consolidation and widespread use of homogenization methods in lubrication applications [176–181]. The homogenization method proposed by Almqvist has been used to investigate the effect of roughness and surface texture on the tribological performance of different machine elements, such as piston-ring cylinder liner contact and rotating devices [179,182–187].

More recently, Rom and Muller [188] proposed a reduced basis method to accurately solve and speed-up the solution of the homogenized Reynolds equation in a finite element framework. This method replaces the computationally expensive solution of the full texture cell problems (micro-scale) with a reduced basis problem of much smaller dimension, which provides a significantly accelerated solution strategy. After the solution of the texture cell problem for a range of film thicknesses, the homogenized finite element matrix and vector are computed to assemble the homogenized problem. The effectiveness of the combined use of both the homogenization method and reduced basis technique is evaluated for textured journal bearings.

A novel homogenized approach was proposed by Scaraggi et al. [189,190] to study the mixed lubrication behavior of steady sliding contacts of elastically soft solids. The coupled effects of asperity-asperity and asperity-fluid interactions have been considered through a mean field theory based upon a perturbation treatment. The results demonstrated how the asperity flattening induced by the fluid-asperity interactions, as well as the local percolation effects and roughness anisotropic deformation govern the fluid flow at the interface. It was also remarked that the lubrication regime is generally not uniform at the interface due to different local average separations. Furthermore, the potential occurrence of an apparent (elasto)-hydrodynamic regime for those lubrication conditions characterized by values of $\bar{h}/h_{rms} \leq 1$ at the macroscopic level was discussed, for which the lubricant is expected to have a negligible influence on the frictional stresses. This effective transition from boundary to (elasto)-hydrodynamic regime for a given value of $\bar{h}/h_{rms}$ occurs due to the increase of the defined sliding parameter $U\eta/(E^*h_{rms})$, which determines a transition from a constant boundary stress value to a power-law shear stress. It was further noticed that this transition disappears for very small values of $\bar{h}/h_{rms}$, when percolation takes place and the average fluid flow vanishes. Afterwards, Scaraggi [191–193] also presented a homogenized method based upon the application of the Bruggeman effective medium to the Reynolds equation to investigate the average effect of textured surfaces on the macroscopic hydrodynamic characteristics of the interface. The method allows for the assessment of generic texture shape, distribution and area density, and was applied to practical cases involving 1D and 2D thrust bearing geometries.

A heterogeneous multi-scale method has been proposed by Gao and Hewson [194] to analyze micro-EHL with small-scale topographical features. The small-scale problem was solved using full CFD simulations including local elastic deformations and coupled to the global scale via scattered data interpolation method. It has been demonstrated that the proposed multi-scale framework successfully modeled the global pressure and film thickness for a textured bearing while maintaining the small-scale modeling features. Later on, Gao et al. [195] extended this multi-scale framework by incorporating the micro-cavitation and local fluid shear thinning properties.

The homogenized Reynolds equation, which simultaneously considers surface roughness and turbulent flow effects, was proposed by Lahmar et al. [196]. A plain journal bearing with periodic isotropic roughness patterns operating under turbulent conditions was used as a case study. The homogenized results agreed well with results obtained from deterministic simulations, showing that the proposed homogenization approach is suitable to study problems with rough surfaces and turbulent conditions.

A promising computational engineering framework was developed by Waseem et al. [197,198] to support the design process of optimized surface textures for hydrodynamic lubrication. The proposed framework makes use of a combination of a two-scale homogenization method and topology optimization schemes. Another important aspect of the developed multi-scale framework is the consideration of the temporal (squeeze-film effect) and spatial (wedge effect) variations in film thickness in the constitutive tensors, which characterize the homogenized response of the surface texture in terms of fluid pressure generation and load carrying capacity on the macroscopic scale. Although temporal and spatial variations in the film thickness are important for the generation of hydrodynamic pressure within the fluid at the interface, they are not always simultaneously considered in homogenization methods applied to lubrication problems.

More recently, Yildiran et al. [199] investigated the lubrication response of conventional textures (i.e., textures with well-defined, smooth geometries, such as dimples, squares, ellipsoidals, V-shapes, etc.) and representative modern re-entrant textures (i.e., textures with more complex geometries, such as trapezoidal and T-shaped features) based upon the homogenization scheme proposed by Bayada and Chambat [164]. After a comprehensive review of the literature on homogenization techniques applied to lubricated contacts and their limitations, the transition between three microscopic lubrication regimes has been demonstrated for conventional and re-entrant textures. In this work, the difference between Reynolds and Stokes roughness is also discussed. In all the above-mentioned references, except [199], the local roughness slope is always assumed small (Reynolds roughness), so that the flow equations at the microscopic scale are well described by the Reynolds equation (i.e., the local inertia effects can be neglected without significant loss of accuracy). When larger local roughness slope is present (Stokes roughness), the local inertia effects need to be taken into account in the analysis, for instance, through the solution of Stokes equations.

4.2.3. Semi-Deterministic Methods

A semi-deterministic modeling strategy was adopted by several authors [30,200–206] to study the combined effect of surface texture and roughness (multi-scale effects) under hydrodynamic and mixed lubrication conditions by solving the averaged Reynolds equation based on the Patir and Cheng model and mass-conservative cavitation. In these works, surface roughness effects (micro-scale) were treated through stochastic models (Patir and Cheng method for the lubricant flow and Greenwood-Williamson based models for asperity contact), while the surface textures (macro-scale) were considered deterministically through proper fine mesh discretization. Figure 16 summarizes the main aspects of different semi-deterministic models proposed in the literature to simulate rough textured surfaces.

Qiu and Khonsari [206] used a mass-conservative cavitation model to investigate the performance of textured dimples in seals and thrust bearings under mixed lubrication conditions. The authors verified the beneficial but minor effect of the surface roughness on the load carrying capacity of dimpled surfaces. It was also concluded that it exists an optimum dimple-to-diameter ratio and dimple density depending on the rotational speed for which the load carrying capacity is maximum. Moreover, it was verified that large dimple depths and increased roughness contribute towards higher seal leakage and that the friction force is decreased due to cavitation over the dimples. Similarly, Brunetière and Tournerie [207] showed that for smooth dimpled surfaces applied to mechanical seals, it is not possible to generate sufficient force to separate the surfaces, whereas a rough dimpled surface can significantly reduce friction.

The semi-deterministic model proposed by Profito et al. [30,203] was validated using experimental results obtained from a reciprocating test with groove surface texture. The same model was used to explain the mechanisms associated with the transient effects induced by moving textures and their influence on the frictional response and film thickness variation in different lubrication regimes. Particularly, with respect to the boundary and mixed lubrication, it was shown that the interplay between inlet suction, asperity contact, cavitation, and fluid squeeze out all contribute to the frictional response and their relative contribution may differ depending on the operating regime. Furthermore, it was also discussed that under certain working conditions in mixed lubrication, as the textures move through the interface, the net effect of inlet suction and the subsequent fluid pressure boosting promoted by the fluid squeeze out tend to increase the film thickness and hence decrease the overall friction.

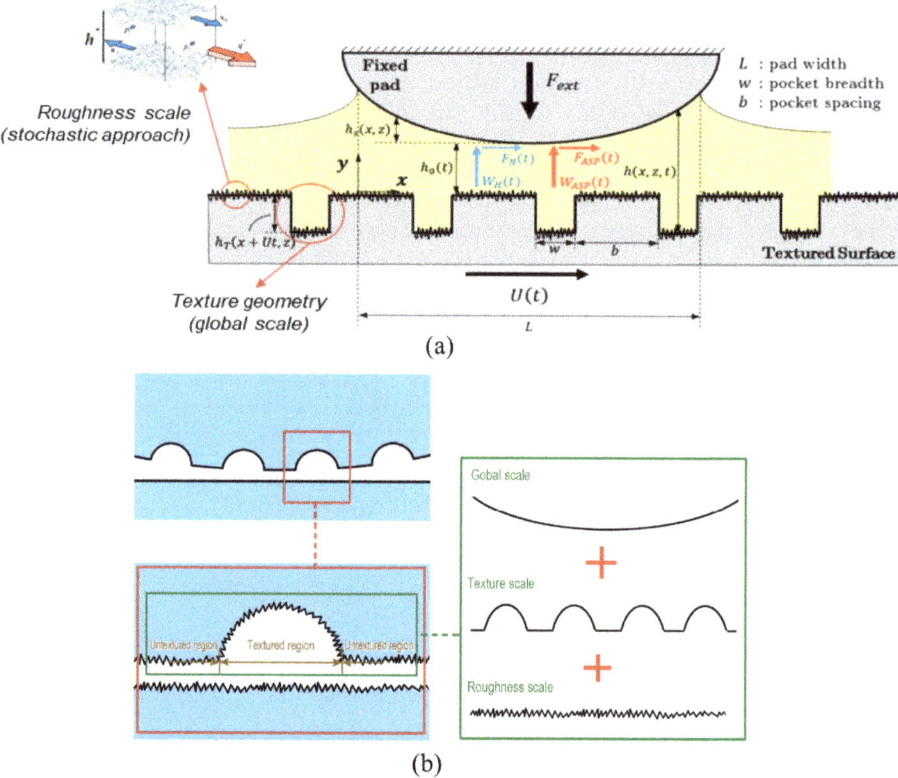

Figure 16. Examples of semi-deterministic models applied to simulate rough textured surfaces. In this modeling strategy, surface textures are considered deterministically, while surface roughness effects are taken into account through averaging (or homogenization) methods for lubricant flow and stochastic models for asperity contact. (**a**) Schematic of the rough contact interface with groove texture used by Profito et al. [30] to investigate the influence of the transient effects induced by moving textures on the frictional response and film thickness in different lubrication regimes. (**b**) Different scales of rough textured surfaces considered by Gu et al. [200] to study mixed lubrication problems in the presence of textures.

4.3. Numerical Multi-Scale Modeling

Finally, the third modeling approach which can be used to model multi-scale surface textures is based upon numerical multi-scale methods. It is important to emphasize that multi-scale modeling is not just about developing analytical models that use explicitly multi-physics coupling based upon the multi-scale expansion of the governing equations, it is also about developing numerical discretization schemes and algorithms. Numerical multi-scale methods can be divided into two main classes [143]. The first class are algorithms conceived to efficiently solve the details of the problem, including the small-scale behavior. Examples of these methods applied to solve tribological problems are the multi-grid [208–213] and adaptive mesh refinements [214–218] methods. In fact, these are linear scaling algorithms, which implies that their computational complexity scales linearly with the number of degrees of freedom necessary to represent the detailed micro-scale solution. The second class is denoted domain decomposition, which provides a platform on which multi-scale methods can be constructed. In this case, the computational domain is divided into sub-domains and a simulation strategy is adopted based upon solving the given problem on each sub-domain, thus making sure that the solutions on different sub-domains match [143]. An important subclass of numerical multi-scale methods are numerical discretization schemes (e.g., finite element, finite volume or finite difference methods), which modify the finite discretization space to consider explicitly the micro-scale features of the problem. In other words, the finite discretization space is adapted by including functions with the proper micro-scale characteristics or by creating multi-scale basis functions that relate the micro- to macro-scale simulations [219–223]. A more in-depth understanding of numerical multi-scale methods for micro-meso-macroscopic scales coupling can be found in [143,144,224,225].

The use of multi-scale domain decomposition methods and multi-scale numerical discretization schemes to solve lubrication-related problems is not widely explored in the literature and has received attention only recently. However, especially due to the inherent sub-domain strategy of these methods, they are convenient and potentially powerful to deal with multi-scale surface textures. For instance, Pei et al. [226] developed a new finite cell method for modeling surface textures in hydrodynamic lubrication. This technique uses a matrix transformation reduction strategy in which the computational domain is divided into a fine-scale domain with texture cells and a coarser-scale domain without texture cells, as can be seen in Figure 17. The proposed methodology was compared with several test cases involving FEM, CFD, and existing theoretical and experimental data, thus demonstrating that both computing time and storage were significantly reduced. Afterwards, the same authors extended their multi-scale method to lubrication problems with rough surfaces considering parallel computation to speed-up the overall numerical solution. The results showed that the method can be used to predict the average mixed lubrication effects on the global scale (coarser mesh) from deterministic calculations in the small-scale, and to accurately recover the deterministic small-scale effects from the global scale results [227]. The same multi-scale methodology was then applied to investigate the influence of surface texture on the lubrication performance of floating ring bearings including thermal effects. Nine different texture patterns were analyzed, and the results verified that textures significantly increased the side leakage and reduced the temperature rise [228].

A multi-scale approach combining a micro-deterministic mixed lubrication model for small-scales and a macro-scale model was proposed by Nyemeck et al. [229,230] to predict the hydrodynamic load carrying capacity with nominally parallel surfaces. In this model, the mass flow conservation is ensured at the boundaries of micro-cells through the calculation of pressure variations at the macro-cell boundaries obtained from a micro-deterministic model with mass-conservative cavitation and asperity contact.

The classical homogenization approach applied to model mixed lubrication was extended by Pérez-Ràfols et al. [231] to study small flows by coupling two scales with a stochastic element. The proposed stochastic element is established using a two-scale formulation based upon the framework of the heterogeneous multi-scale method.

Two important advantages of this model is that (i) the periodic repetition of the topography is not assumed as in conventional homogenization methods, which allows using much smaller micro-scale domains, and (ii) the prediction of more realistic flow patterns compared with conventional homogenization models for similar small-scale domain size is possible. The multi-scale framework proposed by Gao and Hewson [194] was extended by de Boer et al. [232] to 3D micro-scale simulations and more accurate lubricant behavior. Particularly, a two-scale method using a heterogeneous multi-scale approach to study the EHL and micro-EHL effects in tilted-pad bearings were developed. The micro-scale problem was solved by CFD simulations including surface elastic deformation, and a method for the homogenization of the micro-scale results was proposed and coupled to the macro-scale via pressure gradient-mass flow rate relationship.

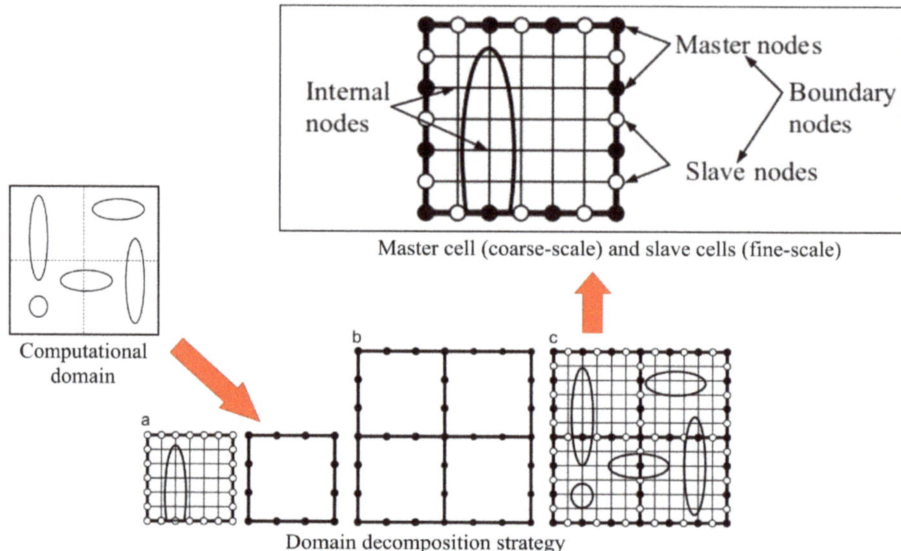

Figure 17. Overview of the finite cell method proposed by [227] for modeling surface textures in different lubrication regimes. (**a**) Illustration of the main domain decomposition steps and nomenclature of the proposed method.

More recently, Brunetière and Francisco [233] presented a multi-scale finite element method applied to the simulation of hydrodynamic lubrication of large rough contact surface. As illustrated in Figure 18a,b, the method is based on dividing the computational domain (macro-scale) into sub-domains (micro-scale) connected by a coarser mesh. The pressure distribution at the macro-scale is used as boundary conditions for the micro-scale problem, and then these boundary pressures are adjusted to guarantee the global mass flow conservation between contiguous sub-domains. A comparison between full deterministic and top-scale results with different values is shown in Figure 18c.

Finally, Costagliola et al. [234–237] proposed a spring-block modeling approach to investigate the fundamental mechanisms of dry friction between textured surfaces and how multi-scale surface textures influence static and dynamic friction. The model was used to show how the intricate surface geometry and local material properties on different length scales strongly affect the macroscopic friction force. Furthermore, it was also demonstrated how global friction properties can be tuned and optimized by designing composite surfaces with varying roughness features or local stiffness values.

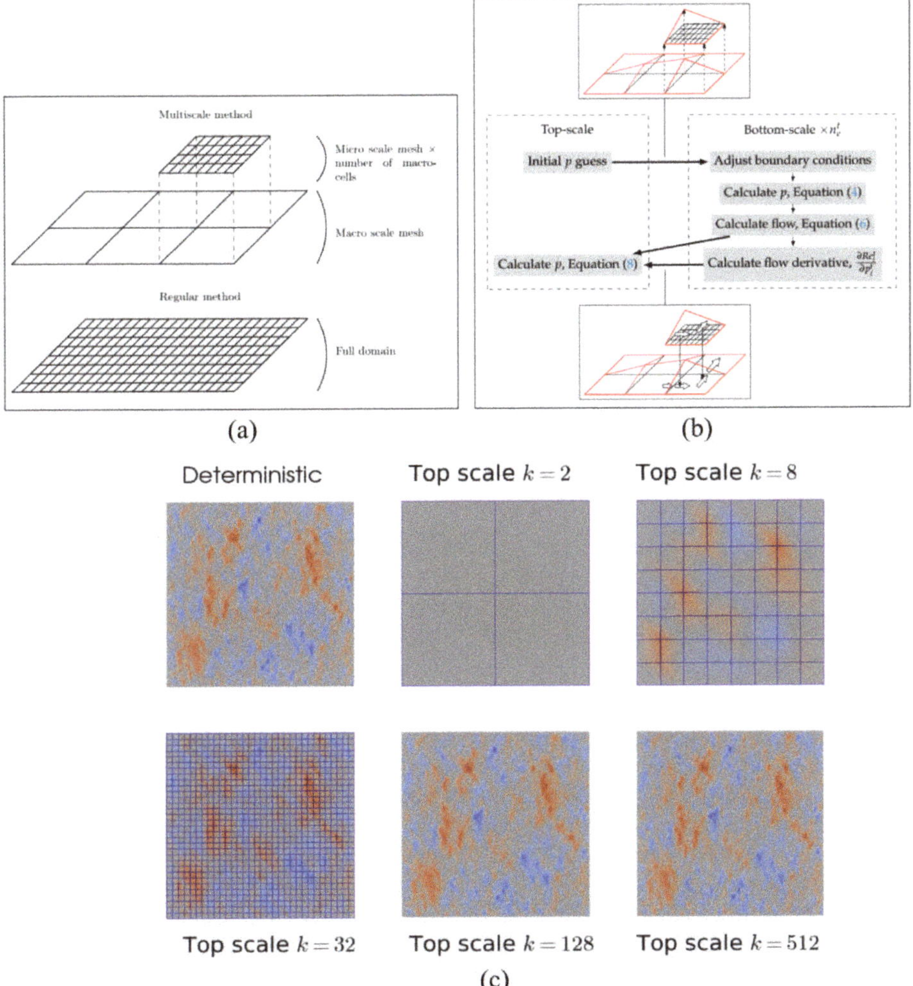

Figure 18. Overview of the multi-scale finite element method proposed by [233] to simulate hydrodynamic lubrication of large rough contact surfaces, which can be extended to deal with multi-scale textures. (**a**) Multi-scale mesh. (**b**) Multi-scale solution procedure. (**c**) Comparison between full deterministic and top-scale pressure distribution with different values.

5. Summary and Future Trends

This section summarizes the current knowledge and presents future tends regarding multi-scale surface textures applied to tribological problems. Even though several experimental and numerical works have tried to address the effects of multi-scale textures on friction and wear, the mechanisms responsible for the observed friction and wear reduction has not been fully identified yet. Therefore, we intend to derive some hypotheses regarding the underlying mechanisms for the improved tribological performance.

As already described in the introduction and depicted in Figure 19, surface textures may contribute to (i) increase the hydrodynamic pressure thus improving the load-carrying capacity (micro-hydrodynamic bearing effect) and reduce the shear-strain rate in the oil over the texture,

(ii) draw additional lubricant into the contact area (inlet-suction effect), (iii) store lubricant and supply it to the interface (oil reservoir effect), (iv) trap wear particles (debris trapping effect), and (v) reduce the real contact area thus reducing friction. Furthermore, under boundary and severe mixed lubrication conditions, surface textures may also affect sealing performance and percolation effects.

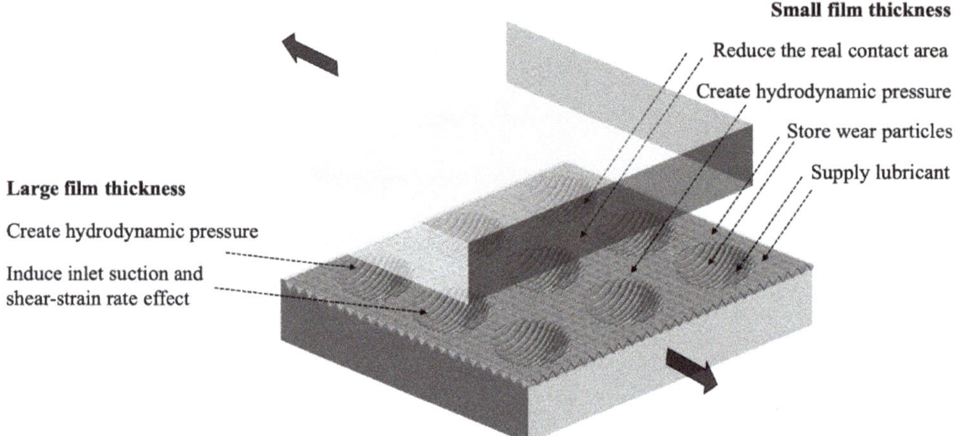

Figure 19. Schematic illustration of the possible mechanisms responsible for improved tribological performance of multi-scale surface textures. In this example, larger dimples with superimposed smaller cross-like textures are shown.

All the above-mentioned contributions are well known and accepted for single-scale textures. However, when taking multi-scale surface textures into consideration, it can be expected that a combination of different contributions may be responsible for superior friction and wear behavior. In this context, it is particularly important to simultaneously consider the type of contact (conformal or non-conformal), the operating conditions and the associated lubrication regime for the design of effective surface textures, since different mechanisms and design strategies need to be taken into consideration depending on the respective contact characteristics. Thereby, it must be emphasized that surface textures, which improve the tribological performance for specific conditions such as a certain range of film thicknesses or a specific lubrication regime, are not necessarily beneficial for all operating conditions [55]. It has been shown that beneficial surface textures optimized for a specific lubrication regime may even induce detrimental effects when used in another regime thus increasing friction and/or wear. In this sense, multi-scale surface textures pursue the goal to extend the range of operations conditions, in which a specific set of surface textures leads to superior tribological performance. This improves the general applicability of surface textures for industrial applications, since many machine elements operate under different lubrication regimes, sometimes even over a single operating cycle, as in piston and piston-ring cylinder liner contacts, connecting-rod bearings, gear meshing, cam-tapped systems among other.

For high-load and low-speed conditions inducing rather small film thicknesses (boundary lubrication), all texture types irrespective of the scale contribute to an improved frictional performance by reducing the real area of contact, supplying additional lubricant to the contact and trapping wear particles (reservoir effect). In this sense, multi-scale surfaces offer the advantage of having a greater surface area covered with textures in which a potentially higher volume of lubricant and wear particles can be stored. However, the combination of low film thickness and small features can also weaken surface integrity due to induced edge effects and stress raisers thus accelerating wear processes. [238,239]. At intermediate load and speed conditions (mixed lubrication), the lubricant film starts to partially carry the applied load. For these conditions, especially shallow textures with rather low depths tend to increase the hydrodynamic pressure and therefore improve the overall frictional behavior. It has been numerically and experimentally shown that when the structural depth is in the same range as the resulting lubricant film thickness, the greatest effects in terms of an additional pressure build-up can be expected [21,34,36,240]. In this regard, under mixed lubrication conditions, multi-scale surfaces can increase the hydrodynamic pressure due to small texture features and, additionally, offer the advantage to provide a great reservoir volume for lubricant and wear debris. Finally, at high lubricant film thicknesses for which the load is mostly carried by the oil, mainly larger texture features improve the frictional performance by increasing the hydrodynamic pressure (inlet suction mechanism and reducing of oil shear-strain rate) thus reducing the transition speed from mixed to hydrodynamic lubrication [93]. Furthermore, it can be assumed that smaller textures have a negligible effect in this lubrication regime. Hence, while not improving the tribological properties compared to single-scale textures, in this case, multi-scale surfaces can still perform well in this lubrication regime.

Summarizing, it can be stated that multi-scale surfaces can have a synergetic effect when suitably combined thus reducing friction and wear over a broader range of operating conditions. To fully use the advantages of multi-scale surface textures, they must be carefully and properly designed. Under hydrodynamic lubrication, this comprises maximizing the hydrodynamic pressure and/or reduce leakage flow without inducing negative effects due to pronounced cavitation or flow circulation problems inside bigger texture features. For mixed lubrication with a significant solid-solid contact (small film thicknesses), textures should be designed in a way that the advantages of an additional hydrodynamic pressure induced by the shallow features overcompensate potential negative effects induced by pronounced cavitation, flow circulation or edge effects (stress raisers) [21, 241,242]. Additionally, smaller and more densely distributed textures may also lead to improved wetting behavior and a better lubricant's distribution in the contact zone, which can help to reduce cavitation and flow circulation thus ultimately improving the load-carrying capacity. Nevertheless, it must be stressed that more fundamental studies will be needed to properly evaluate important parameters in multi-scale textures. In this context, it has already been demonstrated that certain geometrical parameters, such as the aspect ratio, the area density, and their ratio to the texture size, determine the tribological behavior of single-scale surface textures. A comprehensive overview of beneficial single-scale texture geometries under different speed and load conditions can be found in the review presented by Gachot et al. [22]. Moreover, the relation between the acting oil thickness and the involved scales (depths) in multi-scale textures needs to be investigated systematically.

Numerical modeling of the tribological behavior of multi-scale textures is considered to be a powerful tool to optimize the design of multi-scale textures since this approach is more time-efficient and less costly thus reducing the well-practiced trial-and-error methodology. However, it should be emphasized that it is always desirable to cross-correlate the obtained numerical results with experimental data to validate and further improve the mathematical models and optimize the overall design process of multi-scale textures. Since cavitation and the lubricant's flow and distribution in the contact zone play an important role, one interesting approach would be to design tribological experiments with multi-scale textured samples allowing the imaging of the contact zone (in- or ex-situ) and/or the measurement of the lubricant film thickness, the fluid flow velocity, and temperature distributions.

Moreover, modeling and simulation should be further integrated into the design process. In this context, the future trends regarding modeling and simulation of surface textures reside on the continuous development and consolidation of virtual simulation tools, which smartly combine:

(i)   the modeling approaches previously described to accurately predict the tribological behavior of textured surfaces under different lubrication regimes,

(ii)   optimization techniques and, potentially, the use of machine learning algorithms to speed-up the determination of optimum texture designs for friction and wear reduction in different applications,

(iii) the exploration of nature-inspired multi-scale textures for tribological applications.

The development of this computational surface engineering framework can be structured according to the following aspects:

(1) The effect of surface texture on the global, component-size scale would be determined on the basis of well-posed analytical multi-scale methods. Good candidates are the Patir and Cheng average flow model [151,152], the micro-macro method of de Kraker et al. [155,156], as well as the homogenization techniques proposed by Almqvist [176–185], Scaraggi [189–193] and the extension of Wassem et at. [197,198] and Yildiran et al. [199] models to mixed lubrication.

(2) Proper calculation of flow factor (or homogenization) tensors using full deterministic simulations. In this case, an extension of the Hu and Zhu mixed-TEHL model [111,114,115,117,121,126] by including mass-conservative cavitation and advanced rheological models for better predictions of the shear flow factor tensors would be a good alternative.

(3) Use of powerful computational techniques and algorithms, such as multi-grid methods and parallel computing, to speed up deterministic simulations for the determination of flow factor tensors.

(4) Explore the use of numerical multi-scale methods, such as domain decomposition methods and numerical discretization schemes, to model and simulate lubrication problems with multi-scale surface textures.

(5) Development of a computational surface engineering simulation framework as an open source project available to the entire tribology community. Furthermore, the construction of an open source library for storing the flow factors and other important simulation parameters obtained from previously simulated surface textures would be helpful, which, in turn, could be used to accelerate the innovative texture design and optimization for a wide range of surface textures.

Summarizing, numerical methods and experiments should be suitably combined to further push the development and design of multi-scale surfaces to enable lower friction and wear over a broader range of tribological conditions. By doing so, optimized designs such as new texture geometries and multi-scale textures being comprised of features on more than two scales can be tested. Additionally, texturing techniques, which enable the fabrication of multi-scale surfaces should be subject to further investigation. In this context, techniques with many degrees of freedom to create new interesting shapes and efficient techniques, thus paving the way to mass production, are interesting. Gaining more knowledge about the potential tribological effects of multi-scale surface texture combined with the possibility to fabricate more sophisticated texture arrangement can also boost the application side in the future. Having the beneficial effects of multi-scale textures in journal bearings in mind, as outlined in Section 3, it can be imagined that this can be just the beginning of the journey. Multi-scale textures seem to be very promising to improve the friction and wear characteristics in the piston ring cylinder liner contacts. In this regard, textures on different scales could be designed appropriately to enable beneficial frictional properties along the entire stroke irrespective of the sliding velocity. Other machine components, which could significantly benefit from the usage of multi-scale surface textures, are cam followers, rolling element bearings, thrust and sliding bearings among others.

Author Contributions: P.G.G., F.J.P. and A.R. contributed equally regarding the manuscript's conception, the literature review, the manuscript's writing as well as its proof-reading.

Funding: A. Rosenkranz gratefully acknowledges the financial support given by CONICYT in the framework of the project (Fondecyt Iniciacion 11180121). In addition, A. Rosenkranz would like to acknowledge the VID for the financial support given in the project UI 013/2018.

Conflicts of Interest: The authors declare no conflict of interest.

References

1. Dowson, D. *History of Tribology*; Longman: London, UK, 1979; ISBN 978-0-582-44766-0.
2. Williams, J. *Engineering Tribology*; Cambridge University Press: Cambridge, UK, 2005; ISBN 9780511805905.
3. Amontons, G. De la résistance causée dans les machines. *Mémoires l'Académie R. A* **1699**, 257–282.

4. Gao, J.; Luedtke, W.D.; Gourdon, D.; Ruths, M.; Israelachvili, J.N.; Landman, U. Frictional forces and Amontons' law: From the molecular to the macroscopic scale. *J. Phys. Chem. B* **2004**, *108*, 3410–3425. [CrossRef]

5. Bowden, F.P.; Tabor, D. Mechanism of metallic friction. *Nature* **1942**, *150*, 197–199. [CrossRef]

6. Archard, J.F. Contact and rubbing of flat surfaces. *J. Appl. Phys.* **1953**, *24*, 981–988. [CrossRef]

7. Greenwood, J.A.; Williamson, J.B.P. Contact of Nominally Flat Surfaces. *Proc. R. Soc. A Math. Phys. Eng. Sci.* **1966**, *295*, 300–319.

8. Nosonovsky, M.; Bhushan, B. *Multiscale Dissipative Mechanisms and Hierarchical Surfaces*; NanoScience and Technology; Springer: Berlin/Heidelberg, Germany, 2008; ISBN 978-3-540-78424-1.

9. Liu, Z.; Yin, W.; Tao, D.; Tian, Y. A glimpse of superb tribological designs in nature. *Biotribology* **2015**, *1*, 11–23. [CrossRef]

10. Malshe, A.P.; Bapat, S.; Rajurkar, K.P.; Haitjema, H. Bio-inspired textures for functional applications. *CIRP Ann.* **2018**, *67*, 627–650. [CrossRef]

11. Guo, Y.; Zhang, Z.; Zhang, S. Advances in the application of biomimetic surface engineering in the oil and gas industry. *Friction* **2019**, *7*, 289–306. [CrossRef]

12. Dean, B.; Bhushan, B. Shark-skin surfaces for fluid-drag reduction in turbulent flow: A review. *Philos. Trans. R. Soc. A Math. Phys. Eng. Sci.* **2010**, *368*, 4775–4806. [CrossRef]

13. Bixler, G.D.; Bhushan, B. Fluid drag reduction with shark-skin riblet inspired microstructured surfaces. *Adv. Funct. Mater.* **2013**, *23*, 4507–4528. [CrossRef]

14. Greiner, C.; Arzt, E.; Del Campo, A. Hierarchical gecko-like adhesives. *Adv. Mater.* **2009**, *21*, 479–482. [CrossRef]

15. Murphy, M.P.; Kim, S.; Sitti, M. Enhanced adhesion by gecko-inspired hierarchical fibrillar adhesives. *ACS Appl. Mater. Interfaces* **2009**, *1*, 849–855. [CrossRef] [PubMed]

16. Sun, J.; Bhushan, B. Nanomanufacturing of bioinspired surfaces. *Tribol. Int.* **2019**, *129*, 67–74. [CrossRef]

17. Koch, K.; Bhushan, B.; Jung, Y.C.; Barthlott, W. Fabrication of artificial Lotus leaves and significance of hierarchical structure for superhydrophobicity and low adhesion. *Soft Matter* **2009**, *5*, 1386. [CrossRef]

18. Lin, J.; Cai, Y.; Wang, X.; Ding, B.; Yu, J.; Wang, M. Fabrication of biomimetic superhydrophobic surfaces inspired by lotus leaf and silver ragwort leaf. *Nanoscale* **2011**, *3*, 1258–1262. [CrossRef]

19. Etsion, I. Improving tribological performance of mechanical components by laser surface texturing. *Tribol. Lett.* **2004**, *17*, 733–737. [CrossRef]

20. Etsion, I. State of the Art in Laser Surface Texturing. *J. Tribol.* **2005**, *127*, 248–253. [CrossRef]

21. Gropper, D.; Wang, L.; Harvey, T.J. Hydrodynamic lubrication of textured surfaces: A review of modeling techniques and key findings. *Tribol. Int.* **2016**, *94*, 509–529. [CrossRef]

22. Gachot, C.; Rosenkranz, A.; Hsu, S.M.; Costa, H.L. A critical assessment of surface texturing for friction and wear improvement. *Wear* **2017**, *372–373*, 21–41. [CrossRef]

23. Rosenkranz, A.; Grützmacher, P.G.; Gachot, C.; Costa, H.L. Surface Texturing in Machine Elements—A Critical Discussion for Rolling and Sliding Contacts. *Adv. Eng. Mater.* **2019**, *21*, 1900194. [CrossRef]

24. Sudeep, U.; Tandon, N.; Pandey, R.K. Performance of Lubricated Rolling/Sliding Concentrated Contacts With Surface Textures: A Review. *J. Tribol.* **2015**, *137*, 031501. [CrossRef]

25. Rosenkranz, A.; Costa, H.L.; Profito, F.; Gachot, C.; Medina, S.; Dini, D. Influence of surface texturing on hydrodynamic friction in plane converging bearings—An experimental and numerical approach. *Tribol. Int.* **2019**, *134*, 190–204. [CrossRef]

26. Costa, H.L.; Hutchings, I.M. Hydrodynamic lubrication of textured steel surfaces under reciprocating sliding conditions. *Tribol. Int.* **2007**, *40*, 1227–1238. [CrossRef]

27. Fowell, M.; Olver, A.V.; Gosman, A.D.; Spikes, H.A.; Pegg, I. Entrainment and Inlet Suction: Two Mechanisms of Hydrodynamic Lubrication in Textured Bearings. *J. Tribol.* **2007**, *129*, 336. [CrossRef]

28. Fowell, M.T.; Medina, S.; Olver, A.V.; Spikes, H.A.; Pegg, I.G. Parametric study of texturing in convergent bearings. *Tribol. Int.* **2012**, *52*, 7–16. [CrossRef]

29. Vlădescu, S.C.; Ciniero, A.; Tufail, K.; Gangopadhyay, A.; Reddyhoff, T. Looking into a laser textured piston ring-liner contact. *Tribol. Int.* **2017**, *115*, 140–153. [CrossRef]

30. Profito, F.J.; Vlădescu, S.C.; Reddyhoff, T.; Dini, D. Transient experimental and modelling studies of laser-textured micro-grooved surfaces with a focus on piston-ring cylinder liner contacts. *Tribol. Int.* **2017**, *113*, 125–136. [CrossRef]

31. Dobrica, M.B.; Fillon, M.; Pascovici, M.D.; Cicone, T. Optimizing surface texture for hydrodynamic lubricated contacts using a mass-conserving numerical approach. *Proc. Inst. Mech. Eng. Part J J. Eng. Tribol.* **2010**, *224*, 737–750. [CrossRef]

32. Erdemir, A. Review of engineered tribological interfaces for improved boundary lubrication. *Tribol. Int.* **2005**, *38*, 249–256. [CrossRef]

33. Pettersson, U.; Jacobson, S. Influence of surface texture on boundary lubricated sliding contacts. *Tribol. Int.* **2003**, *36*, 857–864. [CrossRef]

34. Rosenkranz, A.; Heib, T.; Gachot, C.; Mücklich, F. Oil film lifetime and wear particle analysis of laser-patterned stainless steel surfaces. *Wear* **2015**, *334–335*, 1–12. [CrossRef]

35. Braun, D.; Greiner, C.; Schneider, J.; Gumbsch, P. Efficiency of laser surface texturing in the reduction of friction under mixed lubrication. *Tribol. Int.* **2014**, *77*, 142–147. [CrossRef]

36. Ramesh, A.; Akram, W.; Mishra, S.P.; Cannon, A.H.; Polycarpou, A.A.; King, W.P. Friction characteristics of microtextured surfaces under mixed and hydrodynamic lubrication. *Tribol. Int.* **2013**, *57*, 170–176. [CrossRef]

37. Gachot, C.; Rosenkranz, A.; Reinert, L.; Ramos-Moore, E.; Souza, N.; Müser, M.H.; Mücklich, F. Dry friction between laser-patterned surfaces: Role of alignment, structural wavelength and surface chemistry. *Tribol. Lett.* **2013**, *49*, 193–202. [CrossRef]

38. Rosenkranz, A.; Reinert, L.; Gachot, C.; Mücklich, F. Alignment and wear debris effects between laser-patterned steel surfaces under dry sliding conditions. *Wear* **2014**, *318*, 49–61. [CrossRef]

39. Prodanov, N.; Gachot, C.; Rosenkranz, A.; Mücklich, F.; Müser, M.H. Contact mechanics of laser-textured surfaces: Correlating contact area and friction. *Tribol. Lett.* **2013**, *50*, 41–48. [CrossRef]

40. Abdel-Aal, H.A. Functional surfaces for tribological applications: Inspiration and design. *Surf. Topogr. Metrol. Prop.* **2016**, *4*, 1–37. [CrossRef]

41. Costa, H.L.L.; Hutchings, I.M. Some innovative surface texturing techniques for tribological purposes. *Proc. Inst. Mech. Eng. Part J J. Eng. Tribol.* **2015**, *229*, 429–448. [CrossRef]

42. Coblas, D.G.; Fatu, A.; Maoui, A.; Hajjam, M. Manufacturing textured surfaces: State of art and recent developments. *Proc. Inst. Mech. Eng. Part J J. Eng. Tribol.* **2015**, *229*, 3–29. [CrossRef]

43. Abdel-Aal, H.A.; El Mansori, M. Tribological analysis of the ventral scale structure in a Python regius in relation to laser textured surfaces. *Surf. Topogr. Metrol. Prop.* **2013**, *1*. [CrossRef]

44. Wang, Y.; Yang, J.; Guo, X.; Zhang, Q.; Wang, J.; Ding, J.; Yuan, N. Fabrication and tribological properties of superhydrophobic nickel films with positive and negative biomimetic microtextures. *Friction* **2014**, *2*, 287–294. [CrossRef]

45. Shafiei, M.; Alpas, A.T. Fabrication of biotextured nanocrystalline nickel films for the reduction and control of friction. *Mater. Sci. Eng. C* **2008**, *28*, 1340–1346. [CrossRef]

46. Shafiei, M.; Alpas, A.T. Nanocrystalline nickel films with lotus leaf texture for superhydrophobic and low friction surfaces. *Appl. Surf. Sci.* **2009**, *256*, 710–719. [CrossRef]

47. Resendiz, J.; Egberts, P.; Park, S.S. Tribological Behavior of Multi-scaled Patterned Surfaces Machined Through Inclined End Milling and Micro Shot Blasting. *Tribol. Lett.* **2018**, *66*, 132. [CrossRef]

48. Grützmacher, P.G.; Rosenkranz, A.; Szurdak, A.; Gachot, C.; Hirt, G.; Mücklich, F. Effects of Multi-Scale Patterning on the Run-In Behavior of Steel-Alumina Pairings under Lubricated Conditions. *Adv. Eng. Mater.* **2018**, *20*, 1700521. [CrossRef]

49. Grützmacher, P.G.; Rosenkranz, A.; Szurdak, A.; Gachot, C.; Hirt, G.; Mücklich, F. Lubricant migration on stainless steel induced by bio-inspired multi-scale surface patterns. *Mater. Des.* **2018**, *150*, 55–63. [CrossRef]

50. Rukosuyev, M.V.; Lee, J.; Cho, S.J.; Lim, G.; Jun, M.B.G.G. One-step fabrication of superhydrophobic hierarchical structures by femtosecond laser ablation. *Appl. Surf. Sci.* **2014**, *313*, 411–417. [CrossRef]

51. Jagdheesh, R.; García-Ballesteros, J.J.; Ocaña, J.L. One-step fabrication of near superhydrophobic aluminum surface by nanosecond laser ablation. *Appl. Surf. Sci.* **2015**, *374*, 2–11. [CrossRef]

52. Martínez-Calderon, M.; Rodríguez, A.; Dias-Ponte, A.; Morant-Miñana, M.C.C.; Gómez-Aranzadi, M.; Olaizola, S.M.M. Femtosecond laser fabrication of highly hydrophobic stainless steel surface with hierarchical structures fabricated by combining ordered microstructures and LIPSS. *Appl. Surf. Sci.* **2016**, *374*, 81–89. [CrossRef]

53. Long, J.; Fan, P.; Zhong, M.; Zhang, H.; Xie, Y.; Lin, C. Superhydrophobic and colorful copper surfaces fabricated by picosecond laser induced periodic nanostructures. *Appl. Surf. Sci.* **2014**, *311*, 461–467. [CrossRef]

54. Gachot, C.; Rosenkranz, A.; Wietbrock, B.; Hirt, G.; Mücklich, F. Advanced Design of Hierarchical Topographies in Metallic Surfaces by Combining Micro-Coining and Laser Interference Patterning. *Adv. Eng. Mater.* **2013**, *15*, 503–509. [CrossRef]
55. Hsu, S.M.; Jing, Y.; Zhao, F. Self-adaptive surface texture design for friction reduction across the lubrication regimes. *Surf. Topogr. Metrol. Prop.* **2016**, *4*, 014004. [CrossRef]
56. Sugioka, K.; Meunier, M.; Piqué, A. (Eds.) *Laser Precision Microfabrication*; Springer Series in Materials Science; Springer: Berlin/Heidelberg, Germany, 2010; Volume 135, ISBN 978-3-642-10522-7.
57. Vorobyev, A.Y.; Guo, C. Direct femtosecond laser surface nano/microstructuring and its applications. *Laser Photonics Rev.* **2013**, *7*, 385–407. [CrossRef]
58. Marczak, J. Micromachining And Pattering In Micro/Nano Scale On Macroscopic Areas. *Arch. Metall. Mater.* **2015**, *60*, 2221–2234. [CrossRef]
59. Kirner, S.V.; Hermens, U.; Mimidis, A.; Skoulas, E.; Florian, C.; Hischen, F.; Plamadeala, C.; Baumgartner, W.; Winands, K.; Mescheder, H.; et al. Mimicking bug-like surface structures and their fluid transport produced by ultrashort laser pulse irradiation of steel. *Appl. Phys. A Mater. Sci. Process.* **2017**, *123*, 1–13. [CrossRef]
60. Hermens, U.; Kirner, S.V.; Emonts, C.; Comanns, P.; Skoulas, E.; Mimidis, A.; Mescheder, H.; Winands, K.; Krüger, J.; Stratakis, E.; et al. Mimicking lizard-like surface structures upon ultrashort laser pulse irradiation of inorganic materials. *Appl. Surf. Sci.* **2017**, *418*, 499–507. [CrossRef]
61. Neuenschwander, B.; Jaeggi, B.; Schmid, M.; Hennig, G. Surface Structuring with Ultra-short Laser Pulses: Basics, Limitations and Needs for High Throughput. *Phys. Procedia* **2014**, *56*, 1047–1058. [CrossRef]
62. Mücklich, F.; Lasagni, A.F.; Daniel, C. Laser interference metallurgy—Using interference as a tool for micro/nano structuring. *Int. J. Mater. Res.* **2006**, *97*, 1337–1344. [CrossRef]
63. Leitz, K.-H.; Redlingshöfer, B.; Reg, Y.; Otto, A.; Schmidt, M. Metal Ablation with Short and Ultrashort Laser Pulses. *Phys. Procedia* **2011**, *12*, 230–238. [CrossRef]
64. Segu, D.Z.; Hwang, P. Effectiveness of multi-shape laser surface texturing in the reduction of friction under lubrication regime. *Ind. Lubr. Tribol.* **2016**, *68*, 116–124. [CrossRef]
65. Segu, D.Z.; Choi, S.G.; Choi, J.H.; Kim, S.S. The effect of multi-scale laser textured surface on lubrication regime. *Appl. Surf. Sci.* **2013**, *270*, 58–63. [CrossRef]
66. Zenebe Segu, D.; Hwang, P. Friction control by multi-shape textured surface under pin-on-disc test. *Tribol. Int.* **2015**, *91*, 111–117. [CrossRef]
67. Kwon, Y.; Patankar, N.; Choi, J.; Lee, J. Design of surface hierarchy for extreme hydrophobicity. *Langmuir* **2009**, *25*, 6129–6136. [CrossRef] [PubMed]
68. Rupp, F.; Scheideler, L.; Rehbein, D.; Axmann, D.; Geis-Gerstorfer, J. Roughness induced dynamic changes of wettability of acid etched titanium implant modifications. *Biomaterials* **2004**, *25*, 1429–1438. [CrossRef] [PubMed]
69. Rosenkranz, A.; Grützmacher, P.G.; Szurdak, A.; Gachot, C.; Hirt, G.; Mücklich, F.; Muecklich, F. Synergetic effect of laser patterning and micro coining for controlled lubricant propagation. *Surf. Topogr. Metrol. Prop.* **2016**, *4*, 034008. [CrossRef]
70. Rosenkranz, A.; Grützmacher, P.G.; Murzyn, K.; Mathieu, C.; Mücklich, F. Multi-scale surface patterning to tune friction under mixed lubricated conditions. *Appl. Nanosci.* **2019**. [CrossRef]
71. Grützmacher, P.G.; Rosenkranz, A.; Szurdak, A.; Grüber, M.; Gachot, C.; Hirt, G.; Mücklich, F. Multi-scale surface patterning—An approach to control friction and lubricant migration in lubricated systems. *Ind. Lubr. Tribol.* **2019**, *71*, 1007–1016. [CrossRef]
72. Grützmacher, P.G.; Rosenkranz, A.; Szurdak, A.; König, F.; Jacobs, G.; Hirt, G.; Mücklich, F. From lab to application—Improved frictional performance of journal bearings induced by single- and multi-scale surface patterns. *Tribol. Int.* **2018**, *127*, 500–508. [CrossRef]
73. Kim, M.; Lee, S.M.; Lee, S.J.; Kim, Y.W.; Lee, D.W. Effect on friction reduction of micro/nano hierarchical patterns on sapphire wafers. *Int. J. Precis. Eng. Manuf.-Green Technol.* **2017**, *4*, 27–35. [CrossRef]
74. Akkan, C.K.; Hammadeh, M.E.; May, A.; Park, H.W.; Abdul-Khaliq, H.; Strunskus, T.; Aktas, O.C. Surface topography and wetting modifications of PEEK for implant applications. *Lasers Med. Sci.* **2014**, *29*, 1633–1639. [CrossRef]
75. Fürstner, R.; Barthlott, W.; Neinhuis, C.; Walzel, P. Wetting and self-cleaning properties of artificial superhydrophobic surfaces. *Langmuir* **2005**, *21*, 956–961. [CrossRef] [PubMed]

76. Cardoso, J.T.; Aguilar-Morales, A.I.; Alamri, S.; Huerta-Murillo, D.; Cordovilla, F.; Lasagni, A.F.; Ocaña, J.L. Superhydrophobicity on hierarchical periodic surface structures fabricated via direct laser writing and direct laser interference patterning on an aluminium alloy. *Opt. Lasers Eng.* **2018**, *111*, 193–200. [CrossRef]

77. Huerta-Murillo, D.; Aguilar-Morales, A.I.; Alamri, S.; Cardoso, J.T.; Jagdheesh, R.; Lasagni, A.F.; Ocaña, J.L. Fabrication of multi-scale periodic surface structures on Ti-6Al-4V by direct laser writing and direct laser interference patterning for modified wettability applications. *Opt. Lasers Eng.* **2017**, *98*, 134–142. [CrossRef]

78. Zhang, K.; Deng, J.; Guo, X.; Sun, L.; Lei, S. Study on the adhesion and tribological behavior of PVD TiAlN coatings with a multi-scale textured substrate surface. *Int. J. Refract. Met. Hard Mater.* **2018**, *72*, 292–305. [CrossRef]

79. Bhushan, B.; Koch, K.; Jung, Y.C. Nanostructures for superhydrophobicity and low adhesion. *Soft Matter* **2008**, *4*, 1799. [CrossRef]

80. Zhang, F.; Low, H.Y. Anisotropic wettability on imprinted hierarchical structures. *Langmuir* **2007**, *23*, 7793–7798. [CrossRef]

81. Tanvir Ahmmed, K.M.; Kietzig, A.M. Drag reduction on laser-patterned hierarchical superhydrophobic surfaces. *Soft Matter* **2016**, *12*, 4912–4922. [CrossRef]

82. Qiu, R.; Wang, P.; Zhang, D.; Wu, J. One-step preparation of hierarchical cobalt structure with inborn superhydrophobic effect. *Colloids Surf. A Physicochem. Eng. Asp.* **2011**, *377*, 144–149. [CrossRef]

83. Wu, B.; Zhou, M.; Li, J.; Ye, X.; Li, G.; Cai, L. Superhydrophobic surfaces fabricated by microstructuring of stainless steel using a femtosecond laser. *Appl. Surf. Sci.* **2009**, *256*, 61–66. [CrossRef]

84. Wang, Y.; Wang, L.; Wang, S.; Wood, R.J.K.; Xue, Q. From natural lotus leaf to highly hard-flexible diamond-like carbon surface with superhydrophobic and good tribological performance. *Surf. Coat. Technol.* **2012**, *206*, 2258–2264. [CrossRef]

85. Abdel-Aal, H.A.; El Mansori, M.; Mezghani, S. Multi-scale investigation of surface topography of ball python (python regius) shed skin in comparison to human skin. *Tribol. Lett.* **2010**, *37*, 517–527. [CrossRef]

86. Abdel-Aal, H.A.; Vargiolu, R.; Zahouani, H.; El Mansori, M. A study on the frictional response of reptilian shed skin. *J. Phys. Conf. Ser.* **2011**, *311*, 012016. [CrossRef]

87. Cuervo, P.; López, D.A.; Cano, J.P.; Sánchez, J.C.; Rudas, S.; Estupiñán, H.; Toro, A.; Abdel-Aal, H.A. Development of low friction snake-inspired deterministic textured surfaces. *Surf. Topogr. Metrol. Prop.* **2016**, *4*, 1–17. [CrossRef]

88. Greiner, C.; Schäfer, M. Bio-inspired scale-like surface textures and their tribological properties. *Bioinspiration Biomim.* **2015**, *10*, 44001. [CrossRef] [PubMed]

89. Schneider, J.; Djamiykov, V.; Greiner, C. Friction reduction through biologically inspired scale-like laser surface textures. *Beilstein J. Nanotechnol.* **2018**, *9*, 2561–2572. [CrossRef]

90. Wang, X.; Kato, K.; Adachi, K.; Aizawa, K. Loads carrying capacity map for the surface texture design of SiC thrust bearing sliding in water. *Tribol. Int.* **2003**, *36*, 189–197. [CrossRef]

91. Wang, X.; Adachi, K.; Otsuka, K.; Kato, K. Optimization of the surface texture for silicon carbide sliding in water. *Appl. Surf. Sci.* **2006**, *253*, 1282–1286. [CrossRef]

92. Yu, H.; Huang, W.; Wang, X. Dimple patterns design for different circumstances. *Lubr. Sci.* **2013**, *25*, 67–78. [CrossRef]

93. Hsu, S.M.; Jing, Y.; Hua, D.; Zhang, H. Friction reduction using discrete surface textures: Principle and design. *J. Phys. D. Appl. Phys.* **2014**, *47*, 335307. [CrossRef]

94. Rosenkranz, A.; Szurdak, A.; Gachot, C.; Hirt, G.; Mücklich, F. Friction reduction under mixed and full film EHL induced by hot micro-coined surface patterns. *Tribol. Int.* **2016**, *95*, 290–297. [CrossRef]

95. Choi, H.; Moin, P.; Kim, J. Direct numerical simulation of turbulent flow over riblets. *J. Fluid Mech.* **1993**, *255*, 503. [CrossRef]

96. Choi, H.; Moin, P.; Kim, J. Active turbulence control for drag reduction in wall-bounded flows. *J. Fluid Mech.* **1994**, *262*, 75–110. [CrossRef]

97. Zhang, D.; Luo, Y.; Li, X.; Chen, H. Numerical Simulation and Experimental Study of Drag-Reducing Surface of a Real Shark Skin. *J. Hydrodyn.* **2011**, *23*, 204–211. [CrossRef]

98. Luo, Y.; Liu, Y.; Zhang, D.; NG, E.Y.K. Influence of morphology for drag reduction effect of sharkskin surface. *J. Mech. Med. Biol.* **2014**, *14*, 1450029. [CrossRef]

99. Luo, Y.; Yuan, L.; Li, J.; Wang, J. Boundary layer drag reduction research hypotheses derived from bio-inspired surface and recent advanced applications. *Micron* **2015**, *79*, 59–73. [CrossRef] [PubMed]

100. Martin, S.; Bhushan, B. Modeling and optimization of shark-inspired riblet geometries for low drag applications. *J. Colloid Interface Sci.* **2016**, *474*, 206–215. [CrossRef] [PubMed]

101. Fu, Y.F.; Yuan, C.Q.; Bai, X.Q. Marine drag reduction of shark skin inspired riblet surfaces. *Biosurface Biotribol.* **2017**, *3*, 11–24. [CrossRef]

102. Belhadjamor, M.; Belghith, S.; Mezlini, S.; El Mansori, M. Effect of the surface texturing scale on the self-clean function: Correlation between mechanical response and wetting behavior. *Tribol. Int.* **2017**, *111*, 91–99. [CrossRef]

103. Brajdic-Mitidieri, P.; Gosman, A.D.; Ioannides, E.; Spikes, H.A. CFD Analysis of a Low Friction Pocketed Pad Bearing. *J. Tribol.* **2005**, *127*, 803. [CrossRef]

104. Vakilian, M.; Gandjalikhan Nassab, S.A.; Kheirandish, Z. CFD-Based Thermohydrodynamic Analysis of Rayleigh Step Bearings Considering an Inertia Effect. *Tribol. Trans.* **2014**, *57*, 123–133. [CrossRef]

105. Ai, X.; Cheng, H.S.; Zheng, L. A Transient Model for Micro-Elastohydrodynamic Lubrication With Three-Dimensional Irregularities. *J. Tribol.* **1993**, *115*, 102. [CrossRef]

106. Chang, L.; Cusano, C.; Conry, T.F. Effects of Lubricant Rheology and Kinematic Conditions on Micro-Elastohydrodynamic Lubrication. *J. Tribol.* **1989**, *111*, 344. [CrossRef]

107. Chang, L.; Webster, M.N.; Jackson, A. A Line-Contact Micro-EHL Model With Three-Dimensional Surface Topography. *J. Tribol.* **1994**, *116*, 21. [CrossRef]

108. Kweh, C.C.; Evans, H.P.; Snidle, R.W. Micro-Elastohydrodynamic Lubrication of an Elliptical Contact With Transverse and Three-Dimensional Sinusoidal Roughness. *J. Tribol.* **1989**, *111*, 577. [CrossRef]

109. Lubrecht, A.A.; Ten Napel, W.E.; Bosma, R. The Influence of Longitudinal and Transverse Roughness on the Elastohydrodynamic Lubrication of Circular Contacts. *J. Tribol.* **1988**, *110*, 421. [CrossRef]

110. Kweh, C.C.; Patching, M.J.; Evans, H.P.; Snidle, R.W. Simulation of Elastohydrodynamic Contacts Between Rough Surfaces. *J. Tribol.* **1992**, *114*, 412. [CrossRef]

111. Venner, C.H.; ten Napel, W.E. Surface Roughness Effects in an EHL Line Contact. *J. Tribol.* **1992**, *114*, 616. [CrossRef]

112. Zhu, D.; Ai, X. Point Contact EHL Based on Optically Measured Three-Dimensional Rough Surfaces. *J. Tribol.* **1997**, *119*, 375. [CrossRef]

113. Xu, G.; Sadeghi, F. Thermal EHL Analysis of Circular Contacts With Measured Surface Roughness. *J. Tribol.* **1996**, *118*, 473. [CrossRef]

114. Hu, Y.-Z.; Zhu, D. A Full Numerical Solution to the Mixed Lubrication in Point Contacts. *J. Tribol.* **2000**, *122*, 1. [CrossRef]

115. Zhu, D.; Hu, Y.-Z. A Computer Program Package for the Prediction of EHL and Mixed Lubrication Characteristics, Friction, Subsurface Stresses and Flash Temperatures Based on Measured 3-D Surface Roughness. *Tribol. Trans.* **2001**, *44*, 383–390. [CrossRef]

116. Zhu, D.; Jane Wang, Q. Effect of Roughness Orientation on the Elastohydrodynamic Lubrication Film Thickness. *J. Tribol.* **2013**, *135*, 031501. [CrossRef]

117. Wang, X.; Liu, Y.; Zhu, D. Numerical Solution of Mixed Thermal Elastohydrodynamic Lubrication in Point Contacts With Three-Dimensional Surface Roughness. *J. Tribol.* **2016**, *139*, 011501. [CrossRef]

118. Ren, N.; Nanbu, T.; Yasuda, Y.; Zhu, D.; Wang, Q. Micro Textures in Concentrated-Conformal-Contact Lubrication: Effect of Distribution Patterns. *Tribol. Lett.* **2007**, *28*, 275–285. [CrossRef]

119. Wang, Q.J.; Zhu, D.; Zhou, R.; Hashimoto, F. Investigating the Effect of Surface Finish on Mixed EHL in Rolling and Rolling-Sliding Contacts. *Tribol. Trans.* **2008**, *51*, 748–761. [CrossRef]

120. Zhu, D.; Hu, Y.-Z. Effects of Rough Surface Topography and Orientation on the Characteristics of EHD and Mixed Lubrication in Both Circular and Elliptical Contacts. *Tribol. Trans.* **2001**, *44*, 391–398. [CrossRef]

121. Wang, Q.J.; Zhu, D. Virtual Texturing: Modeling the Performance of Lubricated Contacts of Engineered Surfaces. *J. Tribol.* **2005**, *127*, 722. [CrossRef]

122. Liu, Y.; Wang, Q.J.; Wang, W.; Hu, Y.; Zhu, D. Effects of Differential Scheme and Mesh Density on EHL Film Thickness in Point Contacts. *J. Tribol.* **2006**, *128*, 641. [CrossRef]

123. Zhu, D. On some aspects of numerical solutions of thin-film and mixed elastohydrodynamic lubrication. *Proc. Inst. Mech. Eng. Part J J. Eng. Tribol.* **2007**, *221*, 561–579. [CrossRef]

124. Ren, N.; Zhu, D.; Chen, W.W.; Liu, Y.; Wang, Q.J. A Three-Dimensional Deterministic Model for Rough Surface Line-Contact EHL Problems. *J. Tribol.* **2008**, *131*, 011501. [CrossRef]

125. Zhu, D.; Jane Wang, Q. Elastohydrodynamic Lubrication: A Gateway to Interfacial Mechanics—Review and Prospect. *J. Tribol.* **2011**, *133*, 041001. [CrossRef]

126. Zhu, D.; Nanbu, T.; Ren, N.; Yasuda, Y.; Wang, Q.J. Model-based virtual surface texturing for concentrated conformal-contact lubrication. *Proc. Inst. Mech. Eng. Part J J. Eng. Tribol.* **2010**, *224*, 685–696. [CrossRef]

127. Demirci, I.; Mezghani, S.; Yousfi, M.; Zahouani, H.; Mansori, M. El The Scale Effect of Roughness on Hydrodynamic Contact Friction. *Tribol. Trans.* **2012**, *55*, 705–712. [CrossRef]

128. Mezghani, S.; Demirci, I.; Zahouani, H.; El Mansori, M. The effect of groove texture patterns on piston-ring pack friction. *Precis. Eng.* **2012**, *36*, 210–217. [CrossRef]

129. Demirci, I.; Mezghani, S.; Yousfi, M.; El Mansori, M. Multiscale Analysis of the Roughness Effect on Lubricated Rough Contact. *J. Tribol.* **2013**, *136*, 011501. [CrossRef]

130. Profito, F.J.; Tomanik, E.; Zachariadis, D.C. Effect of cylinder liner wear on the mixed lubrication regime of TLOCRs. *Tribol. Int.* **2016**, *93*, 723–732. [CrossRef]

131. Li, Y. *Multiphase Oil Transport at Complex Micro Geometry*; Massachusetts Institute of Technology: Cambridge, MA, USA, 2011.

132. Li, Y.; Chen, H.; Tian, T. A Deterministic Model for Lubricant Transport within Complex Geometry under Sliding Contact and its Application in the Interaction between the Oil Control Ring and Rough Liner in Internal Combustion Engines. In *Proceedings of the SAE Technical Paper Series*; SAE International: Warrendale, PA, USA, 2010; Volume 1.

133. Chen, H. *Modeling the Lubrication of the Piston Ring Pack of Internal Combustion Engines Using Deterministic Method*; Massachusetts Institute of Technology: Cambridge, MA, USA, 2011.

134. Chen, H.; Li, Y.; Tian, T. A Novel Approach to Model the Lubrication and Friction between the Twin-Land Oil Control Ring and Liner with Consideration of Micro Structure of the Liner Surface Finish in Internal Combustion Engines. In *Proceedings of the SAE Technical Paper Series*; SAE International: Warrendale, PA, USA, 2010; Volume 1.

135. Liao, K.; Liu, Y.; Kim, D.; Urzua, P.; Tian, T. Practical challenges in determining piston ring friction. *Proc. Inst. Mech. Eng. Part J J. Eng. Tribol.* **2013**, *227*, 112–125. [CrossRef]

136. Tomanik, E.; El Mansori, M.; Souza, R.; Profito, F. Effect of waviness and roughness on cylinder liner friction. *Tribol. Int.* **2018**, *120*, 547–555. [CrossRef]

137. Biboulet, N.; Bouassida, H.; Lubrecht, A.A. Cross hatched texture influence on the load carrying capacity of oil control rings. *Tribol. Int.* **2015**, *82*, 12–19. [CrossRef]

138. Noutary, M.-P.; Biboulet, N.; Lubrecht, A.A. Dimple influence on load carrying capacity of parallel surfaces. *Tribol. Int.* **2018**. [CrossRef]

139. Noutary, M.-P.; Biboulet, N.; Lubrecht, A.A. A robust piston ring lubrication solver: Influence of liner groove shape, depth and density. *Tribol. Int.* **2016**, *100*, 35–40. [CrossRef]

140. Minet, C.; Brunetière, N.; Tournerie, B. A Deterministic Mixed Lubrication Model for Mechanical Seals. *J. Tribol.* **2011**, *133*, 042203. [CrossRef]

141. Pavliotis, G.A.; Stuart, A.M. *Multiscale Methods: Averaging and Homogenization*; Texts Applied in Mathematics; Springer: New York, NY, USA, 2008; ISBN 9788578110796.

142. Cioranescu, D.; Donato, P. *An Introduction to Homogenization*; Oxford University Press: Oxford, UK, 1999.

143. Weinan, E. *Principles of Multiscale Modeling*; Cambridge University Press: Cambridge, UK, 2011; ISBN 9781107096547.

144. Horstemeyer, M.F. Multiscale Modeling: A Review. In *Practical Aspects of Computational Chemistry*; Springer: Dordrecht, The Netherlands, 2009; pp. 87–135. ISBN 9789048126866.

145. Vakis, A.I.; Yastrebov, V.A.; Scheibert, J.; Nicola, L.; Dini, D.; Minfray, C.; Almqvist, A.; Paggi, M.; Lee, S.; Limbert, G.; et al. Modeling and simulation in tribology across scales: An overview. *Tribol. Int.* **2018**, *125*, 169–199. [CrossRef]

146. Tzeng, S.T.; Saibel, E. Surface roughness effect on slider bearing lubrication. *ASLE Trans.* **1967**, *10*, 334–348. [CrossRef]

147. Christensen, H. Stochastic Models for Hydrodynamic Lubrication of Rough Surfaces. *Proc. Inst. Mech. Eng.* **1969**, *184*, 1013–1026. [CrossRef]

148. Christensen, H. Some aspects of the functional influence of surface roughness in lubrication. *Wear* **1971**, *17*, 149–162. [CrossRef]

149. Christensen, H. A Theory of Mixed Lubrication. *Proc. Inst. Mech. Eng.* **1972**, *186*, 421–430. [CrossRef]

150. Chow, L.S.H.; Cheng, H.S. The Effect of Surface Roughness on the Average Film Thickness Between Lubricated Rollers. *J. Lubr. Technol.* **1976**, *98*, 117. [CrossRef]
151. Patir, N.; Cheng, H.S. Application of Average Flow Model to Lubrication Between Rough Sliding Surfaces. *J. Lubr. Technol.* **1979**, *101*, 220. [CrossRef]
152. Patir, N.; Cheng, H.S. An Average Flow Model for Determining Effects of Three-Dimensional Roughness on Partial Hydrodynamic Lubrication. *J. Lubr. Technol.* **1978**, *100*, 12. [CrossRef]
153. Elrod, H.G. A General Theory for Laminar Lubrication With Reynolds Roughness. *J. Lubr. Technol.* **1979**, *101*, 8. [CrossRef]
154. Tripp, J.H. Surface Roughness Effects in Hydrodynamic Lubrication: The Flow Factor Method. *J. Lubr. Technol.* **1983**, *105*, 458. [CrossRef]
155. de Kraker, A.; van Ostayen, R.A.J.; Rixen, D.J. Development of a texture averaged Reynolds equation. *Tribol. Int.* **2010**, *43*, 2100–2109. [CrossRef]
156. de Kraker, A.; van Ostayen, R.A.J.; van Beek, A.; Rixen, D.J. A Multiscale Method Modeling Surface Texture Effects. *J. Tribol.* **2007**, *129*, 221. [CrossRef]
157. Jocsak, J.; Li, Y.; Tian, T.; Wong, V.W. Analyzing the Effects of Three-Dimensional Cylinder Liner Surface Texture on Ring-Pack Performance with a Focus on Honing Groove Cross-Hatch Angle. In Proceedings of the ASME 2005 Internal Combustion Engine Division Fall Technical Conference (ICEF2005), Ottawa, ON, Canada, 11–14 September 2005; pp. 621–632.
158. Leighton, M.; Rahmani, R.; Rahnejat, H. Surface-specific flow factors for prediction of friction of cross-hatched surfaces. *Surf. Topogr. Metrol. Prop.* **2016**, *4*, 025002. [CrossRef]
159. Bartel, D.; Bobach, L.; Illner, T.; Deters, L. Simulating transient wear characteristics of journal bearings subjected to mixed friction. *Proc. Inst. Mech. Eng. Part J J. Eng. Tribol.* **2012**, *226*, 1095–1108. [CrossRef]
160. Kim, T.W.; Cho, Y.J. The Flow Factors Considering the Elastic Deformation for the Rough Surface with a Non-Gaussian Height Distribution. *Tribol. Trans.* **2008**, *51*, 213–220. [CrossRef]
161. Peeken, H.J.; Knoll, G.; Rienäcker, A.; Lang, J.; Schönen, R. On the Numerical Determination of Flow Factors. *J. Tribol.* **1997**, *119*, 259. [CrossRef]
162. Pusterhofer, M.; Bergmann, P.; Summer, F.; Grün, F.; Brand, C. A Novel Approach for Modeling Surface Effects in Hydrodynamic Lubrication. *Lubricants* **2018**, *6*, 27. [CrossRef]
163. Dobrica, M.B.; Fillon, M.; Maspeyrot, P. Mixed Elastohydrodynamic Lubrication in a Partial Journal Bearing—Comparison Between Deterministic and Stochastic Models. *J. Tribol.* **2006**, *128*, 778. [CrossRef]
164. Bayada, G.; Chambat, M. New Models in the Theory of the Hydrodynamic Lubrication of Rough Surfaces. *J. Tribol.* **1988**, *110*, 402. [CrossRef]
165. Bayada, G.; Cid, B.; Vázquez, C. Two-scale homogenization study of a Reynolds-Rod elastohydrodynamic model. *Math. Model. Methods Appl. Sci.* **2003**, *13*, 259–293. [CrossRef]
166. Bayada, G.; Faure, J.B. A Double Scale Analysis Approach of the Reynolds Roughness Comments and Application to the Journal Bearing. *J. Tribol.* **1989**, *111*, 323. [CrossRef]
167. Bayada, G.; Martin, S.; Vázquez, C. Two-scale homogenization of a hydrodynamic Elrod–Adams model. *Asymptot. Anal.* **2005**, *44*, 75–110.
168. Bayada, G.; Martin, S.; Vázquez, C. An Average Flow Model of the Reynolds Roughness Including a Mass-Flow Preserving Cavitation Model. *J. Tribol.* **2005**, *127*, 793. [CrossRef]
169. Jai, M. Homogenization and two-scale convergence of the compressible Reynolds lubrication equation modelling the flying characteristics of a rough magnetic head over a rough rigid-disk surface. *ESAIM Math. Model. Numer. Anal.* **1995**, *29*, 199–233. [CrossRef]
170. Jai, M.; Bou-Saïd, B. A Comparison of Homogenization and Averaging Techniques for the Treatment of Roughness in Slip-Flow-Modified Reynolds Equation. *J. Tribol.* **2002**, *124*, 327. [CrossRef]
171. Kane, M.; Bou-Saïd, B. Comparison of Homogenization and Direct Techniques for the Treatment of Roughness in Incompressible Lubrication. *J. Tribol.* **2004**, *126*, 733. [CrossRef]
172. Buscaglia, G.C.; Jai, M. Homogenization of the Generalized Reynolds Equation for Ultra-Thin Gas Films and Its Resolution by FEM. *J. Tribol.* **2004**, *126*, 547. [CrossRef]
173. Buscaglia, G.; Ciuperca, I.; Jai, M. Homogenization of the transient Reynolds equation. *Asymptot. Anal.* **2002**, *32*, 131–152.
174. Buscaglia, G.C.; Jai, M. A new numerical scheme for non uniform homogenized problems: Application to the non linear Reynolds compressible equation. *Math. Probl. Eng.* **2001**, *7*, 355–378. [CrossRef]

175. Prat, M.; Plouraboué, F.; Letalleur, N. Averaged Reynolds equation for flows between rough surfaces in sliding motion. *Transp. Porous Media* **2002**, *48*, 291–313. [CrossRef]

176. Almqvist, A.; Dasht, J. The homogenization process of the Reynolds equation describing compressible liquid flow. *Tribol. Int.* **2006**, *39*, 994–1002. [CrossRef]

177. Almqvist, A.; Essel, E.K.; Fabricius, J.; Wall, P. Reiterated homogenization applied in hydrodynamic lubrication. *Proc. Inst. Mech. Eng. Part J J. Eng. Tribol.* **2008**, *222*, 827–841. [CrossRef]

178. Almqvist, A.; Essel, E.K.; Persson, L.-E.; Wall, P. Homogenization of the unstationary incompressible Reynolds equation. *Tribol. Int.* **2007**, *40*, 1344–1350. [CrossRef]

179. Almqvist, A. Homogenization of the Reynolds Equation Governing Hydrodynamic Flow in a Rotating Device. *J. Tribol.* **2011**, *133*, 021705. [CrossRef]

180. Almqvist, A.; Fabricius, J.; Spencer, A.; Wall, P. Similarities and Differences Between the Flow Factor Method by Patir and Cheng and Homogenization. *J. Tribol.* **2011**, *133*, 031702. [CrossRef]

181. Almqvist, A.; Lukkassen, D.; Meidell, A.; Wall, P. New concepts of homogenization applied in rough surface hydrodynamic lubrication. *Int. J. Eng. Sci.* **2007**, *45*, 139–154. [CrossRef]

182. Spencer, A.; Almqvist, A.; Larsson, R. A semi-deterministic texture-roughness model of the piston ring–cylinder liner contact. *Proc. Inst. Mech. Eng. Part J J. Eng. Tribol.* **2011**, *225*, 325–333. [CrossRef]

183. Söderfjäll, M.; Almqvist, A.; Larsson, R. A model for twin land oil control rings. *Tribol. Int.* **2016**, *95*, 475–482. [CrossRef]

184. Sahlin, F.; Larsson, R.; Almqvist, A.; Lugt, P.M.; Marklund, P. A mixed lubrication model incorporating measured surface topography. Part 1: Theory of flow factors. *Proc. Inst. Mech. Eng. Part J J. Eng. Tribol.* **2010**, *224*, 335–351. [CrossRef]

185. Sahlin, F.; Larsson, R.; Marklund, P.; Almqvist, A.; Lugt, P.M. A mixed lubrication model incorporating measured surface topography. Part 2: Roughness treatment, model validation, and simulation. *Proc. Inst. Mech. Eng. Part J J. Eng. Tribol.* **2010**, *224*, 353–365. [CrossRef]

186. Hu, Y.; Meng, X.; Xie, Y. A computationally efficient mass-conservation-based, two-scale approach to modeling cylinder liner topography changes during running-in. *Wear* **2017**, *386*, 139–156. [CrossRef]

187. Hu, Y.; Meng, X.; Xie, Y. A new efficient flow continuity lubrication model for the piston ring-pack with consideration of oil storage of the cross-hatched texture. *Tribol. Int.* **2018**, *119*, 443–463. [CrossRef]

188. Rom, M.; Muller, S. A Reduced Basis Method for the Homogenized Reynolds Equation Applied to Textured Surfaces. *Commun. Comput. Phys.* **2018**, *24*, 481–509. [CrossRef]

189. Scaraggi, M.; Carbone, G.; Dini, D. Lubrication in soft rough contacts: A novel homogenized approach. Part II—Discussion. *Soft Matter* **2011**, *7*, 10407. [CrossRef]

190. Scaraggi, M.; Carbone, G.; Persson, B.N.J.; Dini, D. Lubrication in soft rough contacts: A novel homogenized approach. Part I-Theory. *Soft Matter* **2011**, *7*, 10395. [CrossRef]

191. Scaraggi, M. Optimal Textures for Increasing the Load Support in a Thrust Bearing Pad Geometry. *Tribol. Lett.* **2014**, *53*, 127–143. [CrossRef]

192. Scaraggi, M. Textured surface hydrodynamic lubrication: Discussion. *Tribol. Lett.* **2012**, *48*, 375–391. [CrossRef]

193. Scaraggi, M. Lubrication of textured surfaces: A general theory for flow and shear stress factors. *Phys. Rev. E* **2012**, *86*, 026314. [CrossRef]

194. Gao, L.; Hewson, R. A Multiscale Framework for EHL and Micro-EHL. *Tribol. Trans.* **2012**, *55*, 713–722. [CrossRef]

195. Gao, L.; de Boer, G.; Hewson, R. The role of micro-cavitation on EHL: A study using a multiscale mass conserving approach. *Tribol. Int.* **2015**, *90*, 324–331. [CrossRef]

196. Lahmar, M.; Bou-Saïd, B.; Tichy, J. The homogenization method of roughness analysis in turbulent lubrication. *Proc. Inst. Mech. Eng. Part J J. Eng. Tribol.* **2013**, *227*, 1090–1100. [CrossRef]

197. Waseem, A.; Temizer, İ.; Kato, J.; Terada, K. Homogenization-based design of surface textures in hydrodynamic lubrication. *Int. J. Numer. Methods Eng.* **2016**, *108*, 1427–1450. [CrossRef]

198. Waseem, A.; Temizer, İ.; Kato, J.; Terada, K. Micro-texture design and optimization in hydrodynamic lubrication via two-scale analysis. *Struct. Multidiscip. Optim.* **2017**, *56*, 227–248. [CrossRef]

199. Yildiran, I.N.; Temizer, I.; Çetin, B. Homogenization in Hydrodynamic Lubrication: Microscopic Regimes and Re-Entrant Textures. *J. Tribol.* **2017**, *140*, 011701. [CrossRef]

200. Gu, C.; Meng, X.; Xie, Y.; Zhang, D. Mixed lubrication problems in the presence of textures: An efficient solution to the cavitation problem with consideration of roughness effects. *Tribol. Int.* **2016**, *103*, 516–528. [CrossRef]

201. Ma, C.; Duan, Y.; Yu, B.; Sun, J.; Tu, Q. The comprehensive effect of surface texture and roughness under hydrodynamic and mixed lubrication conditions. *Proc. Inst. Mech. Eng. Part J J. Eng. Tribol.* **2017**, *231*, 1307–1319. [CrossRef]

202. Xie, Y.; Li, Y.; Suo, S.; Liu, X.; Li, J.; Wang, Y. A mass-conservative average flow model based on finite element method for complex textured surfaces. *Sci. China Physics Mech. Astron.* **2013**, *56*, 1909–1919. [CrossRef]

203. Profito, F.J.; Vladescu, S.-C.; Reddyhoff, T.; Dini, D. Experimental Validation of a Mixed-Lubrication Regime Model for Textured Piston-Ring-Liner Contacts. *Mater. Perform. Charact.* **2017**, *6*, 112–129. [CrossRef]

204. Zhang, H.; Hua, M.; Dong, G.; Zhang, D.; Chin, K.-S. A mixed lubrication model for studying tribological behaviors of surface texturing. *Tribol. Int.* **2016**, *93*, 583–592. [CrossRef]

205. Meng, X.; Gu, C.; Xie, Y. Elasto-plastic contact of rough surfaces: A mixed-lubrication model for the textured surface analysis. *Meccanica* **2017**, *52*, 1541–1559. [CrossRef]

206. Qiu, Y.; Khonsari, M.M. Performance Analysis of Full-Film Textured Surfaces With Consideration of Roughness Effects. *J. Tribol.* **2011**, *133*, 021704. [CrossRef]

207. Brunetière, N.; Tournerie, B. Numerical analysis of a surface-textured mechanical seal operating in mixed lubrication regime. *Tribol. Int.* **2012**, *49*, 80–89. [CrossRef]

208. Venner, C.H.; Lubrecht, A.A. (Eds.) *Multi-Level Methods in Lubrication*; Elsevier: Amsterdam, The Netherlands, 2000.

209. Zhao, J.; Vollebregt, E.A.H.; Oosterlee, C.W. A full multigrid method for linear complementary problems arising from elastic normal contact problems. *Math. Model. Anal.* **2014**, *19*, 216–240. [CrossRef]

210. Zhang, B.; Boffy, H.; Venner, C.H. Multigrid solution of 2D and 3D stress fields in contact mechanics of anisotropic inhomogeneous materials. *Tribol. Int.* **2019**. [CrossRef]

211. Checo, H.M.; Jai, M.; Buscaglia, G.C. An Improved Fixed-Point Algorithm to Solve the Lubrication Problem with Cavitation. *Mecánica Comput.* **2016**, *34*, 1973–1987.

212. Boffy, H.; Venner, C.H. Multigrid solution of the 3D stress field in strongly heterogeneous materials. *Tribol. Int.* **2014**, *74*, 121–129. [CrossRef]

213. Goodyer, C.E.; Berzins, M. Parallelization and scalability issues of a multilevel elastohydrodynamic lubrication solver. *Concurr. Comput. Pract. Exp.* **2007**, *19*, 369–396. [CrossRef]

214. Li, L.; Bogy, D.B. A Local Adaptive Multigrid Control Volume Method for the Air Bearing Problem in Hard Disk Drives. *J. Tribol.* **2013**, *135*, 032001. [CrossRef]

215. Lu, H.; Berzins, M.; Goodyer, C.E.; Jimack, P.K.; Walkley, M. Adaptive high-order finite element solution of transient elastohydrodynamic lubrication problems. *Proc. Inst. Mech. Eng. Part J J. Eng. Tribol.* **2006**, *220*, 215–225. [CrossRef]

216. Nilsson, B.; Hansbo, P. Adaptive finite element methods for hydrodynamic lubrication with cavitation. *Int. J. Numer. Methods Eng.* **2007**, *72*, 1584–1604. [CrossRef]

217. Lu, H.; Berzins, M.; Goodyer, C.E.; Jimack, P.K. Adaptive high-order Discontinuous Galerkin solution of elastohydrodynamic lubrication point contact problems. *Adv. Eng. Softw.* **2012**, *45*, 313–324. [CrossRef]

218. Ahmed, S.; Goodyer, C.E.; Jimack, P.K. An adaptive finite element procedure for fully-coupled point contact elastohydrodynamic lubrication problems. *Comput. Methods Appl. Mech. Eng.* **2014**, *282*, 1–21. [CrossRef]

219. Babuška, I.; Osborn, J.E. Generalized Finite Element Methods: Their Performance and Their Relation to Mixed Methods. *SIAM J. Numer. Anal.* **1983**, *20*, 510–536. [CrossRef]

220. Efendiev, Y.; Hou, T. *Multiscale Finite Element Methods*; Springer: New York, NY, USA, 2009; ISBN 978-0-387-09495-3.

221. Moyner, O.; Lie, K.-A. The Multiscale Finite Volume Method on Unstructured Grids. In Proceedings of the SPE Reservoir Simulation Symposium, The Woodlands, TX, USA, 18–20 February 2013; Society of Petroleum Engineers: Richardson, TX, USA, 2013.

222. Jiang, L.; Mishev, I.D. Mixed Multiscale Finite Volume Methods for Elliptic Problems in Two-Phase Flow Simulations. *Commun. Comput. Phys.* **2012**, *11*, 19–47. [CrossRef]

223. Jenny, P.; Lee, S.; Tchelepi, H. Multi-scale finite-volume method for elliptic problems in subsurface flow simulation. *J. Comput. Phys.* **2003**, *187*, 47–67. [CrossRef]

224. Weinan, E.; Engquist, B.; Li, X.; Ren, W.; Vanden-Eijnden, E. Heterogeneous Multiscale Methods: A Review. *Commun. Comput. Phys.* **2007**, *2*, 367–450.

225. Steinhauser, M.O. *Computational Multiscale Modeling of Fluids and Solids*; Springer: Berlin/Heidelberg, Germany, 2017; ISBN 978-3-662-53222-5.

226. Pei, S.; Ma, S.; Xu, H.; Wang, F.; Zhang, Y. A multiscale method of modeling surface texture in hydrodynamic regime. *Tribol. Int.* **2011**, *44*, 1810–1818. [CrossRef]

227. Pei, S.; Xu, H.; Shi, F. A deterministic multiscale computation method for rough surface lubrication. *Tribol. Int.* **2016**, *94*, 502–508. [CrossRef]

228. Pei, S.; Xu, H.; Yun, M.; Shi, F.; Hong, J. Effects of surface texture on the lubrication performance of the floating ring bearing. *Tribol. Int.* **2016**, *102*, 143–153. [CrossRef]

229. Nyemeck, A.P.; Brunetière, N.; Tournerie, B. A Multiscale Approach to the Mixed Lubrication Regime: Application to Mechanical Seals. *Tribol. Lett.* **2012**, *47*, 417–429. [CrossRef]

230. Nyemeck, A.P.; Brunetiere, N.; Tournerie, B. A Mixed Thermoelastohydrodynamic Lubrication Analysis of Mechanical Face Seals by a Multiscale Approach. *Tribol. Trans.* **2015**, *58*, 836–848. [CrossRef]

231. Pérez-Ràfols, F.; Larsson, R.; Lundström, S.; Wall, P.; Almqvist, A. A stochastic two-scale model for pressure-driven flow between rough surfaces. *Proc. R. Soc. A Math. Phys. Eng. Sci.* **2016**, *472*, 20160069. [CrossRef] [PubMed]

232. De Boer, G.N.; Hewson, R.W.; Thompson, H.M.; Gao, L.; Toropov, V.V. Two-scale EHL: Three-dimensional topography in tilted-pad bearings. *Tribol. Int.* **2014**, *79*, 111–125. [CrossRef]

233. Brunetière, N.; Francisco, A. Multiscale Modeling Applied to the Hydrodynamic Lubrication of Rough Surfaces for Computation Time Reduction. *Lubricants* **2018**, *6*, 83. [CrossRef]

234. Costagliola, G.; Bosia, F.; Pugno, N.M. Hierarchical Spring-Block Model for Multiscale Friction Problems. *ACS Biomater. Sci. Eng.* **2017**, *3*, 2845–2852. [CrossRef]

235. Costagliola, G.; Bosia, F.; Pugno, N.M. A 2-D model for friction of complex anisotropic surfaces. *J. Mech. Phys. Solids* **2018**, *112*, 50–65. [CrossRef]

236. Costagliola, G.; Bosia, F.; Pugno, N.M. Static and dynamic friction of hierarchical surfaces. *Phys. Rev. E* **2016**, *94*, 063003. [CrossRef]

237. Costagliola, G.; Bosia, F.; Pugno, N.M. Tuning friction with composite hierarchical surfaces. *Tribol. Int.* **2017**, *115*, 261–267. [CrossRef]

238. Wu, X.; Ge, Z.; Niu, H.; Ruan, J.; Zhang, J. Edge effect factor affecting the tribological properties in water of protrusion surface textures on stainless steel. *Biosurface Biotribol.* **2018**, *4*, 46–49. [CrossRef]

239. Joshi, G.S.; Putignano, C.; Gaudiuso, C.; Stark, T.; Kiedrowski, T.; Ancona, A.; Carbone, G. Effects of the micro surface texturing in lubricated non-conformal point contacts. *Tribol. Int.* **2018**, *127*, 296–301. [CrossRef]

240. Wang, X.; Wang, J.; Zhang, B.; Huang, W. Design principles for the area density of dimple patterns. *Proc. Inst. Mech. Eng. Part J J. Eng. Tribol.* **2014**, *229*, 538–546. [CrossRef]

241. Greiner, C.; Merz, T.; Braun, D.; Codrignani, A.; Magagnato, F. Optimum dimple diameter for friction reduction with laser surface texturing: The effect of velocity gradient. *Surf. Topogr. Metrol. Prop.* **2015**, *3*, 044001. [CrossRef]

242. Sahlin, F.; Glavatskih, S.B.; Almqvist, T.; Larsson, R. Two-Dimensional CFD-Analysis of Micro-Patterned Surfaces in Hydrodynamic Lubrication. *J. Tribol.* **2005**, *127*, 96. [CrossRef]

© 2019 by the authors. Licensee MDPI, Basel, Switzerland. This article is an open access article distributed under the terms and conditions of the Creative Commons Attribution (CC BY) license (http://creativecommons.org/licenses/by/4.0/).

 *lubricants*

Article

High-Rate Laser Surface Texturing for Advanced Tribological Functionality

Jörg Schille [1,*], Lutz Schneider [1], Stefan Mauersberger [1], Sylvia Szokup [2], Sören Höhn [2], Johannes Pötschke [2], Friedemann Reiß [3], Erhard Leidich [3] and Udo Löschner [1]

[1] Laserinstitut Hochschule Mittweida, University of Applied Sciences Mittweida, Technikumplatz 17, 09648 Mittweida, Germany; schneide@hs-mittweida.de (L.S.); mauersbe@hs-mittweida.de (S.M.); loeschne@hs-mittweida.de (U.L.)

[2] Fraunhofer Institute for Ceramic Technologies and Systems IKTS, Winterbergstrasse 28, 01277 Dresden, Germany; sylvia.szokup@ikts.fraunhofer.de (S.S.); soeren.hoehn@ikts.fraunhofer.de (S.H.); johannes.poetschke@ikts.fraunhofer.de (J.P.)

[3] Institute of Design Engineering and Drive Technology IKAT, Chemnitz University of Technology, Reichenhainer Straße 70, 09126 Chemnitz, Germany; friedemann.reiss@mb.tu-chemnitz.de (F.R.); erhard.leidich@mb.tu-chemnitz.de (E.L.)

* Correspondence: schille@hs-mittweida.de; Tel.: +49-3727-581838

Received: 27 February 2020; Accepted: 15 March 2020; Published: 20 March 2020

Abstract: This article features with the enhancement of the static coefficient of friction by laser texturing the contact surfaces of tribological systems tested under dry friction conditions. The high-rate laser technology was applied for surface texturing at unprecedented processing rates, namely using powerful ultrashort pulses lasers in combination with ultrafast polygon-mirror based scan systems. The laser textured surfaces were analyzed by ion beam slope cutting and Raman measurements, showing a crystallographic disordering of the produced microscopic surface features. The laser induced self-organizing periodic surface structures as well as deterministic surface textures were tested regarding their tribological behavior. The highest static coefficient of friction was found of μ_{20} = 0.68 for a laser textured cross pattern that is 126% higher than for a fine grinded reference contact system. The line pattern was textured on a shaft-hub connection where the static coefficient of friction increased up to 75% that demonstrates the high potential of the technology for real-world applications.

Keywords: high-rate; laser texturing; high-power; laser; LIPSS; tribology; coefficient of friction

---

1. Introduction

Laser surface texturing has been established as an effective method to tailor and control the micro-topographic but also physiochemical properties of surfaces. During the last decades, a wide variety of multi-scale surface textures ranging from nanometer to micrometer feature dimensions have been reported that can be produced using pulsed and ultrashort pulsed lasers in surface processing. There is also an increasing trend for applying such microscopic laser textures for functionalization in advanced surface engineering. This is emphasized in Figure 1 on the basis of the SCOPUS abstract and citation database [1], showing a steadily growing number of peer-reviewed journal articles published per year for the search term "Laser surface texturing" (blue bar).

Inspired by nature, a great number of these basic research activities reveal the potential of multi-scale laser textures mimicking natural concepts to be applied for biomimetic surface functionalities, interfaces and products. In this manner, Figure 2 illustrates a variety of characteristic laser textures produced on stainless steel plates: (a) rippled surface texture, (b) multi-scale surface texture and (c) riblets, which can be used, for example, to control optical, wetting or aerodynamic surface behaviors. Thereby, the specific functionality strongly depends on the individual texture feature geometry.

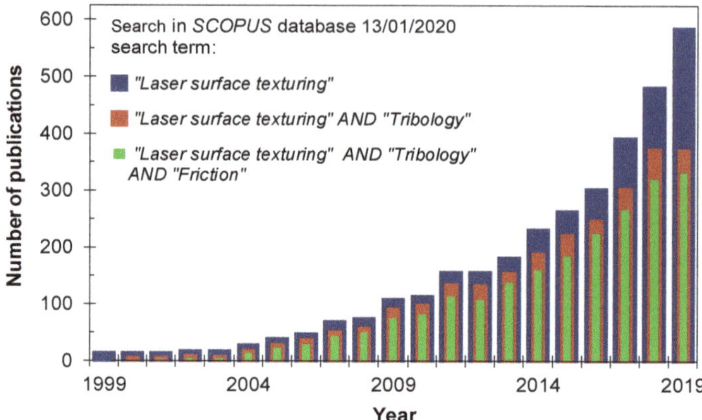

Figure 1. Overview of the number of peer-reviewed journal articles published per year in the SCOPUS abstract and citation database as received for the search term "Laser surface texturing" (blue bar), "Laser surface texturing AND Tribology" (red bar) and "Laser surface texturing AND Tribology AND Friction" (green bar) [1].

Laser surface texturing has also attracted great attention in tribology applications which can be recognized in Figure 1 for the search terms "Laser surface texturing AND Tribology" (red bar) as well as "Laser surface texturing AND Tribology AND Friction" (green bar). The ongoing upward trend of research activities in this field is mainly due to the ability of laser textures to also improve the frictional performance of tribological systems with regard to their efficiency, load capacity, wear resistance and life-time. However, most of these tribological studies report on laser surface texturing for reducing internal torque in the lubricant friction regime and micro-dimpled laser textures were applied to function as micro-hydrodynamic bearings, micro reservoirs for lubricant retention or micro-traps for wear particles of sliding systems, for example, in seals, piston rings and thrust bearings [2–8]. Recent studies also report on the reduction of the coefficient of friction (COF) for ripples tested under lubrication friction conditions [9,10]. However, only little research has been spent so far on the increase of friction forces in dry friction contacts. Therein, it was demonstrated that the static COF of frictional engaged connections (cam, press fittings) can be increased by laser fabricated molten and re-solidified bulged microscopic structures [11,12].

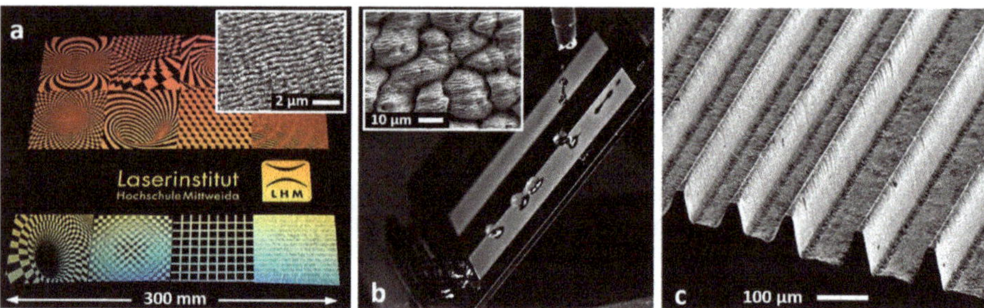

Figure 2. Laser made microscopic surface textures for mimicking natural functionalities on stainless steel: (**a**) rippled surface texture for optical effects; (**b**) multi-scale features imitating the hydrophobic and self-cleaning behavior of the lotus leaf and (**c**) riblets for skin friction drag reduction in turbulent flows inspired by shark's skin micro structure.

In another recent approach, laser made dimples were identified being valuable for COF enhancement in dry friction contacts [13–15]. As a specialty, the microscopic dimples produced on the surface were surrounded by prominent molten and re-solidified walls potentially enhancing the mechanical interconnection along with higher adhesive friction forces for the joint frictional contact. However, the limited processing speed and throughput of these basically investigated processes are actually the major drawback to bring this promising technology to industrial production and manufacturing.

2. High-Rate Laser Processing Technology

The high-rate laser technology was developed at the Laserinstitut Hochschule Mittweida [16–18] with the goal to speed up the achievable processing speeds and throughputs in laser micro machining. In fact, high-rate machining combines powerful laser systems supplying hundreds to thousands Watts average laser powers and ultrafast beam deflection methods. Therewith, the ultrafast laser beam movement is essential for the efficient upscaling of the processing rates when using such unprecedented high laser powers. This is due to the fact that only a few micro joules optical energy at low fluence is needed for most-efficient material removal [19–21]. In consequence, the optimum method for upscaling the processing rate will be increasing the pulse repetition frequency instead of using higher pulse energies. In this case, the laser beam should be deflected at ultrafast speeds thus to avoid strong pulse-overlap effects, for example, laser beam shielding and heat accumulation that will detrimentally affect the machining process [18,22–24]. Therefore, the conventional (galvanometer mirror based) scan systems cannot reach the necessary fast scan speeds of hundreds of meters per second and above. Alternatively, in the last decade rotating cylinders, electro- and acousto-optical deflectors, resonant scanners or fast rotating polygons have been successfully tested for high-speed laser micro machining [25–28]. However, the multi-facet (polygon) mirror based biaxial scan technique seems to be the most promising approach for ultrafast and flexible two-dimensional (2-D) laser beam raster scanning.

Figure 3a depicts a schematic view of the unique polygon scanner architecture engineered at the Laserinstitut Hochschule Mittweida [18,29]. The fast rotating double-reflecting polygon mirror is used to achieve the ultrafast laser beam moving speed along the fast axis. For raster scanning, the laser beam is shifted along the slow axis by the galvanometer scanner when the beam flips from one facet to the other. The moving laser beam can be focused by standard f-theta lenses by what focus spot sizes in the range between 30 μm and 100 μm will be reached that depends on the focal length of the adapted objective. The maximum achievable laser beam moving speed is also determined by the optical scanner configuration, for example, up to 1000 m/s for the 420 mm lens. Therefore, representative laser textures produced on stainless steel metal sheets at ultrafast 950 m/s scan speed can be seen in Figure 3b,c.

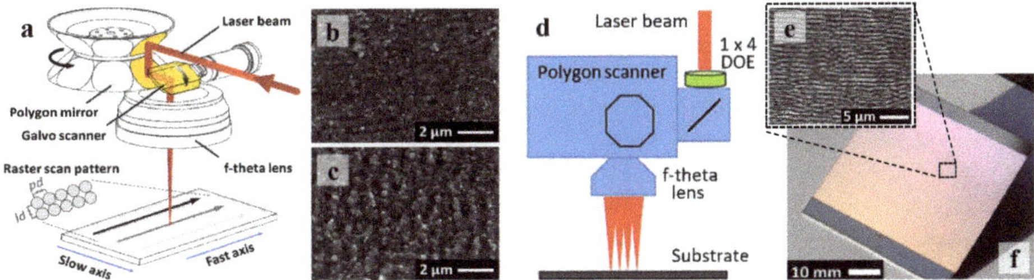

Figure 3. Schematic view of the (**a**) polygon scanner architecture and (**d**) multi-beam high-speed scan technology, *pd* and *ld* represent the spatial pulse distance and line distance of a raster-scan pattern; nano-featured surface textures produced on stainless steel by applying a single pass raster-scan over the substrate at 950 m/s laser beam moving speed and (**b**) 20 μm or (**c**) 10 μm line distance; SEM micrograph (**e**) and optical microscope image (**f**) of a ripple textured stainless steel surface; the effective area processing rate for the given 40×50 mm^2 surface texture was 0.13 m^2/min.

The nano-featured surface textures in Figure 3b were made by a single pass raster-scan over the substrate surface. Therefore, the picosecond laser pulses were placed one after another in fast axis direction by irradiating 263 W average laser power. The pulse distance *pd* was set of about the focus spot diameter and with little overlap in orthogonal direction (slow axis) as the line distance *ld* between the raster scanned lines was half of the spot diameter. In initial studies, such kind of weak surface roughness is considered as inter-pulse feedback mechanism for the self-organizing formation of regular surface patterns as induced upon multi-pulse laser irradiations [30–32]. This material response is also confirmed by the SEM micrograph presented in Figure 3c, showing more prominent surface features originating from the higher accumulated irradiation dose that was due to higher pulse overlap at smaller line distance.

A further technical development is schematically illustrated in Figure 3d, combining the polygon scanner and a four-spot diffractive optical element (DOE) for ultrafast and parallel surface texturing. The feasibility of this groundbreaking technology is exemplified in Figure 3f by the homogeneous ripple texture produced on a stainless steel metal sheet using a femtosecond laser beam of 416 W average power at 560 m/s scan speed. The laser beam was scanned three times over the sample in order to achieve a well-shaped ripple texture, as can be seen in Figure 3e.

The overall processing rate that will be achieved applying high-rate laser technologies in surface texturing can easily be estimated by the following calculations. In theory, the maximum area processing rate APR_{max} for a complete single pass raster-scan pattern over the substrate surface results from the product of the applied scan speed v_S and the chosen line distance *ld* according to Equation (1):

$$APR_{max} = v_S \cdot ld \tag{1}$$

$$APR_{eff} = v_S \cdot ld \cdot \eta \tag{2}$$

From practical point of view, the utilization rate η of the polygon scan system must seriously be taken into consideration to calculate the effective area processing rate APR_{eff} on the substrate. This is described in Equation (2), where the utilization rate is the ratio between the theoretical and effective processible scan length on each polygon mirror facet. Hence, the utilization rate is restricted by the technical polygon scanner architecture and the implemented optical components ranging of 49% or rather 40% for the applied scanner configurations. Table 1 provides an overview of the area processing rates that can be reached for producing the laser textures shown in Figure 3. A maximum area processing rate of 1.14 m²/min or 3.76 m²/min could theoretically be achieved with the applied 263 W picosecond or 416 W femtosecond average laser power. The given effective area processing rates will only be reached when the full length of the accessible scan field is used. This was effectively 0.56 m²/min for the picosecond single spot and 1.51 m²/min for the femtosecond multi spot laser procedure. A further decrease of the area processing rate can be seen for the laser made ripple texture of Figure 3e. The shown highly uniform ripples were produced by three repeated overscans applied on the stainless steel substrate while the effective *APR* reduced to be 0.50 m²/min. Moreover, as the length of the rippled surface texture in Figure 3f was almost 1/4 of the accessible length of the scanning field the duty cycle decreased to 10%. Thus, the effective *APR* was as low as 0.13 m²/min for this specific machining example.

Table 1. Maximum and effective area processing rates *APR* estimated for different processing conditions and by considering a single pass raster-scan procedure applied on the substrate surface at maximal accessible scan length.

| Texture | P_{av} | PRF | H_0 | v_S | pd | ld | APR_{max} | APR_{eff} |
	W	MHz	J/cm^2	m/s	µm	µm	m^2/min	m^2/min
Figure 3b	263	25	1.45	950	38	20	1.14	0.56
Figure 3c	263	25	1.45	950	38	10	0.57	0.28
Figure 3f	416	40 [1]	0.50	560	14	28	3.76	1.51 [1]/0.50 [2]/0.13 [3]

APR_{eff} for [1] 40% facet utilization rate, [2] 3 scan passes and [3] 40 mm field length.

3. Static Friction Analysis Method

A standard analysis procedure developed by the IKAT institute at Chemnitz University of Technology [33,34] was applied for the evaluation of the tribological behavior of the laser textured surfaces. Therewith, both the static coefficient of friction (COF) and the type of frictional performance could be assessed from the recorded slipping curves. The functional principle of this analysis method is shown in Figure 4, displaying the implemented friction test bench assembly (left) and the cylindrical test specimen with a specially defined geometry (top right). The inner and the outer diameter of the ring-shaped contact surface area on top of the cylinder were D_{in} = 15 mm and D_{out} = 30 mm, respectively, and the cylinder height was 65 mm. Two identical cylinders were clamped in the test bench and coaxially fastened with the normal force F_N to achieve a nominal surface pressure of 100 MPa that is a typical contact pressure in tribological systems. Based on Equation (3), the effective friction diameter D_F of the contact area was determined to be 23.3 mm. By rotating the cylinders against each other, friction shear stresses develop in the joint resulting from the frictional torque T_F on the contact area. The frictional torque was measured by the strain gauges as a function of the torsion angle φ, which is indicated in Figure 4.

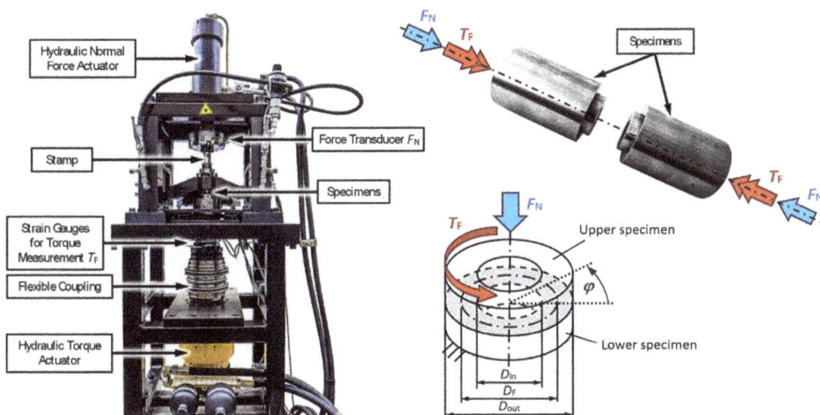

Figure 4. Test bench assembly and standard procedure developed at IKAT (Chemnitz University of Technology) to analyze the performance of frictional contacts [33,34].

The recorded slipping curves are schematically shown in Figure 5 by plotting the torque measurements as function of the relative displacement s_F of the contact areas. In this depiction, the relative displacement s_F calculates according to Equation (4) and by taking into account the torsion angle and the effective friction diameter of the ring-shaped contact surface. As further can be seen in the schematic (bottom, left), three different types of characteristic friction performance could be recognized for the tribological contacts. These different types are distinguished depending on their typical slipping curve progression [35–37] that is briefly described in Figure 5 (right).

$$D_F = \frac{2}{3} \cdot \frac{D_{out}^3 - D_{in}^3}{D_{out}^2 - D_{in}^2} \tag{3}$$

$$s_F = \varphi \cdot \frac{D_F}{2} \tag{4}$$

$$\mu = \frac{2 \cdot T_F}{F_N \cdot D_F} \tag{5}$$

In Coulomb´s law of friction, the static coefficient of friction μ is defined as the constant of proportionality for the frictional contact of two bodies under the action of normal and sheer forces. The static COF states the transition point from sticking/adhesion to sliding and can be understood as the maximum boundary value of the load capacity of a contact which does not yet trigger a relative displacement of the two active partners towards each other. For the specified friction characteristic of Type A, this boundary value is well defined by the first local maximum of the slipping curve. Though, the boundary value is less clear for the friction Types B and C that can be seen by the corresponding slipping curves presented in the schematic of Figure 5 (bottom, left).

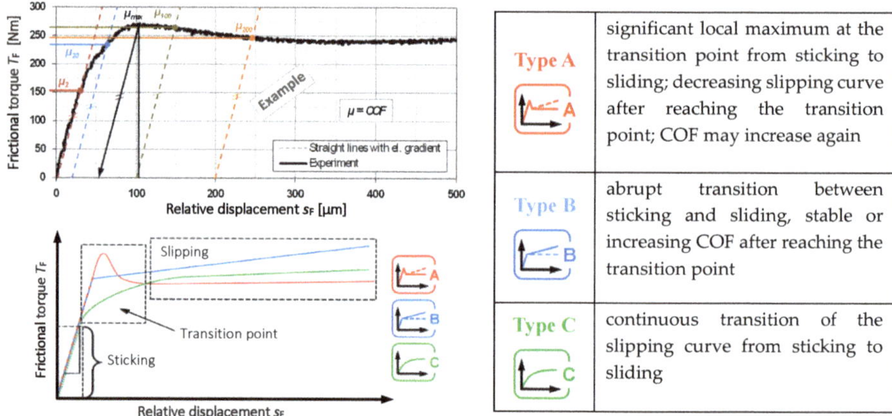

Type A		significant local maximum at the transition point from sticking to sliding; decreasing slipping curve after reaching the transition point; COF may increase again
Type B		abrupt transition between sticking and sliding, stable or increasing COF after reaching the transition point
Type C		continuous transition of the slipping curve from sticking to sliding

Figure 5. Exemplary illustration of the evaluation methods applied to determine the static COF. In the example (top, left), the stiffness lines are indicated by dashed lines to determine the respective COF´s μ_2, μ_{20} etc. Three types of friction can be distinguished depending on the torque progression curve schematically shown bottom, left. A detailed description of the different friction types is provided in the table on the right.

Following the analysis procedure, stiffness lines are included in the plot of the torque vs. relative displacement measurements where their intersection points support the determination of the torque at a specific position along the slipping path. The advantage of this method is to eliminate deformations of the entire tribological system that will affect the nominal value of the torsion angle, such as elastic deformation of the test bench or rather micro-movements in the joint resulting from surface asperities during specimen alignments. In the example of Figure 5 (top, left), the stiffness lines are indicated by the dashed lines. By considering the slope of the slipping curve, the stiffness line is shifted parallel along the abscissa (displacement axis) until its intersection at a desired lateral displacement position, e.g., 2 µm or 20 µm relative displacement between the test specimen contact areas. Then, by taking into account the torque value at the respective intersection of the stiffness line with the slipping curve, the corresponding COFs μ_2, μ_{20}, etc. can be calculated according to Equation (5).

In the following, in order to consider static instead of dynamic friction performance of the laser textured surfaces, the static COFs μ_2 and μ_{20} will be discussed. The static COFs were calculated on the basis of the maximum torque values determined for the torsion angles 0.01° and 0.1° that is equivalent to 2 µm and 20 µm lateral displacements in the frictional contact area. In addition, the maximum value of the COF μ_{max} that could be obtained within the tested range ($\varphi \leq 3°$, $s_F \leq 600$ µm) will be presented. It should be mentioned for the static COF Type A, most commonly the maximum value will be obtained within 20 µm lateral displacement, while for COF Types B and C, the maximum COF value is steadily increasing with increasing relative displacement. In the latter case, where the highest values were obtained at comparably large lateral displacements, the maximum COF´s represent dynamic rather than static frictional performances.

4. Self-Organized Surface Textures for Advanced Tribological Functionality

4.1. Self-Organizing Periodic Surface Features

The development of self-organizing periodic surface structures of microscopic feature size represents a typical phenomenon in laser materials processing. The specific morphology of the originating surface features can be controlled by the irradiation conditions with particular influence of wavelength and polarization state of the laser beam as well as laser peak fluence, number of impinging pulses and respective irradiation dose, the latter quantifying the total optical energy irradiated to the substrates per laser spot area. The great impact of the processing conditions on the characteristic feature properties is exemplified in Figure 6 by a variety of microscopic surface features originating upon femtosecond laser irradiations on mirror-polished stainless steel substrates. Figure 6b illustrates schematically the direct-write laser processing method applied to produce such radial matrixes shown in Figure 6a,c for different laser peak fluence. As indicated in this scheme, the lines 1, 2, 3, … , n to be processed are scanned from center to edge, while the rotating angle α between the individual lines increased steadily from one line to the next. The spatial distance between the pulses within a single scanned line is determined by the chosen pulse repetition frequency and respective scan speed. In Figure 6c, this was set to 200 kHz at 1 m/s and in Figure 6a and 100 kHz and 0.2 m/s, respectively.

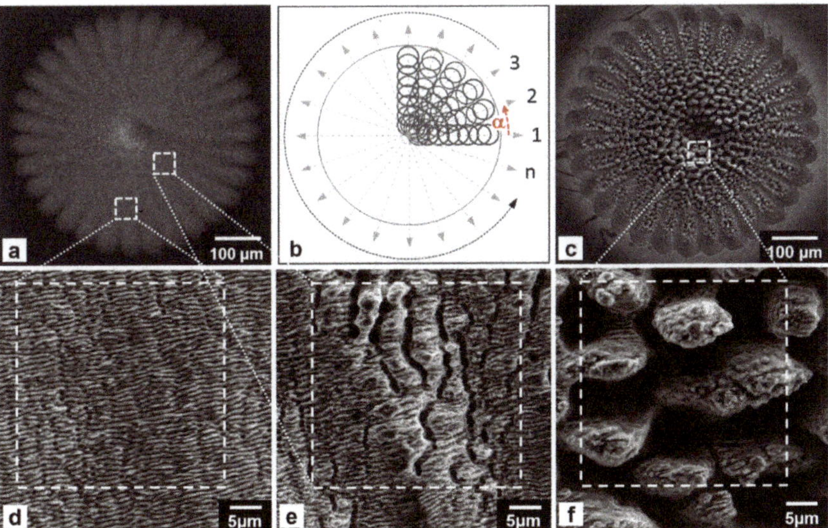

Figure 6. SEM micrographs showing ripple (**a**) and cone-like protrusion (CLP) (**c**) surface textures produced on stainless steel; the direct-write scanning procedure for the radial matrices as used for tightly mapping a variable set of pulse overlaps (**b**); magnified view on low spatial frequency LIPSS (LSFL) (**d**), micro grooves and spikes (**e**) as well as CLPs (**f**) originating at low and high peak fluence pulses of 0.7 J/cm^2 and 3.7 J/cm^2.

In this way, the spatial line distance along the single processed lines enlarges continuously from matrix center to outer edge that is the main advantage of the applied radial matrix method. This in turn, comprehensively overviews the potentially emerging surface features as function of irradiated dose, because the line distance varies from highly overlapping in the center to completely separated pulses at matrix edge. By doing so, a number of complex micro-scale surface morphologies could be identified in Figure 6 that have already been reported for repetitive laser pulse irradiations [38–44]. In fact, the developed self-organizing periodic surface structures can be differentiated as follows:

(i) Ripples, or rather LIPSS, denote laser induced periodic surface structures. The ripples can be distinguished between low spatial frequency LIPSS (LSFL) and high spatial frequency LIPSS (HSFL). As indicated in Figure 6a, LSFLs emerge with spatial periods close to the wavelength of the incident laser beam at comparatively low irradiation dose where pulses of low laser peak fluence, $H_0 = 0.7$ J/cm^2, slightly above ablation threshold at low spatial pulse overlaps were applied. HSFLs originate at lower irradiation dose with the ripple period significantly smaller than the applied laser beam wavelength,

(ii) Undulated surfaces carved with micro-grooves form orthogonal to the ripples at supra-wavelength periods. This kind of surface feature is pointed out in Figure 6e emerging at higher irradiation dose in areas of intermediate line overlaps such as indicated in Figure 6a.

(iii) Micro-spikes start to grow in between the micro-grooves that might be affected by hydrodynamic effects induced by inhomogeneous laser beam absorption at corrugated surfaces [45], as highlighted in Figure 6e.

(iv) Cone-like protrusions (CLPs) originate in regions of highest dose that can be seen in Figure 6c and f for pulses of $H_0 = 3.7$ J/cm^2 impinging at high spatial pulse overlaps.

4.2. Microstructure Characterization

The laser made microscopic surfaces features were investigated regarding their microstructural characteristics in order to assess whether or not the metallographic structure underneath the surface was affected by the laser processing. Therefore, the ion beam slope cutting (IBSC) technique [46,47] was applied with help of the EM TIC 020 Milling System (Leica Microsystems) to prepare the laser textured regions for cross-sectional scanning electron microscope (SEM) and electron back scatter diffraction (EBSD) inspections. The areas of interest were exposed at an angle of 90° to create a slope plane for the representative view on the laser-treated near-surface region. The advantage of the IBSC preparation method over common mechanical preparation techniques was that no additional deformations were introduced to the material [47] and thus the original metallographic structure of the microscopic features was exposed as developed during laser processing.

In the SEM analysis, two detector modes were used for providing different information for the ion beam cut slope planes on the laser processed surfaces. First, the signal detected from the secondary electrons (SE detector mode) is carrying more detailed information from the laser treated surface as it originated from the inelastic interaction between the electron beam and the sample atoms. And second, the signal received from the backscattered electrons (BSE detector mode) resulted from the inelastic interactions between the electron beam and the sample providing in depth information from deeper regions of the laser processed areas. In addition, the EBSD method was applied for a more precise insight on the grain size, crystallographic structure and orientation of the laser affected subsurface regions.

The cross-section SEM micrographs of Figure 7 show two laser textures at different magnifications: LSFL-ripples produced on the steel sample at 1.7 J/cm^2 and 5 scan passes (left, a–e) and CLP formations covering the steel surface that were made with pulses of same laser peak fluence but 20 scan passes (right, f–j). On the one hand, there is a clear difference in the height and period of the developed laser texture features. From Figure 7 left, the height of the LSFL can be estimated of about 0.4 µm and the period is somewhat less than the wavelength of the impinging laser beam of 1.03 µm which is consistent to our previous findings for near-infrared ultrashort laser beams [48]. The height of the CLP structures can be seen in the range between 10 µm and 20 µm and their period of about 10 µm is almost one order of magnitude greater than the ripple period. On the other hand, there is a grain structure variation between the near surface region and the base material thus indicating the laser affected subsurface region. This can clearly be seen in Figure 7c,d as well as in Figure 7h,i where the grain size is much smaller in the molten and re-solidified transition layer than in the bulk metal. According to this, the depth of the laser affected region is approximated by the thickness of the transformed subsurface layer, which is in the range between 2 µm and 5 µm for the ripple textures and from 20 µm up to 30 µm for the CLPs. The formation of such a transition layer is also confirmed in Figure 7e,j showing the results of the EBSD measurements carried out on the laser-treated near-surface region. The re-solidified areas formerly melted by the laser irradiations become visible in the EBSD mappings by slight orientation changes within the larger grains.

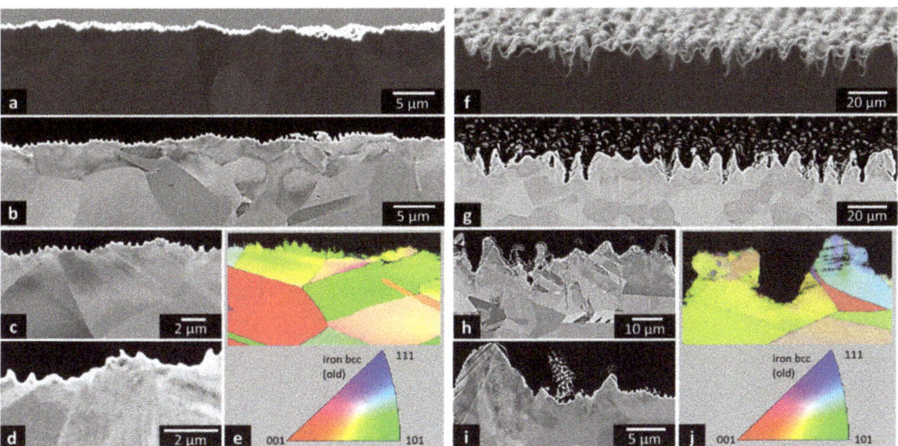

Figure 7. SEM micrographs recorded with the SE (**a,f**) and BSE detector (**b,g**); detailed micrographs (**c,d,h,i**) and electron back scatter diffraction (EBSD) mappings (**e,j**) to illustrate the depth of the laser affected zone.

4.3. Raman Spectroscopy Analysis

The disordering of the crystallographic structure is also validated by Raman spectroscopy measurements on the surface of the laser textured areas using the micro Raman microscope LabRam HR VIS (Horiba Scientific). The excitation wavelength of 473 nm was deliberately chosen to measure with a shallow depth of penetration and thus to determine the superficially changes in the material properties. Figure 8 (left) illustrates the investigated microscopic features by high-resolution field emission scanning electron microscopy (FESEM) images, such as: a) untreated steel surface for reference purposes, as well as ripples and CLP´s produced with $H_0 = 1.7$ J/cm^2 and b) $pd = 2$ µm, $ld = 10$ µm, 5 scans; c) $pd = 2$ µm, $ld = 10$ µm, 20 scans; and d) $pd = 2$ µm, $ld = 2$ µm, 20 scans. The corresponding Raman signals detected from the laser excitation of the substrate surfaces are presented in Figure 8 (right, e–h). The reference spectrum in Figure 8e shows two sharp bands below 800 cm^{-1} which are characteristic for the original (untreated) steel material. On the laser processed surfaces, however, these bands merge and become wider with increasing total energy input. Further information on this signal broadening can be found in the literature reporting broad spectral features appearing on furnace heat treated Fe-Cr-Ni alloy [49]. The spectra broadening was related to spinel-like chromium oxide, disordered FexCr$_3$-xO$_4$ spinel or an amorphous oxide layer forming at the substrate surface at temperatures ranging between 300 °C and 500 °C. From this it can be concluded, the crystal structure of the laser textured surfaces is no longer ordered and becomes more amorphous due to potential deformation or melting during the laser processing.

In Figure 8, the spectral features shown in the range from 600 cm^{-1} to 800 cm^{-1} matches pretty well with FeCr$_2$O$_4$ and Cr$_2$O$_3$ Raman spectra published in the RRUFF™ database [50]. The slight shift of the spectral peaks, for instance 683.8 cm^{-1} versus 679.5 cm^{-1} for the untreated and laser textured substrates, gives a hint for compressive or tensile stresses induced by the laser process. The weak features appearing in the Raman spectra between 1250 cm^{-1} and 1500 cm^{-1} might result from the carbon layer developed on the metal alloy surface without significant effect from the laser texturing. However, the detected Raman signals could not unambiguously be assigned to specific chemical compounds due to overlapping peak positions. In consequence, the Raman analysis carried out here is more suitable for the qualitative assessment of the laser textured substrates than for the exact determination of the chemical composition of the laser treated (sub)surface region.

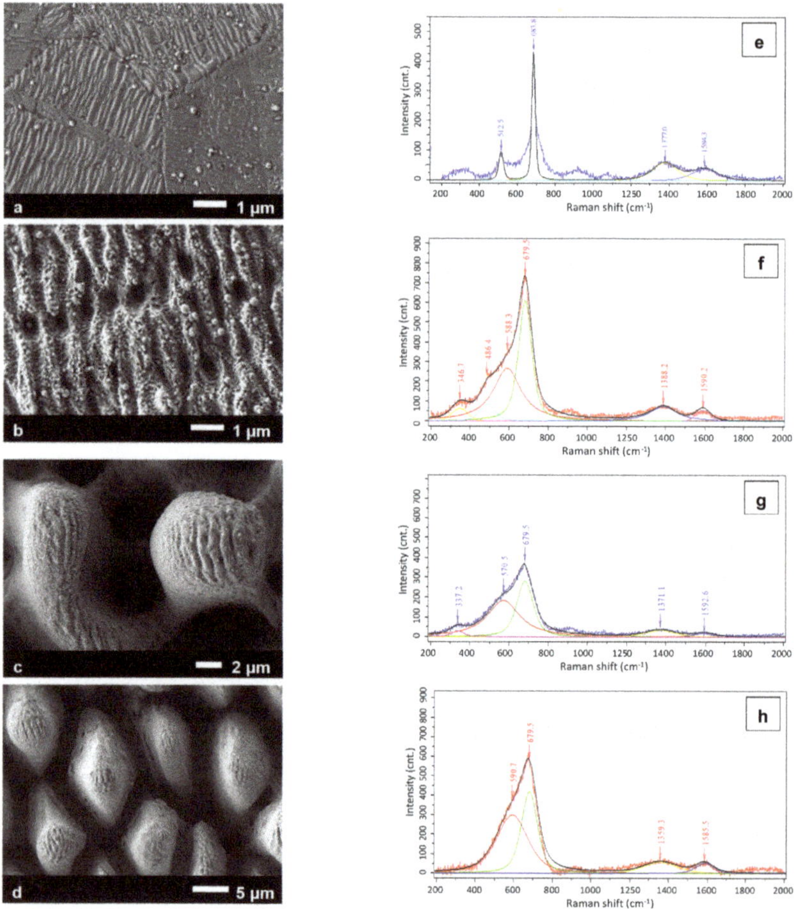

Figure 8. Raman analysis carried out on untreated stainless steel surface for reference purposes (**a**) as well as laser made ripple (**b**) and CLP (**c**,**d**) surface textures. The corresponding Raman signals detected following 473 nm laser excitation are presented on the right side of the figure (**e**–**h**).

4.4. Tribological Performance Test

For tribological performance testing, the self-organizing laser textures were produced on the ring-shaped contact surface of the test specimens described above in Section 3. The test specimens were made of molybdenum alloy steel 42CrMo4 + QT which is a high-grade steel widely used in engineering purposes. Preliminary tests reveal that the microscopic surface textures produced on austenitic chromium-nickel steel 1.4301 metal plates for Raman and cross section analysis develop of similar topographical shape on 42CrMo4 + QT specimen material. The COF values given in the following represent mean values averaged over 3 individual measurements for 100 MPa surface pressure in the contact area under dry friction condition. As a reference, two non-laser treated test specimen with fine grinded surface topography were tested, showing a static COF of $\mu_{max} = 0.30$ and Type A frictional characteristic for 100 MPa contact surface pressure [37]. The corresponding slipping curve for the non-laser textured contact system will be given in the following plots (indicated by a red line) for reference purposes.

The self-organizing CLP texture shown in Figure 9a was made by crossover scanning a near-infrared laser beam of 10 ps pulse duration, 15 W average laser power at 1 MHz pulse repetition frequency and 3.7 J/cm^2 laser peak fluence. The hatch and line distance of the raster-scanned cross grid was kept constant of 10 µm. The effective area processing rate was considerably low of 0.04 cm^2/min that resulted from the high number of scan crossings applied to produce this special shape of CLPs. The maximum height S_Z and the root mean square height S_q as representative surface roughness values were measured of $S_Z = 21.4$ µm and $S_q = 3.46$ µm, respectively.

In the frictional tests, the CLP textures were tested against a counter body with a fine grinded (non-laser treated) contact area of $S_Z = 2.9$ µm and $S_q = 0.28$ µm average surface roughness. The recorded slipping curves presented in Figure 9d indicate a tribological characteristic according to Type A. The summary of the tribological characteristics of the CLP texture is provided in the table of Figure 9 (bottom, right) showing static COFs of $\mu_2 = 0.42$ and $\mu_{20} = 0.44$. The maximum static COF was determined of $\mu_{max} = 0.46$ that is about + 53 % higher than the COF reference value of 0.30. In addition, SEM micrographs taken from the CLP texture and the contact area of the counter body after the friction test are shown in Figure 9b,c. The CLPs appeared squashed after the testing and scratches with a maximum depth of 5.6 µm can be observed on the counter body contact surface.

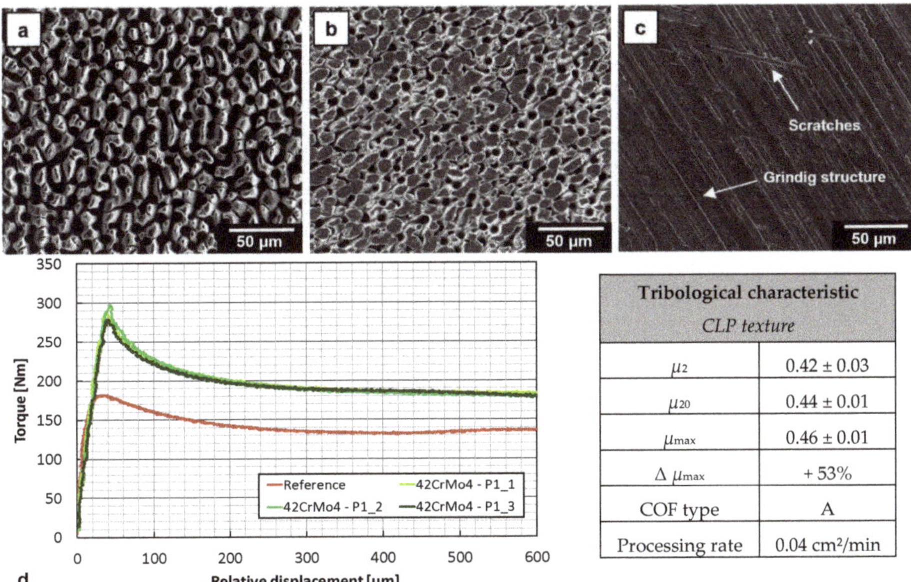

Tribological characteristic	
CLP texture	
μ_2	0.42 ± 0.03
μ_{20}	0.44 ± 0.01
μ_{max}	0.46 ± 0.01
$\Delta \mu_{max}$	+ 53%
COF type	A
Processing rate	0.04 cm²/min

Figure 9. SEM micrographs showing the CLP texture (**a**) before and (**b**) after friction testing and (**c**) the fine grinded counter body surface after the friction test; the recorded slipping curves (**d**) and a summary of the tribological characteristic of CLPs (table bottom, right) are presented.

The LSFL ripples presented in Figure 10a were produced on the fine grinded specimen surface by irradiating ultrashort pulses of 400 fs pulse duration at 1.03 MHz pulse repetition frequency and 7.9 J/cm^2 laser peak fluence. The laser beam was raster-scanned in a single line-scan pattern where the line distance was set of 10 μm. The effective area processing rate was 18 cm^2/min obtained at 10 m/s laser beam moving speed. The height of the ripples could be estimated of about 0.4 μm in Section 4.2. The LSFL ripple textured specimen were tested against another rippled surface textured with the same laser parameter set. The testing against a laser textured counter body was due to the fact that the roughness of the fine grinded surfaces was significantly larger than the ripple height and thus no significant effect on the friction performance was expected for the test of ripple textured surfaces against surfaces of larger roughness. The slipping curve recorded for the ripple versus ripple contact system reveal a clear increase of the static COF of $\mu_2 = 0.41$ at the very beginning until the maximum of $\mu_{max} = 0.42$ was reached. However, with further displacement a rapid drop can be seen in the curve with COF of about the non-laser textured reference value of $\mu_{20} = 0.29$. This result for the stainless steel alloy is a bit contrary to our expectations as higher friction forces were reported for ripple textured silicon [51] as well as the hypothesis of a potential enhancement of the adhesion part of friction resultant from the lower roughness of the ripple textured surfaces. Figure 10a–c shows the LSFL textured surface before and after the friction test, the underlying fine grinded surface structure is still apparent. On both test specimens, the ripple texture smeared, or rather was pushed away during frictional testing. The recorded slipping curve in Figure 10d indicates a COF Type A.

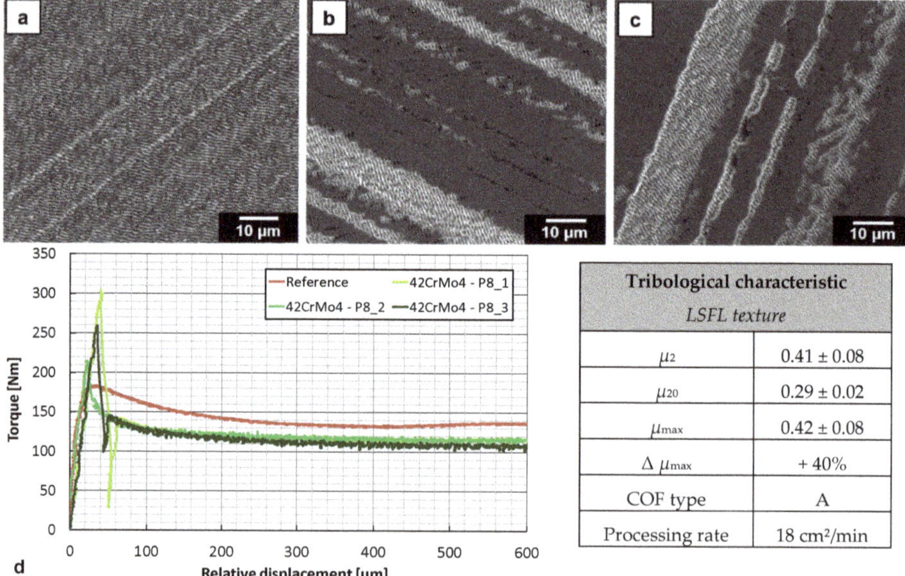

Tribological characteristic	
LSFL texture	
μ_2	0.41 ± 0.08
μ_{20}	0.29 ± 0.02
μ_{max}	0.42 ± 0.08
$\Delta\,\mu_{max}$	$+40\%$
COF type	A
Processing rate	18 cm^2/min

Figure 10. SEM micrographs showing the LSFL ripple texture (**a**) before and (**b**) after friction testing and (**c**) the rippled counter body surface after the friction test; the recorded slipping curves (**d**) and a summary of the tribological characteristic of LSFLs (table bottom, right) are presented.

5. Deterministic Surface Textures for Advanced Tribological Functionality

5.1. Deterministic Surface Textures and Groove Structures

Deterministic surface textures made by direct laser structuring provide another practical method to control the surface functionality in general, and here in particular the frictional performance of tribological systems. Therewith, the surface feature characteristic can easily be adjusted by the chosen processing parameters, for instance, the laser beam spot size in the processing plane, the laser peak fluence of the irradiated pulses and the scan regime by the lateral pulse and line distance as well as the number of scan passes. As the width of the microscopic features can be varied by the pulse/line spacing, the scan number has a greater effect on the feature height. In particular, nanosecond lasers have been used during the past for the fabrication of high aspect-ratio surface features, among others, demonstrated in [52,53]. A recent study reports on the gradually increase of the friction coefficient with higher line density [54]. In any case, the fabrication of deterministic surface features with nanosecond lasers is characterized by strong material melting accompanied by a greater influence of the laser processing on the crystallographic structure, as observed up to 80 µm from the surface for the nanosecond laser regime [53]. This processing behavior can be a serious disadvantage that hinders the use of the cost-effective nanosecond laser technology for desired (micro-)engineering needs. A promising alternative therefore is provided by the ultrashort pulse laser technology, in particular for high-precision engineering applications. As a machining example, Figure 11 presents a number of different surface topographies made with an ultrashort pulse laser beam and a varied beam scanning pattern, such as (a) line pattern, (b) cross pattern and (c) a combination of an alternating line and crosswise scanned laser beam. The feature depth can easily be adjusted by the number of scan passes.

Figure 11. Deterministic surface textures made by ultrashort pulse laser processing under varied laser beam scanning procedures: (**a**) line pattern, (**b**) cross pattern and (**c**) alternating line and crosswise scanned laser beam.

5.2. Tribological Performance

Two different surface textures were investigated regarding their dry friction performance, a line pattern and a cross pattern consisting of 2 perpendicularly crossed lines. The line pattern in Figure 12 was made on the 42CrMo4 + QT specimen contact surface by irradiating 10 ps pulses at 10 MHz and 200 W average laser power and respective 5.0 J/cm^2 laser peak fluence. The lateral distance between the lines in the raster-scanned pattern was 20 µm, the lateral pulse distance within a line was 1 µm. The effective area processing rate for this line pattern textured with a single pass was 34 cm^2/min.

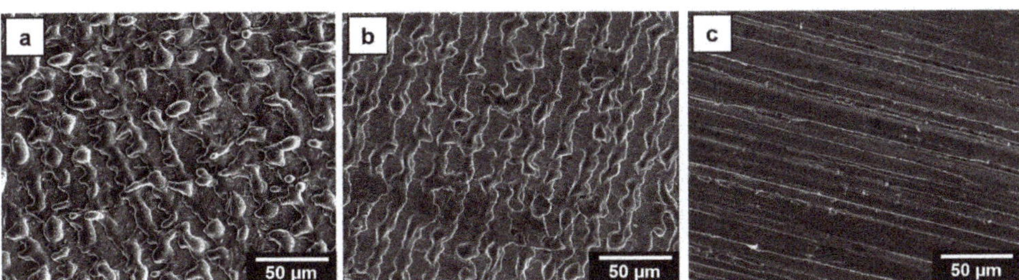

Figure 12. *Cont.*

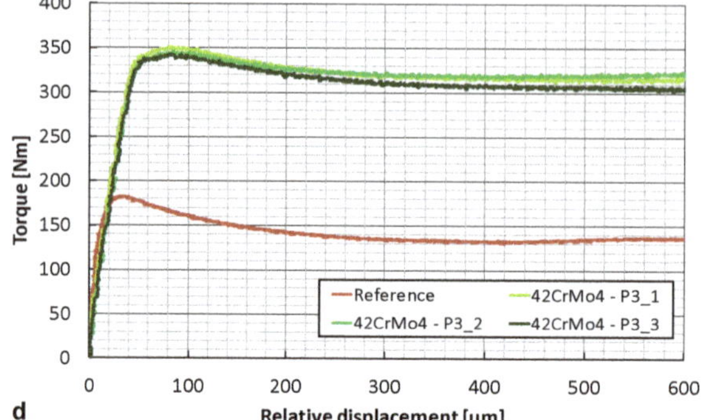

Tribological characteristic	
line pattern	
μ_2	0.51 ± 0.01
μ_{20}	0.55 ± 0.01
μ_{max}	0.56 ± 0.01
$\Delta \mu_{max}$	$+87\%$
COF type	B
Processing rate	34 cm²/min

Figure 12. SEM micrographs showing the line pattern (**a**) before and (**b**) after friction testing and (**c**) the counter body surface after the friction test; the recorded slipping curves (**d**) and a summary of the tribological characteristic of the line pattern texture (table bottom, right) are presented.

In Figure 12a, a high number of molten and re-solidified surface features can be seen following the line-by-line raster scan pattern with a spatial distance of 20 µm. The reason for material melting was the high optical energy deposition onto the material resulting from the high average laser power irradiated at high pulse overlaps within a line. The average surface roughness of the laser textured contact area was measured of $S_Z = 12.8$ µm, $S_a = 1.43$ µm and $S_q = 1.76$ µm. After friction testing, the roughness values have been found almost unchanged of $S_Z = 11.6$ µm, $S_a = 1.57$ µm and $S_q = 1.87$ µm. The SEM inspection of Figure 12b shows slightly compressed features after the friction test while the surface of the counter body appeared corrugated in Figure 12c. A clear indication for this surface corrugation is also given by the surface roughness values where S_Z increased from 2.9 µm to 4.9 µm that was measured before and after the friction test. From the slipping curves presented in Figure 12 d, the static COF was determined of $\mu_{max} = 0.56$ for the line pattern that is about + 87% higher than the reference COF of the non-laser textured contact system. The frictional characteristic of the slipping curve follows Type B.

Aside from the pulse and line distance which were set of 1.5 µm and 40 µm, the laser settings chosen for the line pattern were also applied for the fabrication of the cross pattern shown in Figure 13a. For two scan passes in total the area processing rate was 43 cm²/min. The characteristic topography parameters of the surface features were measured of $S_Z = 14.8$ µm, $S_a = 1.65$ µm and $S_q = 2.05$ µm. After friction testing, the tips seem to be compressed in Figure 13b but the surface roughness parameter were measured of similar dimension as before of $S_Z = 13.2$ µm, $S_a = 1.83$ µm and $S_q = 2.24$ µm. The counter body surface appeared plastically deformed, see Figure 13. The maximum static COF was derived from the Type B slipping curve of $\mu_{max} = 0.66$ that is about + 120% higher than the COF reference value.

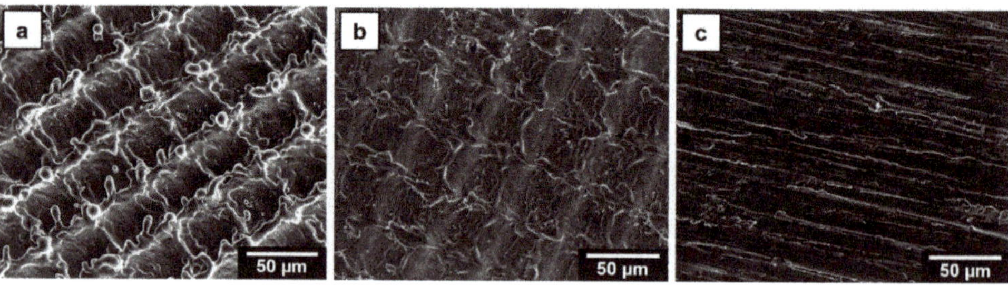

Figure 13. *Cont.*

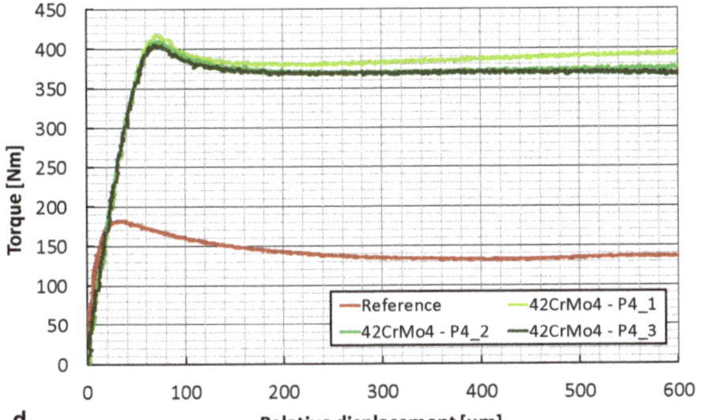

Tribological characteristic	
cross pattern #1	
μ_2	0.56 ± 0.04
μ_{20}	0.66 ± 0.02
μ_{max}	0.66 ± 0.02
$\Delta\,\mu_{max}$	$+ 120\%$
COF type	B
Processing rate	43 cm²/min

Figure 13. SEM micrographs showing the cross pattern made with 2 scan passes (**a**) before and (**b**) after friction testing and (**c**) the counter body surface after the friction test; the recorded slipping curves (**d**) and a summary of the tribological characteristic of the cross pattern texture are presented (table bottom, right).

By further scanning the laser beam, high-aspect ratio surface topographies could be produced. The deterministic surface features presented in Figure 14 have been achieved only by increasing the number of scan passes from 2 to 10. A considerably rougher surface topography was achieved that was measured of $S_Z = 47.3$ µm, $S_a = 8.55$ µm and $S_q = 9.85$ µm. While the processing rate significantly decreased to 9 cm²/min resulting from the higher number of scan passes, only a little higher static COF was determined from the recorded slipping curves of $\mu_{20} = 0.68$ or rather + 127% increase compared to the reference contact. The maximum COF, however, was of $\mu_{max} = 0.86$ implying a substantial COF increase up to + 186%.

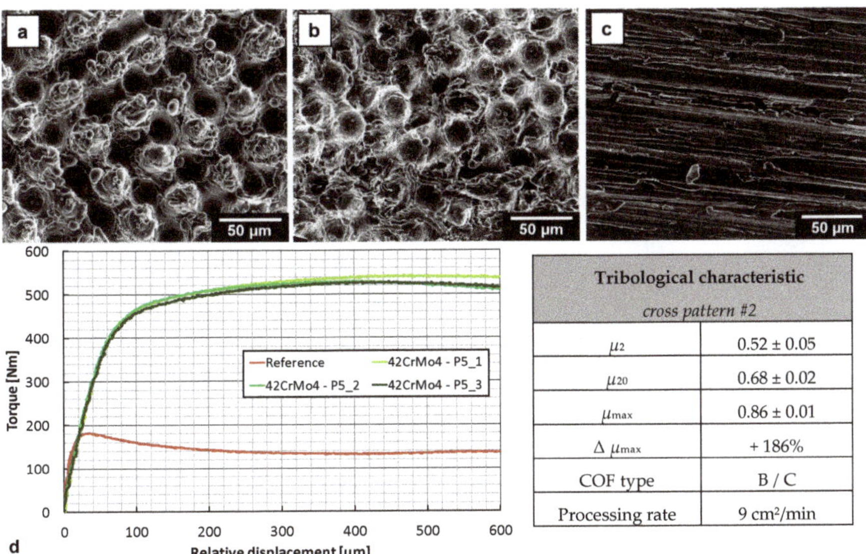

Tribological characteristic	
cross pattern #2	
μ_2	0.52 ± 0.05
μ_{20}	0.68 ± 0.02
μ_{max}	0.86 ± 0.01
$\Delta\,\mu_{max}$	$+ 186\%$
COF type	B / C
Processing rate	9 cm²/min

Figure 14. SEM micrographs showing the cross pattern made with 10 scan passes (**a**) before and (**b**) after friction testing and (**c**) the counter body surface after the friction test; the recorded slipping curves (**d**) and a summary of the tribological characteristic of the cross pattern texture are presented (table bottom, right).

6. Shaft-Hub Connection

The transferability of model-based static COF´s to real components was recently demonstrated for mechanical manufacturing processes [55]. However, in this study a shaft-hub connection was chosen to transfer the advanced tribological performance of laser textured surfaces to a real-life application for the first time. The shaft diameter was 40 mm with H7 press fit for the shaft-hub connection. As a reference, the static COFs were determined of $\mu_{20} = 0.20$ and 0.24 for this specific shaft-hub connection considering a turned or rather fine-grinded surface quality of the contact area [34]. In this very initial test, the width of the laser textured area was 50 mm, as can be seen in the topographical analysis in Figure 15, left. Three different laser textures were analyzed: (i) the LSFL texture of Figure 10, (ii) the line pattern of Figure 12 and for reference purposes (iii) a dimple-shaped micro texture produced with a nanosecond laser as described in a previous study [14]. The frictional performance was analyzed by using a specially designed test rig for joined real components. The surface pressure varied in the measurements that was due to manufacturing tolerances of the shaft-hub assembly. The recorded slipping curves can be seen for the three different laser textures in Figure 15, right. Therefrom, the static and the maximum COFs μ_{20} and μ_{max} as well as the COF type were assessed. It is noteworthy, all COF values determined for the laser textured shaft-hub connection are larger than that for the non-laser textured contact surfaces. Table 2 shows the highest static COF for the line-patterned laser texture, $\mu_{20} = 0.34$ Type A while the maximum COF $\mu_{\mathrm{max}} = 0.4$ was obtained with the dimple textured surface. Thus, the maximum COF increase was found of about + 20% for the LSFL, + 70% for the line pattern and + 100% for the dimple-shaped laser texture, respectively.

Table 2. Summary of the tribological characteristics of the different laser textures tested in a shaft-hub connection.

Tribological characteristic of laser textured shaft-hub connections.			
Laser texture	LSFL	Line pattern	Dimple-shaped texture
Surface pressure	85 ± 2 MPa	92 ± 9 MPa	84 ± 2 MPa
COF type	B	A	A
μ_{20}	0.24 ± 0.01	0.34 ± 0.02	0.32 ± 0.01
$\Delta \mu_{20}$	+ 20%	+ 70%	+ 60%
μ_{max}	0.24 ± 0.01	0.35 ± 0.01	0.40 ± 0.02
$\Delta \mu_{max}$	+ 20%	+ 75%	+ 100%
Processing rate	14.4 cm^2/min	21.0 cm^2/min	14.0 cm^2/min
Topography measurement (before testing)			

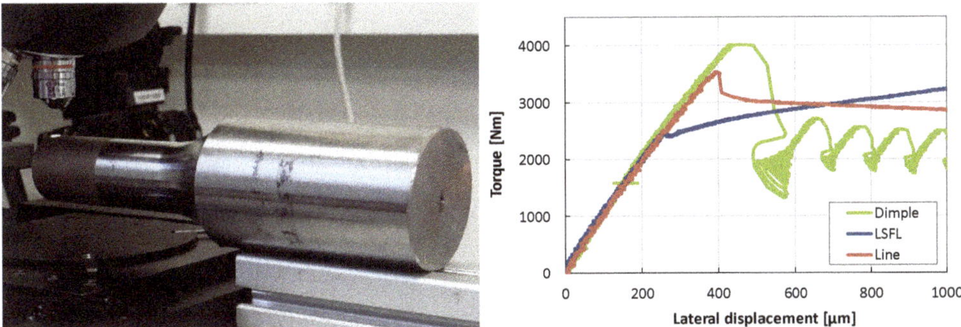

Figure 15. left: Topographical analysis of the shaft, the laser textured area is on the left shaft end; **right:** slipping curves captured in the frictional tests using a test rig for joined real components.

The processing rates given in Table 2 for the LSFL and line pattern, however, are noticeably lower than that presented above in Table 1 for ultrafast laser beam scanning. Therefore, it should be mentioned the processing rates in Table 2 are derived from the total processing time as spent for this specific shaft laser texturing using a galvanometer scanner at this very initial state. By further optimizing the laser texturing implementing polygon-mirror based and multi-beam strategies here, a substantially increase of the processing rates will also be achieved for rotational parts.

7. Summary

High-rate laser texturing was presented as a powerful method for advanced surface engineering and functionalization. By combining high-power laser systems with polygon mirror based scan techniques for ultrafast surface texturing, a maximum area processing rate up to 3.8 m^2/min could be achieved. This unprecedented processing rate was experimentally realized by single pass raster scanning the laser beam at 950 m/s scan speeds producing microscopic surface features over a large area on stainless steel metal sheets. The effective area processing rate decreased noticeably when a higher number of scan passes was required to fabricate a distinct surface topography for a specific surface functionality, as demonstrated by a regular ripple structure where the effective area processing rate was 0.13 m^2/min.

The laser textured surfaces were analyzed regarding their frictional performance by using a standardized friction test method. The highest static COF was found of $\mu_{20} = 0.68$ for the laser textured cross pattern that is about + 126% higher than that obtained for the fine grinded reference contact system. The static COF determined for self-organizing cone-like protrusions was $\mu_{20} = 0.44$ while a little effect on the tribological performance was observed for LSFL textured contact surfaces with $\mu_2 = 0.41$ at the very beginning of the measurement. The effective area processing rates for both self-organizing and deterministic laser surface textures were typically achieved in the range between 0.04 cm^2/min and 43 cm^2/min.

Finally, the ultrashort pulse laser made microscopic surface features were tested for the first time in a shaft-hub connection. For the line pattern textured on the shaft surface, the COF increased to $\mu_{max} = 0.35$ that is + 75% higher than the reference value. Even though further optimization on the surface texture is needed for this special machining example, the high potential of laser textured microscopic surface features could be demonstrated for advanced tribological functionality by a real-world application. Moreover, in addition to the increase of the static coefficient of friction presented here, the laser textured surfaces may also provide great potential to enhance adhesive strength and stability for bonded joints and hard coatings, e.g., on forming and milling tools that will be investigated in the ongoing study.

Author Contributions: Conceptualization, J.S., F.R.; methodology, L.S., J.S., F.R., S.S., S.H., J.P.; investigation, L.S., J.S., F.R., S.S., S.H.; validation, all authors; resources, U.L., J.S., E.L., J.P.; writing—original draft preparation, J.S., F.R., S.S., S.H., S.M.; writing—review and editing, all authors. All authors have read and agreed to the published version of the manuscript.

Funding: The presented study was funded by the German Federal Ministry of Education and Research (grant numbers 03FH037PX4 and 03PSIPT1A) and by the Federal Ministry of Economic Affairs through the German Federation of Industrial Research Associations (AiF grant numbers 17229 BR/1 and 18500 BR/1).

Conflicts of Interest: The authors declare no conflict of interest.

References

1. Scopus Database Document Search. Available online: https://www.scopus.com/search/form.uri?display=basic (accessed on 13 January 2020).

2. Yu, X.Q.; He, S.; Cai, R.L. Frictional characteristics of mechanical seals with a laser-textured seal face. *J. Mater. Process. Technol.* **2002**, *129*, 463–466. [CrossRef]

3. Etsion, I.; Halperin, G.; Brizmer, V.; Kligerman, Y. Experimental investigation of laser surface textured parallel thrust bearings. *Tribol. Lett.* **2004**, *17*, 295–300. [CrossRef]

4. Ryk, G.; Etsion, I. Testing piston rings with partial laser surface texturing for friction reduction. *Wear* **2006**, *261*, 792–796. [CrossRef]

5. Mezzapesa, F.P.; Scaraggi, M.; Carbone, G.; Sorgente, D.; Ancona, A.; Lugarà, P.M. Varying the geometry of laser surface microtexturing to enhance the frictional behavior of lubricated steel surfaces. *Phys. Procedia* **2013**, *41*, 677–682. [CrossRef]

6. Greiner, C.; Merz, T.; Braun, D.; Codrignani, A.; Magagnato, F. Optimum dimple diameter for friction reduction with laser surface texturing: The effect of velocity gradient. *Surf. Topogr. Metrol. Prop.* **2015**, *3*, 044001. [CrossRef]

7. Ma, C.; Gu, W.; Tu, Q.; Sun, J.; Yu, B. Experimental investigation on frictional property of mechanical seals with varying dimple diameter along the radial face. *Adv. Mech. Eng.* **2016**, *8*, 1–9. [CrossRef]

8. Houdková, Š.; Šperka, P.; Repka, M.; Martan, J.; Moskal, D. Shifted laser surface texturing for bearings applications. *J. Phys. Conf. Ser.* **2017**, *843*, 012076. [CrossRef]

9. Bonse, J.; Koter, R.; Hartelt, M.; Spaltmann, D.; Pentzien, S.; Höhm, S.; Rosenfeld, A.; Krüger, J. Tribological performance of femtosecond laser-induced periodic surface structures on titanium and a high toughness. *Appl. Surf. Sci.* **2015**, *336*, 21–27. [CrossRef]

10. Wang, Z.; Zhao, Q.; Wang, C. Reduction of Friction of Metals Using Laser-Induced Periodic Surface Nanostructures. *Micromachines* **2015**, *6*, 1606–1616. [CrossRef]

11. Flores, G.; Wiens, A. Gelaserte Strukturen ersetzen bisherige kostenintensive Lösungen—Teil 1: Herstellung von Funktionsoberflächen mit hoher Haftreibung durch Laserstrukturieren (written in German). *VDI Z* **2016**, *158*, 62–64.

12. Flores, G.; Wiens, A. Gelaserte Strukturen ersetzen bisherige kostenintensive Lösungen - Teil 2: Herstellung von Funktionsoberflächen mit hoher Haftreibung durch Laserstrukturieren (written in German). *VDI Z* **2016**, *158*, 38–40.

13. Dunn, A.; Carstensen, J.V.; Wlodarczyk, K.L.; Hansen, E.B.; Gabzdyl, J.; Harrison, P.M.; Shephard, J.D.; Hand, D.P. Nanosecond laser texturing for high friction applications. *Opt. Lasers Eng.* **2014**, *62*, 9–16. [CrossRef]

14. Schille, J.; Ullmann, F.; Schneider, L.; Gräfensteiner, M.; Schiefer, S.; Gerlach, M.; Leidich, E.; Exner, H. Experimental Study on Laser Surface Texturing for Friction Coefficient Enhancement. *J. Laser Micro Nanoeng.* **2015**, *10*, 245–253. [CrossRef]

15. Dunn, A.; Wlodarczyk, K.L.; Carstensen, J.V.; Hansen, E.B.; Gabzdyl, J.; Harrison, P.M.; Shephard, J.D.; Hand, D.P. Laser surface texturing for high friction contacts. *Appl. Surf. Sci.* **2015**, *357*, 2313–2319. [CrossRef]

16. Exner, H.; Hartwig, L.; Ebert, R.; Klötzer, S.; Streek, A.; Schille, J.; Löschner, U. High rate laser micro processing using high brilliant cw laser radiation. In Proceedings of the 11th International Symposium on Laser Precision Microfabrication 2010, Stuttgart, Germany, 7–10 June 2010.

17. Löschner, U.; Schille, J.; Streek, A.; Knebel, T.; Hartwig, L.; Hillmann, R.; Endisch, C. High-rate laser microprocessing using a polygon scanner system. *J. Laser Appl.* **2015**, *27*, S28007-1.

18. Schille, J.; Schneider, L.; Streek, A.; Klötzer, S.; Löschner, U. High-throughput machining using a high-average power ultrashort pulse laser and high-speed polygon scanner. *Opt. Eng.* **2016**, *55*, 096109-1. [CrossRef]

19. Raciukaitis, G.; Brikas, M.; Gecys, P.; Voisiat, B.; Gedvilas, M. Use of high repetition rate and high power lasers in microfabrication: How to keep the efficiency high? *J. Laser Micro Nanoeng.* **2009**, *4*, 186–191. [CrossRef]

20. Neuenschwander, B.; Jäggi, B.; Schmid, M.; Hennig, G. Surface structuring with ultra-short laser pulses: Basics, limitations and needs for high throughput. *Phys. Procedia* **2014**, *56*, 1047–1058. [CrossRef]

21. Schille, J.; Schneider, L.; Löschner, U. Process optimization in high-average-power ultrashort pulse laser microfabrication: How laser process parameters influence efficiency, throughput and quality. *Appl. Phys. A* **2015**, *120*, 847–855. [CrossRef]

22. Ancona, A.; Röser, F.; Rademaker, K.; Limpert, J.; Nolte, S.; Tünnermann, A. High speed laser drilling of metals using a high repetition rate, high average power ultrafast fiber CPA system. *Opt. Express* **2008**, *16*, 8958–8968. [CrossRef] [PubMed]

23. Finger, J.; Reininghaus, M. Effect of pulse to pulse interactions on ultrashort pulse laser drilling of steel with repetition rates up to 10 MHz. *Opt. Express* **2014**, *22*, 18790–18799. [CrossRef]

24. Jäggi, B.; Remund, S.; Streubel, R.; Gökce, B.; Barcikowski, S.; Neuenschwander, B. Laser Micromachining of Metals with Ultra-Short Pulses: Factors Limiting the Scale-Up Process. *J. Laser Micro Nanoeng.* **2017**, *12*, 267–273.

25. Zeng, S.; Luo, Q.; Li, D.; Lue, X. Femtosecond pulse laser scanning using Acousto-Optic Deflector. *Sci. China Series G Phys. Mech. Astron.* **2009**, *52*, 685–692. [CrossRef]

26. Brüning, S.; Hennig, G.; Eifel, S.; Gillner, A. Utrafast scan techniques for 3D-μm structuring of metal surfaces with high repetitive ps-laser pulses. *Phys. Procedia* **2011**, *12*, 105–115. [CrossRef]

27. De Loor, R. Polygon scanner system for ultra short pulsed laser micro-machining applications. *Phys. Procedia* **2013**, *41*, 544–551. [CrossRef]

28. Römer, G.R.B.E.; Bechtold, P. Electro-optic and Acousto-optic Laser Beam Scanners. *Phys. Procedia* **2014**, *56*, 29–39. [CrossRef]

29. Streek, A.; Klötzer, S. Polygonscantechnik für die Lasermikrobearbeitung (written in German). *Sci. Rep. J. Univ. Appl. Sci. Mittweida* **2015**, *4*, 82–85.

30. Bonse, J.; Krüger, J. Pulse number dependence of laser-induced periodic surface structures for femtosecond laser irradiation of silicon. *J. Appl. Phys.* **2010**, *108*, 034903. [CrossRef]

31. Reif, J.; Varlamova, O.; Ratzke, M.; Schade, M.; Leipner, H.S.; Arguirov, T. Multipulse feedback in self-organized ripples formation upon femtosecond laser ablation from silicon. *Appl. Phys. A* **2010**, *101*, 361–365. [CrossRef]

32. Rudenko, A.; Mauclair, C.; Garrelie, F.; Stoian, R.; Colombier, J.P. Self-organization of surfaces on the nanoscale by topography-mediated selection of quasi-cylindrical and plasmonic waves. *Nanophotonics* **2019**, *8*, 459–465. [CrossRef]

33. Leidich, E.; Vidner, J.; Gräfensteiner, M. Endlich vergleichbare Werte - Standardisiertes Prüfverfahren für Reibungszahlen (written in German). *Antriebstechnik* **2012**, *1*, 32–35.

34. Leidich, E.; Reiß, F.; Gräfensteiner, M. Untersuchungen zur Übertragbarkeit von modellbasierten Haftreibwerten und Reibcharakteristiken auf gefügte Realbauteilgeometrien (written in German). Abschlussbericht, FVV-Heft Nr. 1148, Forschungsvereinigung Verbrennungskraftmaschinen e.V., Frankfurt/Main, **2018**.

35. Boultcouncil—Research Council on Structural Connections. Specification for Structural Joints Using High-Strength Bolts. Chicago. Available online: www.boltcouncil.org (accessed on 11 November 2019).

36. Leidich, E.; Gräfensteiner, M. Ermittlung charakterisierender Kennwerte für reibschlüssige Verbindungen (written in German). *Forsch. Ing.* **2016**, *80*, 71–84. [CrossRef]

37. Leidich, E.; Gräfensteiner, M. GECKO - Analyse und Synthese charakterisierender Kennwerte zur funktionsgerechten Gestaltung von Wirkflächenpaaren in reibschlüssigen Verbindungen (written in German). Abschlussbericht/Teilbericht I, FVV-Heft Nr. 1077, Forschungsvereinigung Verbrennungskraftmaschinen e.V., Frankfurt/Main, **2015**.
38. Varlamova, O.; Costache, F.; Ratzke, M.; Reif, J. Control parameters in pattern formation upon femtosecond laser ablation. *Appl. Surf. Sci.* **2007**, *253*, 7932–7936. [CrossRef]
39. Kurselis, K.; Kiyan, R.; Chichkov, B.N. Formation of corrugated and porous steel surfaces by femtosecond laser irradiation. *Appl. Surf. Sci.* **2012**, *258*, 8845–8852. [CrossRef]
40. Ahmmed, T.K.M.; Grambow, C.; Kietzig, A.M. Fabrication of Micro/Nano Structures on Metals by Femtosecond Laser Micromachining. *Micromachines* **2014**, *5*, 1219–1253. [CrossRef]
41. Ardron, M.; Weston, N.; Hand, D. A practical technique for the generation of highly uniform LIPSS. *Appl. Surf. Sci.* **2014**, *313*, 123–131. [CrossRef]
42. Bonse, J.; Kirner, S.V.; Griepentrog, M.; Spaltmann, D.; Krüger, J. Femtosecond Laser Texturing of Surfaces for Tribological Applications. *Materials* **2018**, *11*, 801. [CrossRef]
43. Fraggelakis, F.; Mincuzzi, G.; Lopez, J.; Manek-Hönninger, I.; Kling, R. Texturing metal surface with MHz ultra-short laser pulses. *Opt. Express* **2017**, *25*, 18131–18139. [CrossRef]
44. Baron, C.F.; Mimidis, A.; Puerto, D.; Skoulas, E.; Stratakis, E.; Solis, J.; Siegel, J. Biomimetic surface structures in steel fabricated with femtosecond laser pulses: Influence of laser rescanning on morphology and wettability. *Beilstein J. Nanotechnol.* **2018**, *9*, 2802–2812. [CrossRef]
45. Tsibidis, G.D.; Fotakis, C.; Stratakis, E. From ripples to spikes: A hydrodynamical mechanism to interpret femtosecond laser-induced self-assembled structures. *Phys. Rev. B* **2015**, *92*, 041405. [CrossRef]
46. Hauffe, W. Vorrichtung zur Vorbereitung der Stoffproben für die Untersuchung, Gepr. Patent zum Ionenstrahl-Böschungsschnitt-Verfahren (written in German). Available online: https://depatisnet.dpma.de/DepatisNet/depatisnet?action=pdf&firstdoc=1&docid=DD000000139670A3 (accessed on 14 November 2019).
47. Krawczyk-Bärsch, E.; Däbritz, S.; Hauffe, W. Combined use of ion beam slope cutting and scanning electron microscopy for the investigation of the 3-dimensional micro-structure of altered mediaeval glass. *Mikrochim. Acta* **1997**, *125*, 89–91. [CrossRef]
48. Schneider, L.; Schille, J.; Mauersberger, S.; Kujawa, K.; Löschner, U. Großflächige und schnelle Funktionalisierung von technischen Oberflächen mittels Ultrakurzpuls-Laserbearbeitung (written in German). *Sci. Rep.* **2017**, *2*, 37–42.
49. Renusch, D.; Veal, B.; Natesan, K.; Grimsditch, M. Transient Oxidation in Fe-Cr-Ni Alloys: A Raman-Scattering Study. *Oxid. Met.* **1996**, *46*, 365–381. [CrossRef]
50. Raman Spectra Search in RRUFF Sample Data Base. Available online: http://rruff.info/ (accessed on 24 February 2019).
51. Eichstädt, J.; Römer, G.R.B.E.; Huis in't Veld, A.J. Towards friction control using laser-induced periodic Surface Structures. *Phys. Procedia* **2011**, *12*, 7–15. [CrossRef]
52. Romero, P.M.; Otero, N.; González, A.; Vázquez, P. High aspect ratio microfeatures with laser texturing in mixed ablative-melting regime. *Proc. SPIE Int. Soc. Opt. Eng.* **2012**. [CrossRef]
53. Gregorčič, P.; Šetina-Batič, B.; Hočevar, M. Controlling the stainless steel surface wettability by nanosecond direct laser texturing at high fluences. *Appl. Phys. A* **2017**, *123*, 1–8. [CrossRef]
54. Conradi, M.; Drnovšek, A.; Gregorčič, P. Wettability and friction control of a stainless steel surface by combining nanosecond laser texturing and adsorption of superhydrophobic nanosilica particles. *Sci. Rep.* **2018**, *8*, 1–9. [CrossRef] [PubMed]
55. Leidich, E.; Reiß, F. Transferability of Model-based Static Coefficients of Friction to Real Components. *MTZ Worldw* **2018**, *79*, 62–67. [CrossRef]

© 2020 by the authors. Licensee MDPI, Basel, Switzerland. This article is an open access article distributed under the terms and conditions of the Creative Commons Attribution (CC BY) license (http://creativecommons.org/licenses/by/4.0/).

 *lubricants* 

Article

Possibilities of Dry and Lubricated Friction Modification Enabled by Different Ultrashort Laser-Based Surface Structuring Methods

Stefan Rung [1,*], Kevin Bokan [1], Frederick Kleinwort [2], Simon Schwarz [1], Peter Simon [2], Jan-Hendrik Klein-Wiele [2], Cemal Esen [3] and Ralf Hellmann [1]

[1] Applied Laser and Photonics Group, University of Applied Sciences Aschaffenburg, Würzburger Straße 45, 63743 Aschaffenburg, Germany; s130088@th-ab.de (K.B.); simon.schwarz@th-ab.de (S.S.); ralf.hellmann@th-ab.de (R.H.)

[2] Laser-Laboratorium Göttingen e.V., Hans-Adolf-Krebs-Weg 1, 37077 Göttingen, Germany; frederick.kleinwort@llg-ev.de (F.K.); peter.simon@llg-ev.de (P.S.); jan-hendrik.klein-wiele@llg-ev.de (J.-H.K.-W.)

[3] Applied Laser Technologies, Ruhr-Universität Bochum, Universitätsstraße 150, 44801 Bochum, Germany; esen@lat.rub.de

* Correspondence: stefan.rung@th-ab.de; Tel.: +49-6022-81-3694

Received: 16 April 2019; Accepted: 13 May 2019; Published: 17 May 2019

Abstract: In this contribution we report on the possibilities of dry and lubricated friction modification introduced by different laser surface texturing methods. We compare the potential of Laser-Induced Periodic Surface Structures and Laser Beam Interference Ablation on 100Cr6 steel in a linear reciprocating ball-on-disc configuration using 100Cr6 steel and tungsten carbide balls with load forces between 50 mN and 1000 mN. For dry friction, we find a possibility to reduce the coefficient of friction and we observe a pronounced direction dependency for surfaces fabricated by Laser Beam Interference Ablation. Furthermore, Laser-Induced Periodic Surface Structures result in a load-dependent friction reduction for lubricated linear reciprocating movements. This work helps to identify the modification behaviour of laser generated micro structures with feature sizes of approximately 1 µm and reveals new possibilities for surface engineering

Keywords: laser-induced periodic surface structures; laser beam interference ablation; dry friction; lubricated friction; laser surface texturing; smart surfaces

---

1. Introduction

Laser surface texturing enables a precise and reliable method to modify surface properties and provides new ways for the design and fabrication of novel surface functionality towards smart surfaces. In this wide field of surface engineering, the controllable modification of friction and wear is of high interest. A recent study reveals that about 23% of the world's energy consumption and 8120 Mt/year of CO_2 emission are caused by these tribological effects, in turn posing the necessity for wear and friction control [1]. While laser-based surface functionalization has already been intensively studied during the last decade, femtosecond laser technology has recently expedited surface structuring on different length scales to advance tribological properties [2–7]. In terms of laser surface texturing, different methods are available to modify the friction behaviour. On the one hand, regular ablation processes using short (pulse duration $\tau > 10$ ps) and ultrashort (pulse duration $\tau < 10$ ps) laser pulses are capable of generating meso- to micro-structures with spatial feature size limitation of the applied laser spot size, typically several tens of micrometers. On the other hand, advanced processing techniques can fabricate surface texturing patterns with feature sizes magnitudes below the applied laser spot size. Out of this research field, we use Laser-Induced Periodic Surface Structures (LIPSS) [8] and Laser

Beam Interference Ablation (LBIA) [9] to fabricate periodic linear patterns on solid surfaces with feature sizes of approx. 1 μm. The general difference between LIPSS and LBIA is their origin and the respective profile depth. While LBIA structures are generated by a special interferometer-based optical setup, LIPSS occur in a self organizing process based on the interference of the laser irradiation and Surface Plasmon Polaritons (SPP) [10]. Ultrashort laser-based periodic surface structures can be generated on all kind of solid materials [11–13]. Both laser-induced surface morphologies, LIPSS and LBIA, can be generated in a single-step process and provide multiple possibilities to modify different surfaces properties. Possible applications of this laser surface texturing methods involve colorization due to diffraction at the periodic structures [14,15], modification of surface wetting properties [16,17], influence on surface cell growth [18,19], and friction management [20–22]. The friction properties of surfaces depend not only on the specific material and several properties of the sliding surface itself, but also on the properties of the counter body, the sliding speed, the environmental conditions, and the nature of any lubricant. Therefore, it is obvious that numerous experimental studies are needed to get fully acquainted with the effects of laser generated periodic surface structures on such tribological properties.

With growing possibilities of flexible methods for laser-based surface functionalization, the question arises, which method leads to the required effects in the field of surface engineering. A line-wise laser surface texturing can be achieved by both methods, LBIA and LIPSS, and they both have the potential to modify friction behaviour. In different tribological test scenarios, LIPSS fabricated surfaces with features in the range of 1 μm and below have shown that it is either possible to increase [23] or to decrease [24] the coefficient of friction (COF). While Gachot et al. [20] report a friction reduction after applying LBIA structures on 100Cr6 while performing a dry linear reciprocating ball-on-disc test, Kasem et al. [25] identify a friction increase on LBIA structures on SAE 1035 steel in a lubricated linear sliding test. Beside the structural modification, the laser-based surface temperature increase, caused by the laser beam absorption, can influence the chemical properties and thus lead to a modification of sliding friction. The important role of the surface chemistry is supported by the studies of Gachot et al. [26] for LBIA and Bonse et al. [7] for LIPSS.

To the best of our knowledge, a direct comparison of the tribological modification behaviour of different laser generated micro structures of approx. 1 μm is not investigated yet. Thus, we use both laser surface texturing methods and demonstrate the possibilities of friction modification and show which method results in which friction behaviour.

2. Materials and Methods

2.1. Laser Surface Texturing

To compare the different laser surface texturing methods, the structures were applied on a chrome steel (100Cr6) substrate with a thickness of 6.6 mm. Before the laser modification is carried out, the base plate was prepared by grinding and polishing with a multi directional polishing using several abrasive grits and afterwards polished with suspensions down to a grain size of 1 μm to achieve a reference surface with a roughness of $R_a = 14$ nm.

2.1.1. LSFL

Generally, LIPSS are categorized into 3 groups, namely low spatial frequency LIPSS (LSFL), high spatial frequency LIPSS (HSFL), and cone-like protrusions (CLP) [8]. In this study, we use LSFL to alter tribological properties. These structures appear upon laser irradiation with linear polarized light. The origin of LSFL is commonly explained by an interference effect of the incident laser light and a surface electro-magnetic wave generated by a laser-induced surface plasmon polarition (SPP) [8]. Due to this origin, LSFL occur with a spatial periodicity in the range of the used laser wavelength and an orientation perpendicular to the polarization of the laser light. Several research groups have shown that LSFL properties like spatial period, orientation, and homogeneity can also be controlled by the applied laser fluence [27,28], pulse to pulse overlap [29], and initial surface roughness of the solid material [30]. For laser surface processing, we used a micro-machining station (MM200 USP, Optec, Frameries, Belgium) equipped with an ultrashort pulsed laser (Pharos 10-600-PP, Light Conversion, Vilnius, Lithuania) having a pulse duration of 220 fs (FWHM) at a repetition rate of 300 kHz. For the generation of the micro- and nano-structures, the fundamental emission wavelength of 1030 nm was utilized. Figure 1a shows the experimental setup for the surface treatment. The energy of the laser was adjusted by an external attenuator based on a rotating wave plate and a polarizer. Using a half wave plate in front of the focusing unit, the linear polarization of the laser beam was rotated orthogonal to the scanning direction. A galvo scanner (RTA AR800, Newson, Dendermonde, Belgium) was used in combination with a telecentric lens (f = 100 mm) to focus the beam onto the sample with a spot diameter of 38 μm ($1/e^2$).

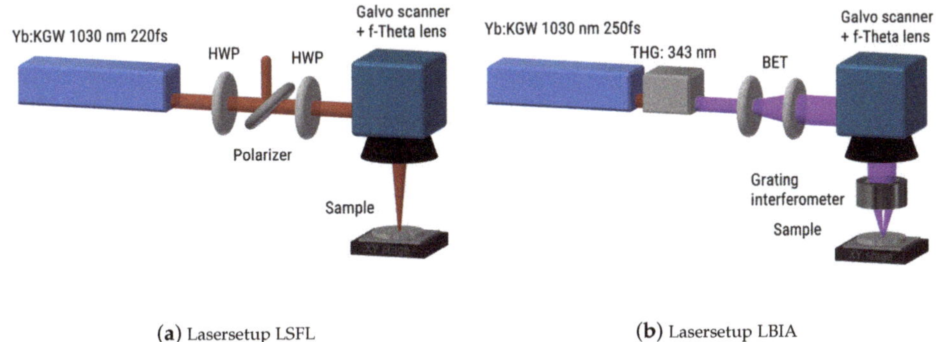

(**a**) Lasersetup LSFL (**b**) Lasersetup LBIA

Figure 1. Methods for laser surface structuring (**a**) low spatial frequency LIPSS (laser-induced periodic surface structures, LSFL) setup (**b**) laser beam interference ablation (LBIA) setup.

2.1.2. LBIA

Beside the LSFL process, the test disc was structered with high definition periodic line gratings of a single orientation across the whole disc. For the fabrication of these surface structures, laser beam interference ablation (LBIA) was used as shown in Figure 1b, applying a special interferometer-based technique developed at LLG. In this approach, the key component is a so-called grating interferometer [9,31], which creates a well-defined, high quality interference pattern, projected onto the sample surface. This interferometer consists of two transmission phase gratings made of fused silica with their grating lines lined up parallel to each other. In the present experiment, the resulting period of the linear grating-like structures projected onto the sample surface was 1.5 μm. A fast scanning system (intelliSCAN$_{de}$ 14, SCANLAB GmbH, Puchheim, Germany) with a focusing optic is applied to ablate arbitrary patterns on the sample, while allowing full coverage of the entire processed area with a highly deterministic periodic structure. The modulation depth of the resulting relief structure can be varied within a wide range by adjusting the laser energy and the scanning speed. In the present study, the modulation depth was adjusted to approx. 1.6 μm on the surface of the test discs, in order to achieve an aspect ratio of roughly 1:1. The Yb:KGW femtosecond laser source (Pharos 20-1000-PP, Light Conversion, Vilnius, Lithuania) used for the ablation process provides pulses at 300 kHz with a duration of 250 fs at a wavelength of 1030 nm. For the presented investigation the laser system was equipped with a third harmonics module which converted the wavelength to 343 nm. Those ultrashort laser pulses are focused to approx. 20 μm and scanned across the sample area.

2.2. Surface Analysis

The generated surface structures were analyzed using atomic force microscopy (XE-150, Park Systems, Suwon, Korea) and scanning electron microscopy (EVO MA10,Carl Zeiss Microscopy GmbH, Jena, Germany). The AFM measurements were used to obtain the 3D surface topology of the different structures in order to determine the depth of the structures. The AFM was operated in non-contact mode using a cantilever with high aspect ratio (>5:1) silicon tip (NANOSENSORS AR5-NCHR, NanoWorld AG, Neuchatel, Switzerland). Images with 512×512 pixels were recorded at a scan rate of 0.2 lines per second and a scan size of $10 \times 10\,\mu m$. The SEM images with 1024×768 pixels were taken at a high voltage of 18–20 kV, a filament current of 2.39–2.44 A, and a probe current of 3–20 pA. The working distance was 5–8 mm and the minimum pixel size 5.86 nm.

2.3. Tribological Analysis

The measurement of the COF was performed in a ball-on-disc (BoD) setup. We used a universal mechanical tester (UMT TriboLab, Bruker Corporation, Billerica, MA, USA) to perform the tribological evaluation. The system is equipped with a linear drive allowing linear reciprocating translation and a second drive unit to apply a specific load of the ball on the base plate. The used force sensor allows measuring the load FL and the lateral force FX up to 5 N. The resulting COF is calculated by FX/FL. We measured the friction behaviour between the 100Cr6 steel substrate and either tribometer balls of chrome steel (100Cr6) or tungsten carbide, both with a diameter of 6.3 mm. The resulting wear on the substrate is captured using laser scanning microscopy (VK-X200, KEYENCE DEUTSCHLAND GmbH, Neu-Isenburg, Germany). Therefore, the track width (point-to point) of the wear tracks on the substrate is measured by a threefold rating. Beside the dry friction test, the COF is measured in a lubricated environment by a thin film of engine oil (5W40).

Due to the elastic deformation of the 100Cr6 steel and tungsten carbide balls, different contact areas appear. Using a Hertzian deformation model of a sphere in contact with a flat sample, the diameter of the contact area is calculated in Table 1.

Table 1. Diameter of contact areas in the specific ball-on-disc configurations.

Tribometer Ball	Load/mN				
	50	100	200	500	1000
100Cr6 steel	20.2 µm	25.6 µm	32.0 µm	43.4 µm	54.7 µm
Tungsten carbide	17.8 µm	22.4 µm	28.3 µm	38.4 µm	48.3 µm

An overview of the applied tribologcial test conditions is given in Table 2 for both sliding regimes, dry and under lubrication. Due to the decreased wear generation using lubrication, higher load forces and numbers of test cycles are possible.

Table 2. Summarized test conditions of tribological evaluation.

Sliding Regime	Test Methode	Substrate	Tribometer Ball	Load Force	Velocity	Cycles	Lubricant
dry	Ball-on-disc	100Cr6	100Cr6 tungsten carbide	50 mN 200 mN	4 mm/s	500	-
lubrication	Ball-on-disc	100Cr6	100Cr6 tungsten carbide	100 mN 500 mN 1000 mN	4 mm/s	1000	5W40

3. Results and Discussion

3.1. Laser Surface Texturing

In the first step, the required laser parameters for continuous surface texturing on 100Cr6 steel are determined. For the LSFL process, the laser fluence and the pulse overlap are optimized to generate a homogeneous LSFL-covered surface. Figure 2a shows the SEM image were the fluence is set to $0.5\,\mathrm{J/cm^2}$ and the pulse to pulse overlap for subsequent laser pulses and scanning lines is 80%. The use of the laser wavelength of 1030 nm and a linear polarization of the laser light results in a linear pattern with an average spatial periodicity of ≈900 nm. Due to the slightly smaller spatial period with respect to the laser wavelength and the perpendicular appearance of the structures, these LIPSS are identified as LSFL. The periodic structures show a uniform periodicity and a straight vertical course. The topographical shape is shown in Figure 2b by a AFM measurement. The height profile shows a homogeneous modulation depth of (201 ± 32) nm.

For the LBIA process the laser parameters and the scanning overlapp of the focused beam were adjusted to produce lines at a depth of approx. 1.6 to 1.7 μm. Due to the projection of lines of the grating interferometer onto the sample surface, the resulting surface structures are highly regular with a strict spatial periodicity of 1.5 μm. The resulting images of the SEM mesurements as well as the measured structural depth from AFM measurement is shown in Figure 2c,d.

The comparision between the LSFL and LBIA structures in Figure 2 shows that the LBIA structures show a higher uniformity and a much more pronounced orientation of the lines. The LSFL structures occur with a 0.6 times smaller spatial periodicity, while the modulation depth of the LBIA structures was about 8 times larger.

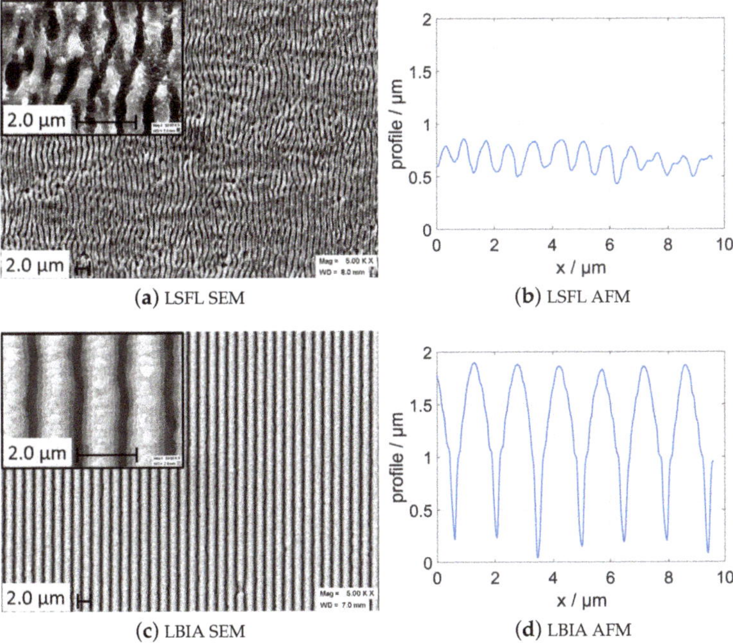

(**a**) LSFL SEM   (**b**) LSFL AFM

(**c**) LBIA SEM   (**d**) LBIA AFM

Figure 2. Laser-based surface modification using the LSFL and LBIA method. SEM image of LSFL (**a**) reveals a mostly vertical orientation of the structes and the corosponding AFM profile (**b**) shows the height profile. Structures fabricated by LBIA show a high homogeneity in the SEM image (**c**) and the AFM profile (**d**).

3.2. Friction Measurement

In the following diagrams (Figures 3 and 5–7), the resulting COF is shown for the reference surface (black solid curve), the LBIA-structured surfaces (red curves) and the LSFL-covered surfaces (blue curves). Furthermore, the relation between the periodic surface structures and the sliding direction is indicated by the line style. For a sliding direction along the periodic grooves, dotted lines and the index "−0" are used. Dashed lines represent a perpendicular sliding movement which is also indicated by the added index "−90". The presented COF values are calcuated by the average of 50 s of steady measurement of the applied force and the induced lateral force. The shown error bars in Figures 3 and 5–7 represent the deviation of the calculated COF for a time range of 50 s.

3.2.1. Dry Test 100Cr6 on 100Cr6

Figure 3 shows the results of the tribological linear reciprocating ball-on-disc evaluation using a 100Cr6 disc and a 100Cr6 triboball for 50 mN and 200 mN load force. The oscillating sliding test is performed for 500 s with a stroke of 2 mm and a frequency of 1 Hz which corresponds a total track distance of 2000 mm. On the reference surface, the COF starts at values below 0.2 and rises slightly with ongoing runtime. After 300 s, the COF rises strongly and approaches a steady state after 450 s with a COF of 0.8. This COF incline can be described by a damage of the natural superficial oxide layer on the surface [32]. With increasing load force, the COF incline starts already within the first 50 s and after a total run time of 150 s, the steady state is nearly reached at a COF of approx. 0.9–1. Surfaces covered by LSFL reveal a comparable high COF directly at the beginning of the linear reciprocating movement. For the movement parallel to the LSFL orientation, the COF starts with a value of 0.65 and increases with time to a value of approx. 0.93. The perpendicular movement on LSFL show higher values for the COF over the entire test duration, starting at 0.81 and with a steady state value of 1. Contrary to the LSFL-covered surface, the LBIA structures introduce only a small COF increase at the beginning of the tribological test. For a perpendicular movement, the COF starts with 0.3 and 0.5 for 50 mN and 200 mN respectively. The break-in effect, which is defined as the initial COF modification by ongoing test duration, is visible for both load forces: For 50 mN the steady state is reached after 400 s and for 200 mN after 250 s. While LSFL-0, LSFL-90, and LBIA-90 result in a COF increase over the entire linear reciprocating ball-on-disc evaluation for both load forces, LBIA structures combined with a parallel movement (LBIA-0) result in a stable COF value at approx. 0.2 for both 50 mN and 200 mN.

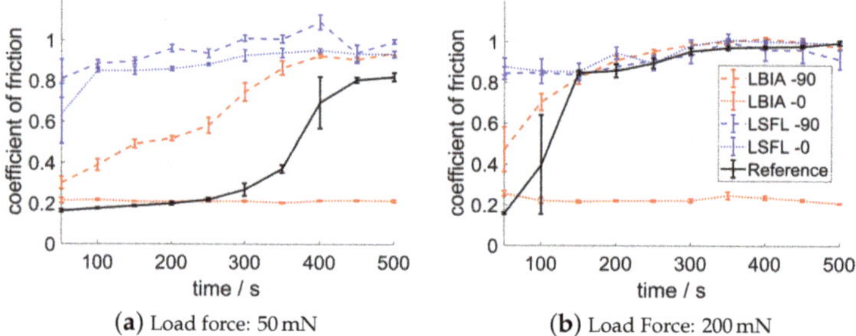

(**a**) Load force: 50 mN   (**b**) Load Force: 200 mN

Figure 3. Temporal evolvement of the coefficient of friction using 100Cr6 triboball without lubrication for a load force of 50 mN (**a**) and 200 mN (**b**).

The differences of the coefficient of friction for LSFL- and LBIA-covered surfaces are also visible in the microscopy image after the linear reciprocating ball-on-disc evaluation. The high COF value for laser surface texturing using LSFL implies a strong interaction between the disc and the triboball. Figure 4a shows the borderline between the initial LSFL and the friction affected area using a parallel movement and a load force of 50 mN. In the sliding track, LSFL vanished completely and big artefacts caused by a strong material adhesion and galling [33] are visible. As indicated by the small COF in

Figure 3a, the wear on the LBIA surface after the parallel movement in Figure 4b is less pronounced. The wear track is visible but the laser surface texturing is still present. Along the complete stroke, no artefacts of material adhesion and galling are visible.

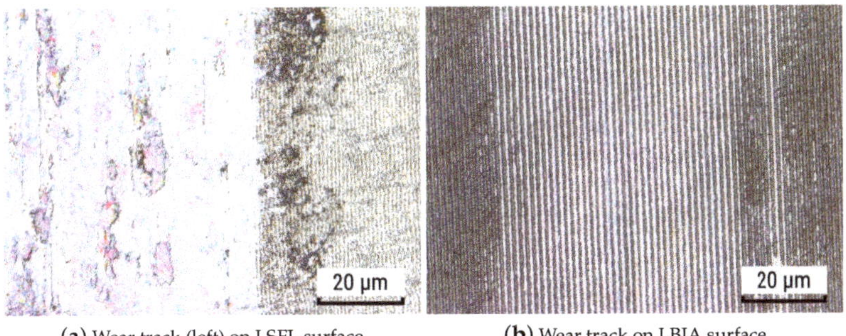

(**a**) Wear track (left) on LSFL surface   (**b**) Wear track on LBIA surface

Figure 4. Wear track after dry linear reciprocating evaluation with a load force of 50 mN and a duration of 500 s on LSFL-0 (**a**) and LBIA-0 (**b**) covered surfaces.

3.2.2. Dry Test Tungsten Carbide on 100Cr6

The results of the tribological linear reciprocating ball-on-disc evaluation using a 100Cr6 disc and a tungsten carbide triboball for 50 mN and 200 mN load force are shown in Figure 5. The general comparison to the pairing 100Cr6/100Cr6 shows that the coefficient of friction is lower for all tested surfaces. On the polished reference surface, the coefficient of friction is 0.18 in the beginning and rises slightly up to 0.25 for a load force of 50 mN. With increasing load (Figure 5b), the COF increases to values of 0.55 at the end of the evaluation time of 500 s. The LSFL-structured surfaces lead to a COF increase over the entire evaluation time. For parallel and perpendicular movements, the COF starts at 0.45 and stays nearly constant for parallel sliding direction. The COF for the perpendicular movement on LSFL is higher (c.f. dry test 100Cr6/100Cr6) and rises to a steady state at 0.5. This behaviour also appears for the higher load force of 200 mN. The COF for LSFL-0 and LSFL-90 is initially 0.57 and increases to 0.6 for a parallel movement and to 0.68 for a perpendicular sliding direction. The LBIA surface introduce a small COF increase at the beginning of the test for both test loads. Using a perpendicular movement, the COF shows a time-based increase and runs from 0.24 to 0.3 for 50 mN and from 0.27 to 0.41 for 200 mN. Contrary to this behaviour, the parallel movement on the LBIA surface reveals a slight COF decrease with ongoing evaluation time. For 50 mN the COF decreases from 0.21 to 0.19. Using a load force of 200 mN, the COF reduces from 0.19 to 0.16.

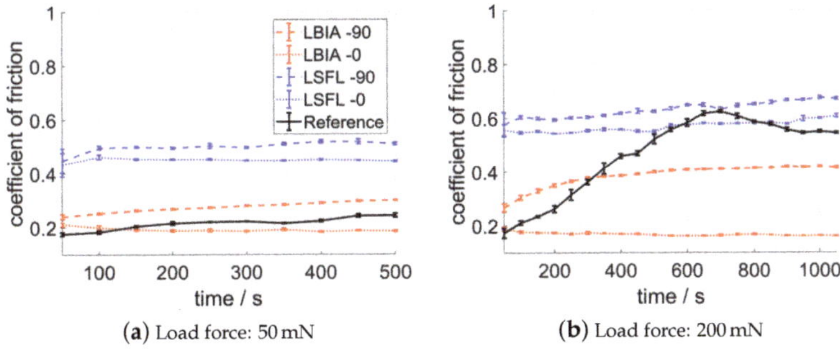

(**a**) Load force: 50 mN   (**b**) Load force: 200 mN

Figure 5. Temporal evolvement of the coefficient of friction using tungsten carbide triboball without lubrication for a load force of 50 mN (**a**) and 200 mN (**b**).

3.2.3. Lubricated Test 100Cr6 on 100Cr6

In Figure 6, the results for the lubricated tribology test using the 100Cr6 triboball are shown. Due to the lubricated friction and the adjunctive COF and wear decrease, it is possible to increase the maximum applied load to 1000 mN and the test duration to 1000 s. The oscillating sliding test is performed with a stroke of 2 mm and a frequency of 1 Hz which corresponds a total track distance of 4000 mm. Figure 6a–c shows the coefficient of friction in a 100Cr6/100Cr6 configuration for 100 mN, 500 mN, and 1000 mN respectively. The COF measurement on the polished reference surface reveals no remarkable influence of the applied load force and is stable at 0.14–0.15. For all lubricated test configurations, the initial COF changes with ongoing test duration. Contrary to the dry friction evaluation, this observed break-in effect causes a COF reduction after a specific test duration. While the applied load does not effect the COF on the reference surface, both LSFL- and LBIA-structured surfaces reveal a load dependency for the COF. For a test load of 100 mN, laser surface structuring in general introduce a COF increase. Figure 6a reveals that the highest friction is measured on the LBIA surface for a parallel movement. Starting with a COF of 0.29, the ongoing COF decrease within the test duration implies that the break-in effect, i.e., the ongoing change of the COF until it reaches a steady state, is not finished after 1000 s. For 100 mN, the influence of sliding direction is well pronounced for both LBIA and LSFL surfaces. The initial friction of LBIA-90 is 0.2 and reduces to 0.18. The COF for LSFL-covered surfaces is smaller; for a perpendicular sliding direction it starts at 0.19 and reduces to 0.16 and for a parallel movement it starts with a COF of 0.17 and decline to 0.15. With increasing load force, a remarkable effect is observed, the COF reduces on structured surfaces with increasing load force. For 500 mN and 1000 mN the COF for a parallel movement is smaller on both LBIA- and LSFL-covered surfaces. At the end of the observation time, LBIA surfaces still show an increased COF. According to the reference COF, perpendicular sliding direction on LBIA introduces a 12% increase and parallel movement a 3% increase. LSFL reveal the potential to decrease the COF. While the perpendicular movement on LSFL introduces a COF reduction of only 1%, the friction reduction for parallel movement is 7%. Figure 6c shows that increasing the load force to 1000 mN lead to a further friction reduction for parallel movement on LSFL. After the break-in (approx. 200 s), a stable COF reduction of 12% according to the reference surface can be measured. On the LBIA-structured surface, a small COF increase of 9% remains for perpendicular movement and 6% for parallel movement.

Figure 6d shows the width of the generated wear track on the disc's surface for the lubricated oscillating sliding test with a applied load of 1000 mN. On laser textured surfaces, the wear track width decreases according to the measured COF values, i.e., the highest COF (LBIA-90) leads to the highest wear width and the lowest COF (LSFL-0) leads to the lowest wear track width. The comparison of the magnitude of the wear track width and the contact diameter between the ball and the flat surface shows a pronounced difference. On LBIA-90 the wear track is more than double for a load force of 1000 mN (cf. Table 1). A friction-based temperature rise in the contact zone could lead to elastic–plastic deformation of the involved interfaces [34,35]. It is worthwhile to mention that the similar COF of the reference and the LSFL-90 surface does not result in a similar wear track width. The laser processing generates a thin oxidized passivation layer on the surface [32], which protects the surface against friction-induced damage.

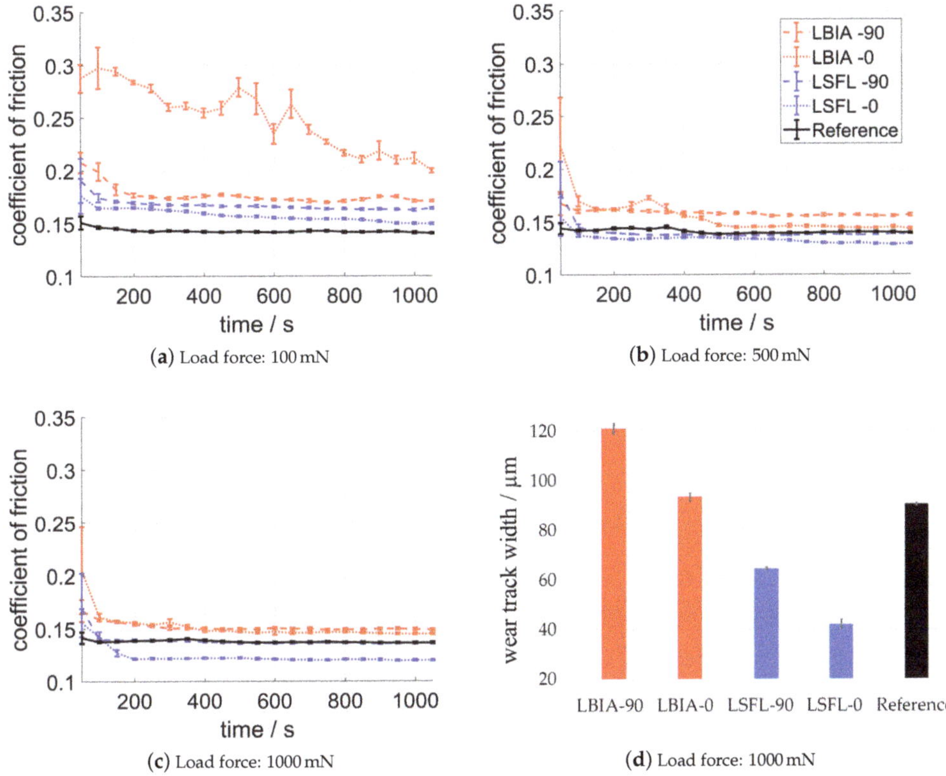

Figure 6. Tribological evaluation using 100Cr6 triboballs under lubrication. Temporal evolvement of the coefficient for a load force of 100 mN (**a**), 500 mN (**b**), and 1000 mN (**c**). Wear track width on the 100Cr6 substrate after after 1000 s test duration using a load force of 1000 mN and 100Cr6 triboballs (**d**).

3.2.4. Lubricated Test Tungsten Carbide on 100Cr6

Figure 7 summarizes the results of the lubricated tribological linear reciprocating ball-on-disc evaluation using a 100Cr6 disc and a tungsten carbide triboball for 100 mN, 500 mN, and 1000 mN load force. Comparable to the results for lubricated 100Cr6/100Cr6 experiments, the COF of the polished reference surface remains nearly unaffected over the test duration. After a short break-in, the COF is stable between 0.13 and 0.14. The slight increase of the COF with increasing load force is attributed to the decreasing film thickness of the lubricant between the WC ball and the polished disc surface [36]. For the smallest investigated load of 100 mN, all laser textured surfaces introduce a COF larger (15%–20%) than the reference surface. With increasing load, both laser textured surfaces are characterized by decreasing COF. According to the research of Ben-David et al. [37] or Ma et al. [38], the nonlinear increase of the degree of surface contact with increasing contact pressure of a rough surface is responsible for this effect. Again, the COF of LBIA surfaces stays above the reference surface. LSFL-covered surfaces introduce a COF decrease for an applied load of 500 mN and 1000 mN (c.f. Figure 7b,c). In both load cases, the direction dependency on LSFL vanishes and a COF reduction of 2% and 9% for 500 mN and 1000 mN can be measured, respectively. The COF for LBIA-0 increases with increasing load force. This behaviour is in contrast to all other lasers structures surfaces in this study, and can be expalined by numerical simulations by Zhu and Wang [34,39], who studied the lubricant film thickness between a tribo ball and a surface with linear surface modulations similiar to those of the LBIA surface. Sliding along the linear features effects a decrease of the film thickness of the lubricant. On the cusps of the LBIA features, however, little or no lubricant film will be spread. Due to the steady contact during sliding, the lubricant locally leaks out leading to an increased friction. The sliding movement transverse to the line shaped structures (LBIA-90) provides more resitance to the lubricant, causing an increase of the average lubrication film thickness between the ball and the rough surface [34,39].

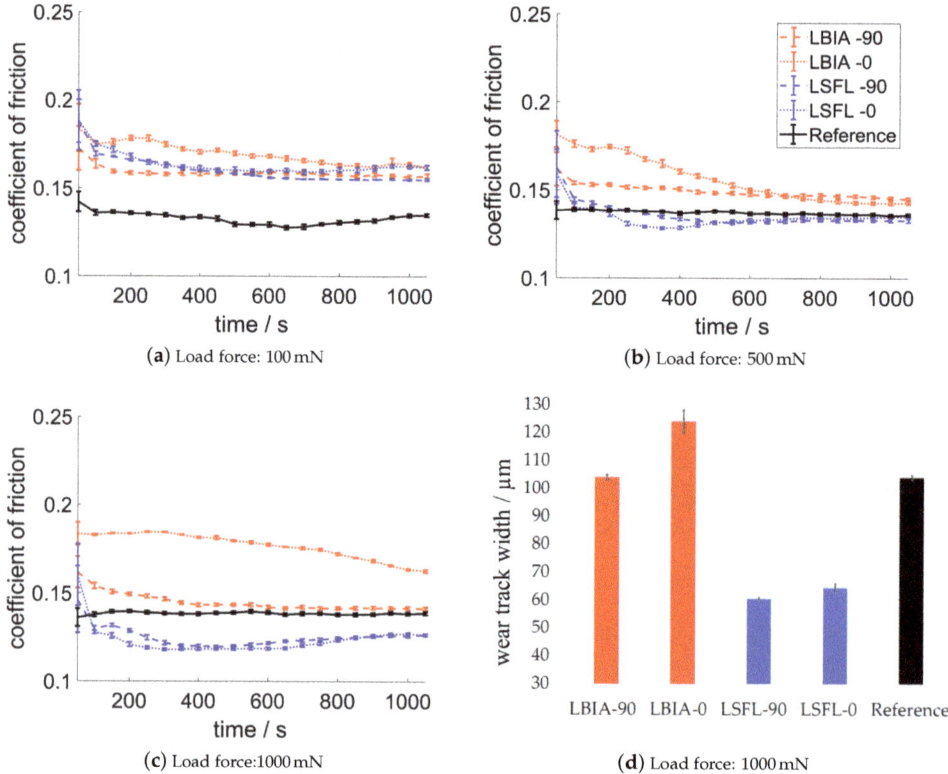

Figure 7. Tribological evaluation using tungsten carbide triboballs in a lubricated enviroment. Temporal evolution of the coefficient for a load force of 100 mN (**a**), 500 mN (**b**), and 1000 mN (**c**). Wear track width on the 100Cr6 substrate after after 1000 s test duration using a load force of 1000 mN and tungsten carbide triboballs (**d**).

The increased and decreased friction caused by LBIA and LSFL also affects the generated wear track width shown in in Figure 7d. As mentioned above, the increased friction on LBIA leads to a temperature increase and thus to an elastic–plastic surface deformation. Figure 7d also reveals that the COF reduction on LSFL-covered surfaces leads to a wear reduction on the 100Cr6 disc. According to the reference surface, a reduction wear track width of up to 42% is possible using LSFL and parallel movement.

4. Conclusions

This research work focused on femtosecond laser surface texturing on 100Cr6 steel using two different methods for the generation of periodic surface structures. Both methods produce a linear pattern on the surface. The LSFL approach fabricates a periodic surface pattern with a spatial period of $\approx$900 nm and a modulation depth of $\approx$200 nm. Using the LBIA technique, a surface modulation with a period of 1.5 µm and a depth of ≈1.6 µm is generated. To compare the potential of friction modification, a tribological linear reciprocating ball-on-disc evaluation is performed using 100Cr6 steel and tungsten carbide (WC) triboballs. The effects of sliding direction and load force are evaluated against a polished reference surface for dry and lubricated conditions. Table 3 summarizes the observed effects with respect to the reference surface, highlighting the potential fields of applications for LBIA and LSFL.

Table 3. Summarized modification possibilities with respect to a polished reference surface indicating no influence (0), increase (+), strong increase (++), decrease (-), and strong decrease (- -) of the coefficient of friction.

| | Dry | | | | Lubricated | | | |
| Ball | LBIA | | LSFL | | LBIA | | LSFL | |
	90	0	90	0	90	0	90	0
Steel	0	- -	+	+	++	+	0	-
WC	-	- -	++	+	+	++	- -	- -

Although LBIA and LSFL generate surface structures with similar periodicity, the overall tribological behaviours are different. For the tribological analysis without lubrication, LSFL causes a COF increase for both the 100Cr6/100Cr6 and 100Cr6/tungsten carbide combinations. The perpendicular movement on LBIA structures generates COF comparable to the reference surface. Parallel sliding direction on LBIA surfaces decreases the friction remarkably for both 100Cr6 and tungsten carbide triboballs. Comparing the lubricated friction behaviour of LSFL and LBIA shows the possibility of COF reduction using LSFL-covered surfaces up to 12% for 100Cr6/100Cr6 and up to 9% for 100Cr6/tungsten carbide configuration. An additional benefit of LSFL-covered surfaces in the introduced test scenario is the reduced wear track width. In contrast to the COF reduction introduced by LSFL for lubricated friction, LBIA lead to a COF increase for both 100Cr6 and tungsten carbide triboballs under lubricated test conditions.

Author Contributions: Conceptualization, S.R. and K.B.; Methodology, S.R., K.B., S.S. and J.-H.K.-W.; Software, S.R.; Validation, K.B. and F.K.; Formal analysis, S.R.; Investigation, S.R., K.B. and F.K.; Resources, P.S. and R.H.; Writing—original draft preparation, S.R., K.B. and F.K.; Writing—review and editing, S.R., F.K., P.S., J.-H.K.-W., C.E. and R.H.; Visualization, S.R., K.B. and F.K.; Supervision, P.S., C.E. and R.H.; Funding acquisition, P.S. and R.H.

Funding: This research received no external funding.

Conflicts of Interest: The authors declare no conflict of interest.

References

1. Holmberg, K.; Erdemir, A. Influence of tribology on global energy consumption, costs and emissions. *Friction* **2017**, *5*, 263–284. [CrossRef]
2. Ancona, A.; Carbone, G.; Scaraggi, M.; Mezzapesa, F.P.; Sorgente, D.; Lugarà, P.M. Laser surface micro-texturing to enhance the frictional behavior of lubricated steel. In Proceedings of the SPIE Laser-Based Micro- and Nanoprocessing VIII, San Francisco, CA, USA, 1–6 February 2014; Volume 8968, p. 896806.
3. Scaraggi, M.; Mezzapesa, F.P.; Carbone, G.; Ancona, A.; Tricarico, L. Friction Properties of Lubricated Laser-MicroTextured-Surfaces: An Experimental Study from Boundary- to Hydrodynamic-Lubrication. *Tribol. Lett.* **2013**, *49*, 117–125. [CrossRef]
4. Mizuno, A.; Honda, T.; Kikuchi, J.; Iwai, Y.; Yasumaru, N.; Miyazaki, K. Friction Properties of the DLC Film with Periodic Structures in Nano-scale. *Tribol. Online* **2006**, *1*, 44–48. [CrossRef]
5. Gnilitskyi, I.; Rotundo, F.; Martini, C.; Pavlov, I.; Ilday, S.; Vovk, E.; Ilday, F.Ö.; Orazi, L. Nano patterning of AISI 316L stainless steel with Nonlinear Laser Lithography: Sliding under dry and oil-lubricated conditions. *Tribol. Int.* **2016**, *99*, 67–76. [CrossRef]
6. Eichstädt, J.; Römer, G.; Huis in'tVeld, A.J. Towards Friction Control using laser-induced periodic Surface Structures. *Phys. Procedia* **2011**, *12*, 7–15. [CrossRef]

7. Bonse, J.; Höhm, S.; Koter, R.; Hartelt, M.; Spaltmann, D.; Pentzien, S.; Rosenfeld, A.; Krüger, J. Tribological performance of sub-100-nm femtosecond laser-induced periodic surface structures on titanium. *Appl. Surf. Sci.* **2016**, *374*, 190–196. [CrossRef]

8. Bonse, J.; Höhm, S.; Kirner, S.V.; Rosenfeld, A.; Kruger, J. Laser-Induced Periodic Surface Structures—A Scientific Evergreen. *IEEE J. Sel. Top. Quantum Electron.* **2017**, *23*, 9000615. [CrossRef]

9. Bekesi, J.; Meinertz, J.; Ihlemann, J.; Simon, P. Fabrication of large-area grating structures through laser ablation. *Appl. Phys. A* **2008**, *93*, 27–31. [CrossRef]

10. Bonse, J.; Rosenfeld, A.; Krüger, J. On the role of surface plasmon polaritons in the formation of laser-induced periodic surface structures upon irradiation of silicon by femtosecond-laser pulses. *J. Appl. Phys.* **2009**, *106*, 104910. [CrossRef]

11. Nayak, B.K.; Gupta, M.C. Self-organized micro/nano structures in metal surfaces by ultrafast laser irradiation. *Opt. Lasers Eng.* **2010**, *48*, 940–949. [CrossRef]

12. Schwarz, S.; Rung, S.; Hellmann, R. Generation of laser-induced periodic surface structures on transparent material-fused silica. *Appl. Phys. Lett.* **2016**, *108*, 181607. [CrossRef]

13. Derrien, T.J.Y.; Krüger, J.; Itina, T.E.; Höhm, S.; Rosenfeld, A.; Bonse, J. Rippled area formed by surface plasmon polaritons upon femtosecond laser double-pulse irradiation of silicon: The role of carrier generation and relaxation processes. *Appl. Phys. A* **2014**, *117*, 77–81. [CrossRef]

14. Vorobyev, A.Y.; Guo, C. Direct femtosecond laser surface nano/microstructuring and its applications. *Laser Photonics Rev.* **2013**, *7*, 385–407. [CrossRef]

15. Li, G.; Li, J.; Hu, Y.; Zhang, C.; Li, X.; Chu, J.; Huang, W. Femtosecond laser color marking stainless steel surface with different wavelengths. *Appl. Phys. A* **2015**, *118*, 1189–1196. [CrossRef]

16. Rung, S.; Schwarz, S.; Götzendorfer, B.; Esen, C.; Hellmann, R. Time Dependence of Wetting Behavior Upon Applying Hierarchic Nano-Micro Periodic Surface Structures on Brass Using Ultra Short Laser Pulses. *Appl. Sci.* **2018**, *8*, 700. [CrossRef]

17. van Ta, D.; Dunn, A.; Wasley, T.J.; Li, J.; Kay, R.W.; Stringer, J.; Smith, P.J.; Esenturk, E.; Connaughton, C.; Shephard, J.D. Laser textured surface gradients. *Appl. Surf. Sci.* **2016**, *371*, 583–589.

18. Raimbault, O.; Benayoun, S.; Anselme, K.; Mauclair, C.; Bourgade, T.; Kietzig, A.M.; Girard-Lauriault, P.L.; Valette, S.; Donnet, C. The effects of femtosecond laser-textured Ti-6Al-4V on wettability and cell response. *Mater. Sci. Eng. C Mater. Biol. Appl.* **2016**, *69*, 311–320. [CrossRef] [PubMed]

19. Wallat, K.; Dörr, D.; Le Harzic, R.; Stracke, F.; Sauer, D.; Neumeier, M.; Kovtun, A.; Zimmermann, H.; Epple, M. Cellular reactions toward nanostructured silicon surfaces created by laser ablation. *J. Laser Appl.* **2012**, *24*, 042016. [CrossRef]

20. Gachot, C.; Rosenkranz, A.; Reinert, L.; Ramos-Moore, E.; Souza, N.; Müser, M.H.; Mücklich, F. Dry Friction Between Laser-Patterned Surfaces: Role of Alignment, Structural Wavelength and Surface Chemistry. *Tribol. Lett.* **2013**, *49*, 193–202. [CrossRef]

21. Ancona, A.; Carbone, G.; de Filippis, M.; Volpe, A.; Lugarà, P.M. Femtosecond laser full and partial texturing of steel surfaces to reduce friction in lubricated contact. *Adv. Opt. Technol.* **2014**, *3*, 539–547. [CrossRef]

22. Rung, S.; Bokan, K.; Rutsch, K.; Schwarz, S.; Esen, C.; Hellmann, R. Laser Induced Periodic Surface Structures on 100Cr6 Steel for Modification of Friction Demonstrated with Stribeck Test. *J. Laser Micro/Nanoeng.* **2018**, *13*, 160–165.

23. Bonse, J.; Kirner, S.V.; Griepentrog, M.; Spaltmann, D.; Krüger, J. Femtosecond Laser Texturing of Surfaces for Tribological Applications. *Materials* **2018**, *11*, 801. [CrossRef] [PubMed]

24. Voyer, J.; Klien, S.; Ausserer, F.; Velkavrh, I.; Ristow, A.; Diem, A. Friction Reduction Through Sub-Micro Laser Surface Modifications. *Tribologie Schmierungstechnik* **2015**, *62*, 13–18.

25. Kasem, H.; Stav, O.; Grützmacher, P.; Gachot, C. Effect of Low Depth Surface Texturing on Friction Reduction in Lubricated Sliding Contact. *Lubricants* **2018**, *6*, 62. [CrossRef]

26. Gachot, C.; Grützmacher, P.; Rosenkranz, A. Laser Surface Texturing of TiAl Multilayer Films—Effects of Microstructure and Topography on Friction and Wear. *Lubricants* **2018**, *6*, 36. [CrossRef]

27. Golosov, E.V.; Emel'yanov, V.I.; Ionin, A.A.; Kolobov, Y.R.; Kudryashov, S.I.; Ligachev, A.E.; Novoselov, Y.N.; Seleznev, L.V.; Sinitsyn, D.V. Femtosecond laser writing of subwave one-dimensional quasiperiodic nanostructures on a titanium surface. *JETP Lett.* **2009**, *90*, 107–110. [CrossRef]

28. Okamuro, K.; Hashida, M.; Miyasaka, Y.; Ikuta, Y.; Tokita, S.; Sakabe, S. Laser fluence dependence of periodic grating structures formed on metal surfaces under femtosecond laser pulse irradiation. *Phys. Rev. B* **2010**, *82*, 165417. [CrossRef]

29. Bonse, J.; Krüger, J. Pulse number dependence of laser-induced periodic surface structures for femtosecond laser irradiation of silicon. *J. Appl. Phys.* **2010**, *108*, 034903. [CrossRef]

30. Preusch, F.; Rung, S.; Hellmann, R. Influence of Polishing Orientation on the Generation of LIPSS on Stainless Steel. *J. Laser Micro/Nanoeng.* **2016**, *11*, 137–142. [CrossRef]

31. Bekesi, J.; Simon, P.; Ihlemann, J. Deterministic sub-micron 2D grating structures on steel by UV-fs-laser interference patterning. *Appl. Phys. A* **2014**, *114*, 69–73. [CrossRef]

32. Kirner, S.V.; Slachciak, N.; Elert, A.M.; Griepentrog, M.; Fischer, D.; Hertwig, A.; Sahre, M.; Dörfel, I.; Sturm, H.; Pentzien, S.; et al. Tribological performance of titanium samples oxidized by fs-laser radiation, thermal heating, or electrochemical anodization. *Appl. Phys. A* **2018**, *124*, 326. [CrossRef]

33. Czichos, H.; Habig, K.H. *Tribologie-Handbuch: Tribometrie, Tribomaterialien, Tribotechnik*, 4th ed.; Springer Fachmedien Wiesbaden: Wiesbaden, Germany, 2015.

34. Zhu, D.; Wang, J.; Wang, Q.J. On the Stribeck Curves for Lubricated Counterformal Contacts of Rough Surfaces. *J. Tribol.* **2015**, *137*, 021501. [CrossRef]

35. Gelinck, E.; Schipper, D.J. Calculation of Stribeck curves for line contacts. *Tribol. Int.* **2000**, *33*, 175–181. [CrossRef]

36. Zhu, D.; Wang, Q.J. On the lambda ratio range of mixed lubrication. *Proc. Inst. Mech. Eng. Part J J. Eng. Tribol.* **2012**, *226*, 1010–1022. [CrossRef]

37. Ben-David, O.; Fineberg, J. Static friction coefficient is not a material constant. *Phys. Rev. Lett.* **2011**, *106*, 254301. [CrossRef]

38. Ma, X.; de Rooij, M.; Schipper, D. A load dependent friction model for fully plastic contact conditions. *Wear* **2010**, *269*, 790–796. [CrossRef]

39. Zhu, D.; Wang, Q.J. Effect of Roughness Orientation on the Elastohydrodynamic Lubrication Film Thickness. *J. Tribol.* **2013**, *135*, 031501. [CrossRef]

© 2019 by the authors. Licensee MDPI, Basel, Switzerland. This article is an open access article distributed under the terms and conditions of the Creative Commons Attribution (CC BY) license (http://creativecommons.org/licenses/by/4.0/).

Article

Avoiding Starvation in Tribocontact Through Active Lubricant Transport in Laser Textured Surfaces

Tobias Stark [1], Thomas Kiedrowski [1,*], Holger Marschall [2] and Andrés Fabián Lasagni [3,4]

1 Robert Bosch GmbH, Robert-Bosch-Campus 1, 71272 Renningen, Germany; tobias.stark@alumni.ethz.ch
2 Computational Multiphase Flow, Department of Mathematics, Technische Universität Darmstadt, Alarich-Weiss Straße 10, 64287 Darmstadt, Germany; marschall@mma.tu-darmstadt.de
3 Institut für Fertigungstechnik, Technische Universität Dresden, George-Bähr-Str. 3c, 01069 Dresden, Germany; andres_fabian.lasagni@tu-dresden.de
4 Fraunhofer-Institut für Werkstoff- und Strahltechnik IWS, Winterbergstr. 28, 01277 Dresden, Germany
* Correspondence: thomas.kiedrowski@de.bosch.com

Received: 11 April 2019; Accepted: 17 June 2019; Published: 25 June 2019

Abstract: Laser texturing is a viable tool to enhance the tribological performance of surfaces. Especially textures created with Direct Laser Interference Patterning (DLIP) show outstanding improvement in terms of reduction of coefficient of friction (COF) as well as the extension of oil film lifetime. However, since DLIP textures have a limited depth, they can be quickly damaged, especially within the tribocontact area, where wear occurs. This study aims at elucidating the fluid dynamical behavior of the lubricant in the surroundings of the tribocontact where channel-like surface textures are left after the abrasion wear inside the tribocontact area. In a first step, numerical investigations of lubricant wetting phenomena are performed applying OpenFOAM®. The results show that narrow channels (width of 10 μm) allow higher spreading than wide channels (width of 30 μm). In a second step, fluid transport inside DLIP textures is investigated experimentally. The results show an anisotropic spreading with the spreading velocity dependent on the period and depth of the laser textures. A mechanism is introduced for how lubricant can be transported out of the channels into the tribocontact. The main conclusion of this study is that active lubricant transport in laser textured surfaces can avoid starvation in the tribocontact.

Keywords: stribeck curve; lubricant transport; laser surface texturing; direct laser interference patterning; phase-field method

---

1. Introduction

Tribology is a wide-spread field with various application fields including the automobile industry. For instance, in passenger cars, friction occurs in the engine, transmission, tires and brakes. According to Holmberg et al. [1], the direct frictional losses, with braking friction excluded, represent 28% of the fuel energy. In total, 21.5% of the fuel energy is used to move the car. Therefore, there is an enormous potential to reduce the global energy consumption by enhancing the tribological performance of engine parts. There are also other fields of application, e.g., bionic engineering where, for example, the sandfish's skin, which is structured in a way that friction is reduced, serves as a model [2].

Laser surface texturing (LST) has shown to be a versatile tool to enhance surface functions. Especially in the field of tribology, it has been used to reduce the coefficient of friction (COF) as well as to extend the oil film lifetime (e.g., [3–8]). Several research studies in the field of LST and tribology deal with the fabrication of microholes or dimples forming closed surface structures. Important to mention is the pioneer work of Etsion [9], who has already given an overview in this field by outlining the variety in shape, size and textured density of dimples and the influence of these parameters on different tribological effects. In particular, three different working principles for dimples have been distinguished, denoting that they can serve as (i) micro-reservoirs for lubricant in cases of starved lubrication conditions; (ii) micro-traps for wear debris in either lubricated or dry sliding conditions and (iii) micro-hydrodynamic bearings in cases of full or mixed lubrication. For the last point, cavitation is an important factor because it is necessary to achieve a positive pressure build-up which permits to reduce friction. Therefore, an overall low pressure level is required to be able to reach cavitation pressure. Further information about the cavitation phenomenon has been described by Braun [10].

An important parameter that significantly affects the tribological properties of surfaces is related to the fact if the textures have an open or closed characteristic. For example, if a lubricant can be displaced through open surface textures, starvation in the tribocontact might occur which results in a higher COF. On the other hand, lubricant can also be transported actively in open surface textures which might act as a supply of lubricant. The following paragraph therefore outlines the tribological effect of open surface textures.

For example, Wahl [5] compared the behavior of dimples (close) and channel-like (open) structures with different orientation with regard to the sliding direction. For the channels, the width was varied from 60 to 300 µm and the depths were 10 and 50 µm. It was found that for high sliding velocities (>0.5 m/s), dimples (close morphology) could reduce the coefficient of friction whereas the cross-like channels (open morphology) show a significant increase of the coefficient of friction, both compared to a polished reference sample. Nonetheless, Stark et al. [11] found that channel-like structures in a cross-like pattern can also reduce the COF, when using smaller feature sizes. In that work, two different laser texturing methods were applied, namely Direct Laser Writing (DLW) and Direct Laser Interference Patterning (DLIP). For the DLW method, the widths of the channels were 12 and 30 µm and the structure depth was varied between 3 and 24 µm. It was found, that the overall reduction for the mixed lubrication regime can be up to 33%. Additionally, it was found that similar results can be obtained applying DLIP method when fabricating channels with a cross-like pattern geometry. With this method the period of the textures were 4 and 8 µm and the depths around 1 µm. DLIP allows higher throughput compared to DLW [12]. A drawback, however, is that, due to their limited depths, DLIP textures are erased quickly inside the tribocontact where wear occurs. During the testcycle for the tribological evaluation it could be shown that the DLIP textures were erased quickly after the start of the cycle but afterwards the COF stayed at a low level. This was the finding of a temporal investigation of the relation between wear picture and reduction of COF (not shown). Since the DLIP textures inside the tribocontact are erased they cannot be responsible for keeping the COF low. However, in the direct surrounding of the contact point the textures are still present. They form a seamless transition into the tribocontact. The motivation of this work is to investigate whether lubricant can be transported actively in the DLIP textures and how this transport can avoid starvation. Starvation may lead to increased friction, wear and could result in the total malfunction of a machine. Starvation in the tribocontact occurs if oil is displaced from the tribocontact due to various effects such as inertial or thermal phenomena. A solution is to guarantee oil presence in the tribocontact.

1.1. Solution Approach

Klima [13] followed the approach of active lubricant transport towards the tribocontact with the objective to avoid starvation. The driving force for the transport in the channel-like textures is the capillary effect. For this effect, Lucas [14] and Washburn [15] derived the so-called Lucas–Washburn equation which describes a square-root dependence of the fluid column length $L_u(t)$ from the time:

$$L_u(t) = \sqrt{\frac{b\gamma\cos(\theta)}{4\eta}t} = \sqrt{W}\sqrt{t},$$ (1)

where W is a factor combining multiple parameters which essentially governs the speed of fluid advancement [13]. Equation (1) was verified experimentally for circular tubes [15]. However, there have also been investigations for microchannels with different cross-sections (e.g., [16,17]). They found a Washburn-like behavior with the same time law $L_u(t) \propto \sqrt{t}$. Gruetzmacher et al. [18] investigated the flow behavior of a typical lubricant in DLIP textures and found an anisotropic spreading behavior with a higher spreading velocity parallel to the surface patterns compared to perpendicular to them. In the study of Klima [13], the textures ended at the transition region into the tribocontact in order to avoid fluid displacement. The difference in this study is that the textures are created throughout the entire sample, except for the last part, were selective structuring is applied. Since the DLIP textures have a low depth and therefore are erased quickly inside the tribocontact, the lubricant transport in the surrounding channel-like textures might be important. Since the DLIP textures are in the lower micrometer-range, microfluidics play an important role. Therefore, an important non dimensional number is introduced. Non dimensional numbers are of the utmost importance to compare the relative importance of different physical phenomena [19]. The Bond number compares the relative importance of gravity and surface tension:

$$\mathrm{Bo} = \frac{\rho g l^2}{\gamma}, \tag{2}$$

where l is a characteristical length, in our case the channel width. When the length $l = l_c$, Bo= 1. l_c is the capillary length which is defined as

$$l_c = \sqrt{\frac{\gamma}{\rho g}}, \tag{3}$$

where γ is the surface tension between the liquid and an adjacent fluid, typically air, and g is the gravitational acceleration. For this study, l_c is around 2 mm, which is much larger than our characteristic length for the DLIP textures which leads to a small Bond number. A small Bond number indicates that surface forces dominate gravity. This is typical for microfluidics and often allows the neglect of gravitational influence.

In order to better understand the microfluidic transport processes, a numerical model is applied. For this model, the open-source platform OpenFOAM® is applied. The two-phase wetting phenomena are described by the phase-field method. In this method, the Cahn-Hilliard equation describing the multiphase system is coupled with the Navier Stokes equations governing the flow phenomena. Due to the diffusive character of the Cahn-Hilliard equation, this method allows a motion of the contact line in combination with a no-slip boundary condition at the solid wall [20]. The first implementation of this method in OpenFOAM® was done by Cai et al. [21] who validated the numerical model for different test cases and investigated the wetting phenomena of a droplet on a flat substrate. Fink et al. [22] investigated the hydrophobicity of micro-textured surfaces and showed good agreement between the numerical model and experiments, at least for the advancing phase.

1.2. Outline of Article

This study aims at elucidating the underlying phenomena of oil transport in DLIP textures and oil transition into the tribocontact. To this end, first, a numerical model is introduced and applied. It allows the investigation of the spreading (advancing phase) of a fluid in channels of different sizes which are typical for the case of laser textured surfaces (widths down to 10 µm). The results of the numerical investigation convey the idea that the capillary effect inside the channels is the driving force for the oil transport in the channels towards the tribocontact (similar to the work from Klima [13]). In a second step, experiments for understanding the fluid transport in DLIP channels are conducted. These are twofold. On the one hand, the fluid transport inside the channel-like textures is examined and on the other hand one possible transition process of the oil out of the channels into the tribocontact is investigated.

2. Materials and Methods

2.1. Materials

For the texturing, the fluid transport evaluation and the tribological experiments, martensitic stainless steel X90CrMoV18 (1.4112) was used. The sheets were cut in rectangular shaped parts of $16\,\text{mm} \times 16\,\text{mm}$ and $16\,\text{mm} \times 6\,\text{mm}$ for the fluid transport and the tribological evaluation, respectively. All samples were polished by lapping in order to obtain a roughness of $R_z = 0.20\,\mu\text{m}$. Before the experiments, the polished samples were treated with the laser except for the reference sample. All samples were cleaned using isopropanol in an ultrasonic bath for $10\,\text{min}$. Afterwards, the samples were dried and then rinsed with petrolether (boiling temperature 40 to $60\,°\text{C}$). This procedure was repeated before the laser process, before the fluid transport experiments and before the tribological evaluation.

As a lubricating oil, a Polyalphaolefin (PAO) with a viscosity of $39.5\,\text{mm}^2/\text{s}$ at $25\,°\text{C}$ (operating condition) was used. It is the base oil of the commonly used grease Isoflex Topas L32. This oil has no additives in order to avoid side effects which are not induced by the surface structures. The properties of this oil were also used for the numerical model. There, it was assumed that the oil has the behavior of a Newtonian fluid.

As a tribological counterbody, a 100Cr6 ball with a diameter of $12.7\,\text{mm}$ was used.

2.2. Numerical Methods

This section introduces the mathematical formulation of the diffuse interface model. In our method the Cahn-Hilliard phase-field equation and the Navier-Stokes equations are coupled.

2.2.1. Phase-Field Approach for Interface Evolution

In the phase-field method, the phase distribution of two phases, liquid (L) and gas (G), is described by an order parameter C. This parameter takes respective values $C_L = 1$ and $C_G = -1$ for the corresponding bulk phases and varies rapidly but smoothly in a transition layer, which is called the diffuse interface. The interface dynamics is governed by an evolution equation for C, the convective Cahn-Hilliard equation. It is important to note that it is the diffusion process of C that governs the motion of the contact line. The Cahn-Hilliard equation reads

$$\frac{\partial C}{\partial t} + \nabla \cdot (\mathbf{u}C) = \kappa \Delta \Phi, \tag{4}$$

where $\mathbf{u}$ is the velocity field, t is the time, κ the mobility and Φ the chemical potential, which is defined as

$$\Phi = \frac{\lambda}{\varepsilon^2}C(C^2 - 1) - \lambda \Delta C. \tag{5}$$

Here, ε is an interfacial thickness parameter and λ is the energy mixing parameter,

$$\lambda = \frac{3\sigma\varepsilon}{2\sqrt{2}}. \tag{6}$$

As can be seen, it depends on the surface tension σ and ε. The Cahn number $Cn = \varepsilon/L_{\text{ref}}$ relates the interfacial width to the macroscopic length scale.

2.2.2. Governing Equations for the Fluid Flow

In this study, two immiscible, incompressible, isothermal Newtonian fluids are considered. Hence, one can describe the two-phase flow by the Navier-Stokes equations in the following form:

$$\nabla \cdot \mathbf{u} = 0 \tag{7}$$

$$\frac{\partial(\rho\mathbf{u})}{\partial t} + \nabla \cdot (\rho\mathbf{u} \otimes \mathbf{u}) = \nabla p - \nabla \cdot \left[\mu(\nabla\mathbf{u} + (\nabla\mathbf{u})^{\mathsf{T}})\right] + \mathbf{f}_\sigma + \rho\mathbf{g} \tag{8}$$

The density and viscosity fields depend on the order parameter as

$$\rho = \frac{1+C}{2}\rho_L + \frac{1-C}{2}\rho_G, \tag{9}$$

$$\mu = \frac{1+C}{2}\mu_L + \frac{1-C}{2}\mu_G, \tag{10}$$

where $\rho_{L,G}$ and $\mu_{L/G}$ are the density and viscosity of the pure phases. In Equation (8), p denotes the pressure field and $\mathbf{f}_\sigma$ the surface tension force which reads

$$\mathbf{f}_\sigma = -C\nabla\Phi. \tag{11}$$

2.2.3. Numerical Aspects

Simliar to the work from Cai et al. [21], OpenFOAM® was applied to solve the above system of equations numerically with a finite-volume method. Spatial derivatives are approximated by a high-resolution scheme (Gamma scheme) and time integration is performed by a second-order two time level backward scheme (Gear's method). The time step is chosen such that the maximum Courant number is 0.1. For further details see [21]. As suggested by Zolper et al. [23], PAO is considered to behave as a Newtonian fluid. In OpenFOAM®, it is also possible to model fluids with non-Newtonian behavior. However, the multiphase character of the problem in this study makes this more challenging, see for example [24]. Nevertheless, Niethammer et al. [25] show a rigorous treatment of a gas–liquid system with a non-Newtonian liquid phase. In this study, focus is laid on the steady state solution which does not depend on the transient viscosity behavior.

2.3. Experimental Methods

2.3.1. Laser Surface Texturing

For the DLIP texturing process, an ultrashort pulsed laser (Trumpf Trumicro 5 × 50) was used, with a wavelength of 1030 nm and a pulse length of 6 ps. The base repetition rate was 400 kHz. A DLIP optical head (Fraunhofer IWS, Dresden, Germany) is installed to automatically create the interference pattern. The setup of the optical head has been already published elsewhere [26]. In this study, the interference of two coherent laser beams was used, which leads to a line-like structure. The objective for the laser texturing is to obtain textures with a period of 4.0 µm and 8.0 µm with two depths for each period and the constraint that one depth should be comparable for both periods. The final period can be adjusted by fixing a pyramid position inside the optical head which influences the optical path. Since the optical head is installed in a fixed position the samples are moved on a linear axis system. They are moved with a speed of 25 mm/s in a meandering shape with an offset given by the hatch distance. This distance needs to be a multiple of the period in order to avoid destruction of the textures. In Table 1, the laser parameters applied to create the DLIP textures are outlined. The laser parameters need to be adapted for the different textures to guarantee good quality because they all influence each other. For example, S2 and S3 should have a period of 8.0 µm and 4.0 µm, respectively, but at the same time the depth should be the same. Therefore, if all other parameters stay the same, the pulse energy must be reduced if the hatch distance for the 4.0 µm period is smaller. As a summary, the most uniform line-like textures can be produced for low pulse energies with small hatch distances and a comparably small pulse overlap. The reservoir and the scale on the fluid transport samples were also textured with an ultrashort pulsed laser (Trumpf Trumicro 5070 Femto Edition).

Table 1. Laser parameters for the structures evaluated in this study (S1–S4).

	Period	Hatch Distance	Frequency	Pulse Energy	Fluence	Pulse Overlap
S1	8.0 µm	16.0 µm	20 kHz	10.3 µJ	0.82 J/cm^2	96.88%
S2	8.0 µm	8.0 µm	10 kHz	3.5 µJ	0.55 J/cm^2	93.75%
S3	4.0 µm	4.0 µm	10 kHz	2.6 µJ	0.41 J/cm^2	93.75%
S4	4.0 µm	4.0 µm	6.7 kHz	3.5 µJ	0.55 J/cm^2	90.63%

2.3.2. Fluid Transport Evaluation

The pictures and videos for the fluid transport evaluation were taken with a Keyence microscope using two different lenses (20×–200× magnification and 100×–1000× magnification). Figure 1a shows the experimental setup for the fluid transport experiments and Figure 1b shows the setup for the fluid transition experiments with the glass lens. 1 µl of the oil was dispensed on the sample inside the textured reservoir with the help of an Eppendorf pipette. The glass lens used as a counterpart in the meniscus experiments was an uncoated plano-convex lens made of N-BK7 with the same radius of curvature as the original tribological counterpart.

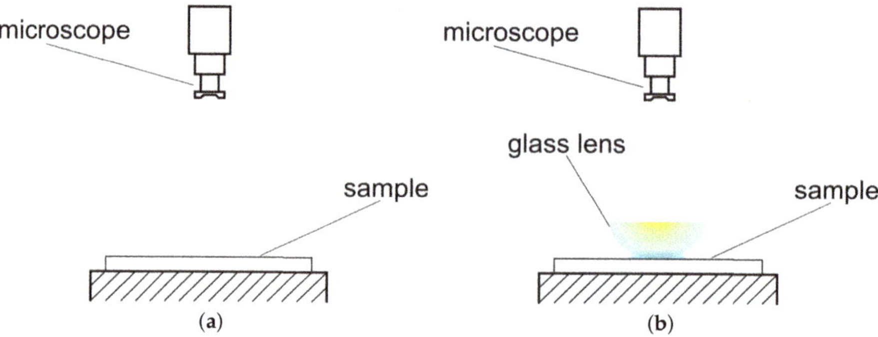

Figure 1. Experimental setup for fluid transport evaluation. In (**a**) the setup for the fluid transport evaluation inside the laser textures is shown. The fluid propagation in the channel-like textures is observed via a microscope from the top. In (**b**) the setup for the lens experiments is illustrated. The transparent glass lens is a substitute for the tribological counterpart and allows fluid flow analysis at the transition region between channels and tribocontact.

2.3.3. Characterization of Tribological Properties

For the tribological evaluation of the surface textures, an Anton Paar Tribometer (MCR301) was used. This device is a ball-on-three-plates tribometer. Figure 2 shows the experimental setup. As a first step, three samples were fixed in the sample holder with an angle of 45° regarding the horizontal plane. Secondly, 1 µL of the PAO was put on each sample with a µL-pipette (Eppendorf). Then, the sample holder was fixed on the device and the ball in the ball-holder was placed above the sample. The ballholder was approached to the sample which makes the oil ousted around the tribocontact. Finally, a normal force of 20 N was applied, which corresponds to 9.43 N per sample. When this force is reached, the ball started to rotate and the measurement cycle began. The measurement cycle consisted of 9 Stribeck curves, which were grouped in 3 sections. Furthermore, it consisted of three endurance parts at the beginning of the cycle and in between the Stribeck sections. Their purpose is to avoid run-in behavior. The total normal force always was 20 N and the rotation speed during the endurance parts was 20 rpm which corresponds to a sliding velocity of 0.009 m/s. The Stribeck curves were recorded using a logarithmic ramp. The normal force per sample of 9.43 N is kept constant while the ball starts to rotate and speeds up to a sliding velocity of 1.4 m/s. The COF is recorded during the entire ramp. Within the test cycle this procedure is repeated nine times such that in the end a mean value and the standard deviation of the COF versus the sliding velocity can be determined.

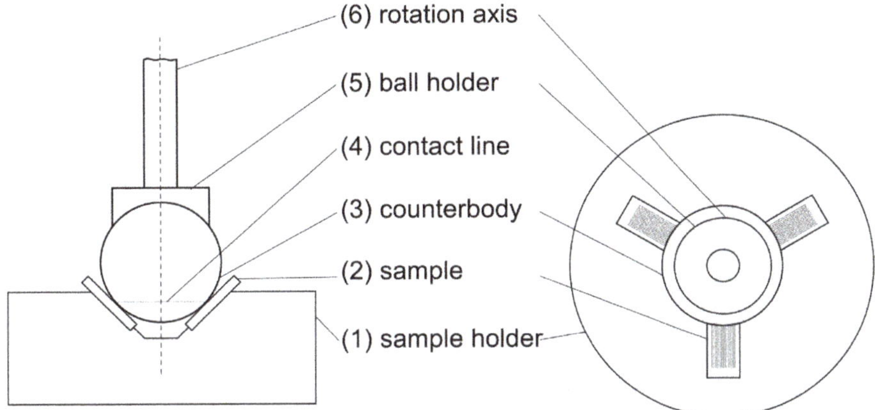

Figure 2. Experimental setup for tribological evaluation. A side view on the left-hand side and a top view on the right-hand side shows the important components of the setup and the contact line (**4**) on the rotating ball (**3**).

2.3.4. Surface Characterization

A Laser Confocal Microscope (Olympus LEXT OLS4000) was used to characterize the surface morphology of the laser treated samples. For postprocessing of the images the software MountainsMap was applied.

3. Results and Discussion

3.1. Numerical Results-Droplet Wetting on Line-Like Textures

In Figure 3, the numerical results for the case of droplet wetting on line-like textures are shown. The wetting process of a droplet with an initial diameter of 140 µm and an impact velocity of 1×10^{-6} m/s is investigated for line-like textures with channel widths of 10 to 30 µm. The depth of the channels is kept constant at 10 µm. The equilibrium contact angle for all cases is 60°. The mesh resolution is the same independent of the channel width and corresponds to a cell length of 1.67 µm in all directions. As a consequence, the minimum number of cells in a channel is 5. This is neccesary for a correct numerical resolution of the gas–liquid interface. In all simulations, the global Cahn number is Cn = 0.02. In Appendix A.2, a full overview of all numerical parameters and settings is given. Two different views are chosen in order to illustrate the spreading behavior along the line-like textures. The first view (left column in Figure 3) shows a cut through the domain at the geometrical center of the droplet. The second view (right column) shows one half of the spreaded droplet filling the channels from the bottom.

Figure 4 shows the quantitative results of the numerical investigations. For this, the spreading factor χ is defined as

$$\chi = \frac{\text{total extension of spreaded droplet in channel}}{\text{initial droplet diameter}}. \tag{12}$$

This quantity is plotted over a relative simulation time t^* which is defined as $t^* = t/t_{\text{total}}$, where $t_{\text{total}} = 0.039$ s for all simulations.

From Equation (12), it is possible to see that the smaller the channel width the higher the spreading factor. The minimal channel width in the simulation is 10 µm. For smaller channels the cell size has to be reduced even further in order to ensure correct representation of the gas–liquid interface. However, smaller cells result in significant higher computational efforts which were not affordable for this study. However, already at a size of 10 μm, the spreading factor reaches a value greater than 2. Since the driving force is the capillary force inside the channels and since this force increases for smaller widths, the demonstrated effect even increases for smaller channel widths.

This phenomenon illustrates the anisotropic flow behavior of line-like textures which can be applied to guide lubricant towards the tribocontact. Therefore, the numerical results can serve as a theoretical demonstration of the possibility for active lubricant transport in laser textures of small sizes and with a line-like pattern.

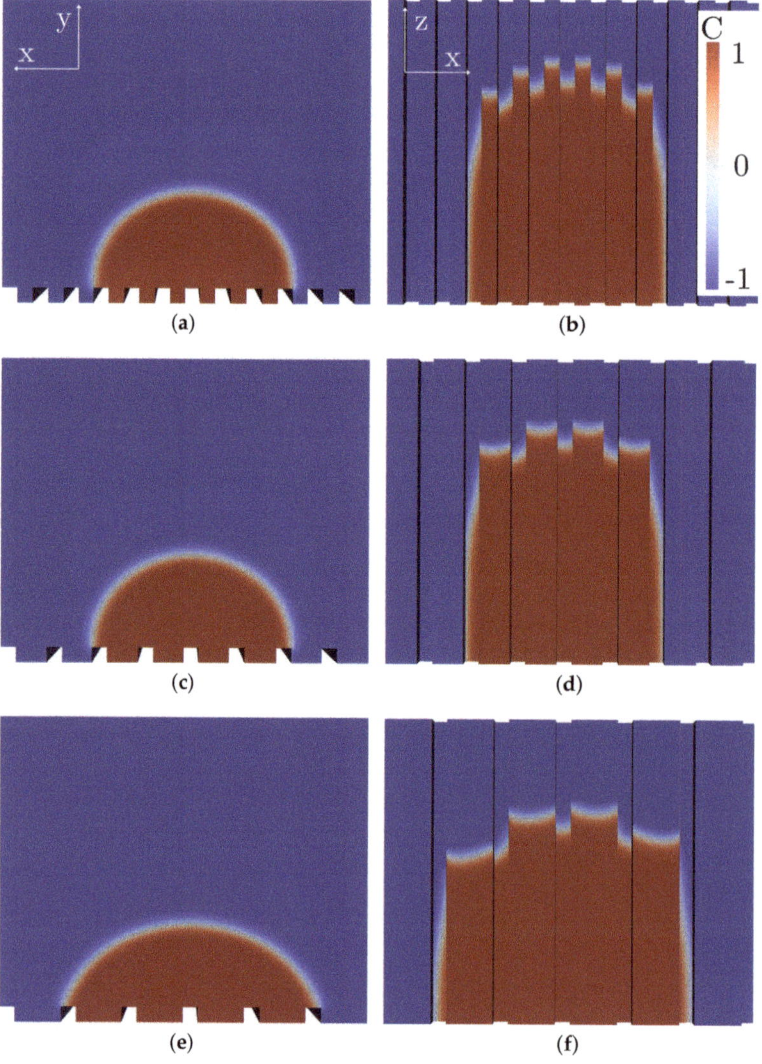

Figure 3. Simulation of oil in channels of different widths at t = 0.039 s. The left column (**a**,**c**,**e**) illustrates a cut through the domain at the geometrical center of the droplet. The right column (**b**,**d**,**f**) shows one half of the spreaded droplet filling the channels from the bottom.

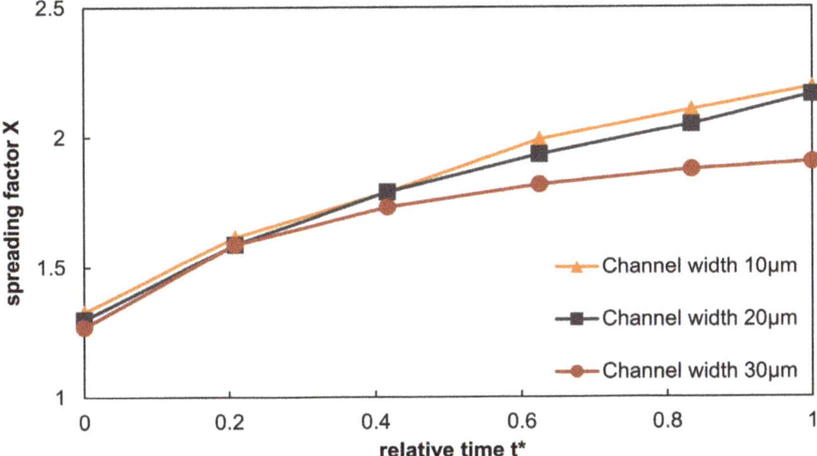

Figure 4. Quantitative evaluation of the numerical results comparing the spreading factor χ for the different channels widths as a function of the relative time t^*.

3.2. Experimental Results

3.2.1. Fabrication of Line-Like Surface Textures Using DLIP

Four different laser textures were created as explained in Section 2.3.1 applying DLIP. For all textures a line-like pattern was chosen. Figure 5 illustrates microscope pictures of all textures which are evaluated in this study Figure 5a–d and the corresponding profiles for these textures Figure 5e–h. Structures S1 and S2 have a period of 8 µm and S3 and S4 a period of 4 µm. Table 2 shows the depths of the textures. The value is an averaged value for the entire surface. Figure 6 shows a side view of the channels for texture S1. This image shows exemplarily the integrity of the produced surface textures. For the fluid dynamical behavior it is of utmost importance that no defects of the channels occur because otherwise the fluid is not transported along the channels.

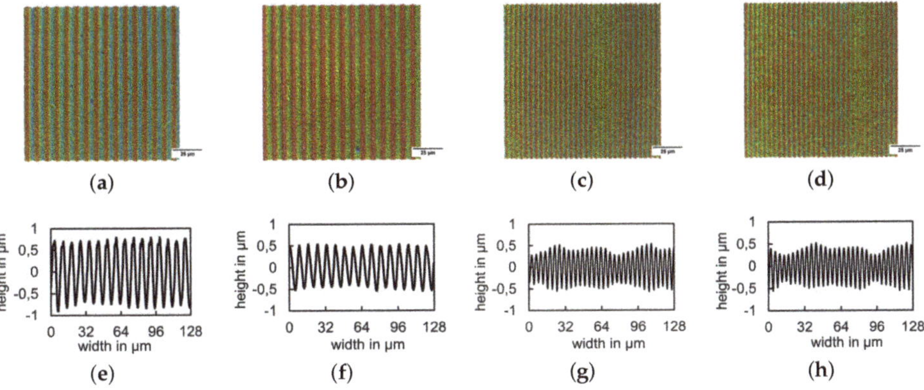

Figure 5. Microscope pictures and profiles of laser textures evaluated in this study. Textures and profiles of S1 (**a**) and (**e**), S2 (**b**) and (**f**), S3 (**c**) and (**g**) and S4 (**d**) and (**h**).

Table 2. Depths of the laser textures S1–S4.

	S1	**S2**	**S3**	**S4**
period	8.0 µm	8.0 µm	4.0 µm	4.0 µm
depth	1.5 µm	1.0 µm	1.0 µm	0.7 µm

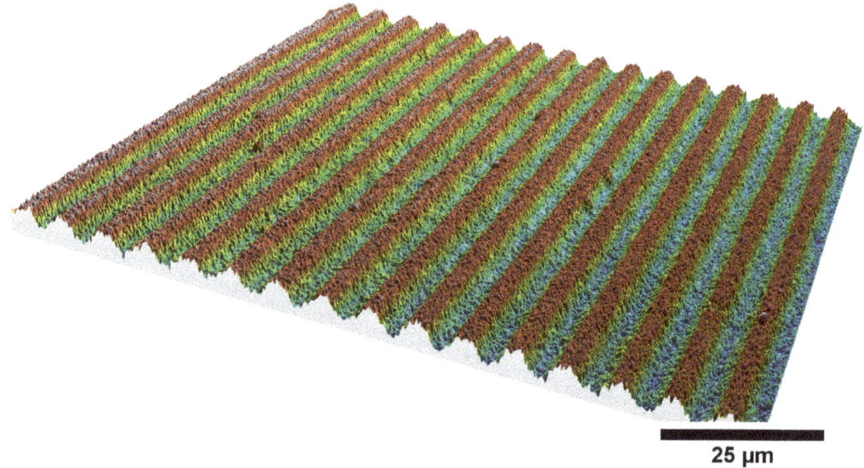

Figure 6. Side view of channels from texture S1. This microscopic image demonstrates in an exemplary manner the integrity of the laser produced textures. It can be seen that there are no discontinuities which lead to open structures.

3.2.2. Fluid Transport Inside Laser Textured Surfaces

For evaluating the fluid transport inside the DLIP textures specially designed samples were used. As can be seen in Figure 7a, the sample has four reservoirs (1) with a capacity of around 1 µl each. The DLIP textures (2) start inside the reservoir. Furthermore, next to the DLIP textures a millimeter-scale (3) is placed in order to measure the traveled distance of the fluid front. One bar at the scale corresponds to 500 µm. This means that for each texture the experiment can be repeated four times. Figure 7b shows a zoom-in of the traveling fluid front at a distance of around 500 µm. The fluid transport in three different DLIP textures was evaluated (S2–S4).

Figure 8 shows the results of the fluid transport experiments. On the vertical axis the traveled distance in millimeter and on the horizontal axis the time in seconds are plotted. All curves show a square-root behavior similar to the Washburn-behavior mentioned in Section 1. This is due to the increasing friction with increasing length of the wetted part of the textures. The texture with a period of 8 µm allows slower transport than textures with a period of 4 µm (smaller channel width). Comparing the depths of the 4 µm textures, a higher depth (1 µm) leads to faster transport than a lower depth (0.7 µm).

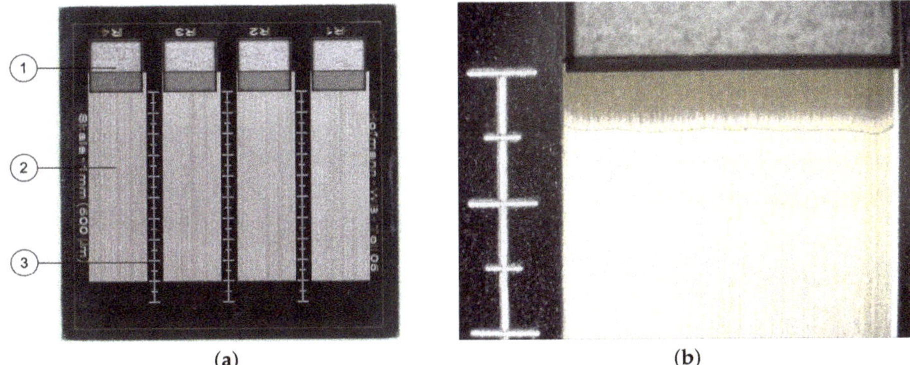

Figure 7. Experimental investigation of fluid transport in laser textured surfaces. (**a**) Specially designed sample with different regions. 1: Reservoir, 2: Laser surface textures, 3: mm-scale (500 µm per bar); (**b**) fluid front propagating in laser textures at a distance of 500 µm.

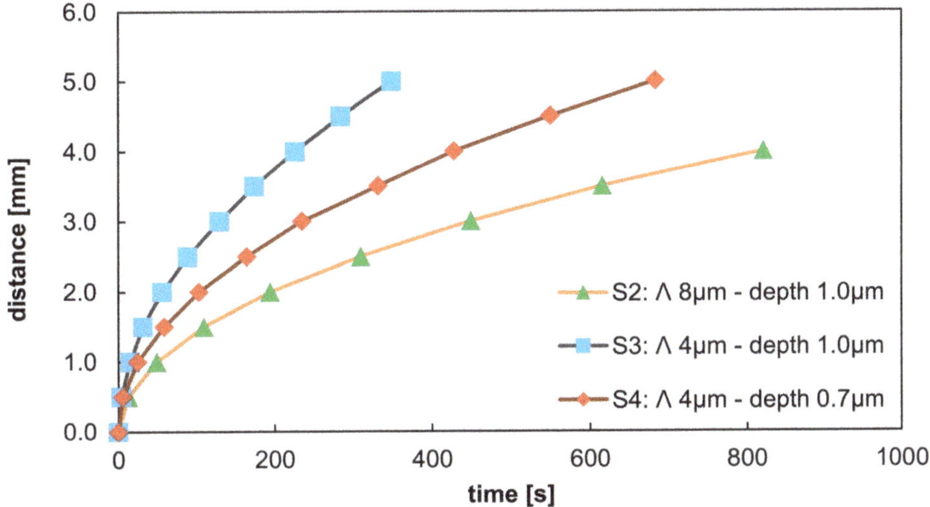

Figure 8. Quantitative results of flow behavior in different Direct Laser Interference Patterning (DLIP) textures (S2–S4). The sample introduced in Figure 7 was used to measure the distance of the fluid column in different channel-like textures as a function of time. The curves show Washburn-like behavior with the fastest transport in channels with smaller period and higher depth.

3.2.3. Evaluation of Tribological Performance

In this section, the results of the tribological evaluation are outlined. The comparison of the Stribeck curves for the different DLIP textures S1–S4 introduced in this study show no difference in terms of coefficient of friction (see Figure A1 in Appendix A.1). Therefore, texture S1 with a period of 8 µm and a depth of around 1.5 µm is chosen for further tribological evaluation. The idea is that three samples with the same parameters as for S1 but with differently structured regions are compared to each other and to an untextured, polished reference sample. The textured samples differ in the area which is textured. The objective of this experiment is to understand whether it makes a difference if the surface textures directly end at the tribocontact or at some distance away from it. As showed in Section 2.3.3, the setup of the tribometer is such that a ball rotates on the samples in a point contact. Therefore, the position of the tribocontact is known a priori. Hence, three differently structured samples are compared: a fully textured sample, a textured sample where only the region of the tribocontact is textured and the opposite case, namely a textured sample where the region of the tribocontact is blank but everything else is textured. The diameter of the tribocontact is around 250 µm such that the region of the tribocontact is chosen to have a width of 500 µm. The introduced textured samples are called "selectively textured samples". One can distinguish between "fully textured", "triboregion only" and "triboregion blank".

Besides the four Stribeck curves, Figure 9 also shows schematics of the samples described and evaluated in this section. It shows typical Stribeck curves for the three cases compared to the one of a polished reference sample. The polished sample has the highest coefficient of friction for all lubricating regimes. The coefficient of friction for the case "triboregion blank" is smaller , especially for the mixed lubrication regime, but the difference is negligible. However, this is not the case for the samples where the region of the tribocontact is textured ("fully textured" or "triboregion only"). Their difference again is negligible but compared to the other two samples, the reduction of the COF, especially for the mixed lubrication regime, is significant. This result might be a bit surprising. In the following section an attempt is made to explain this behavior.

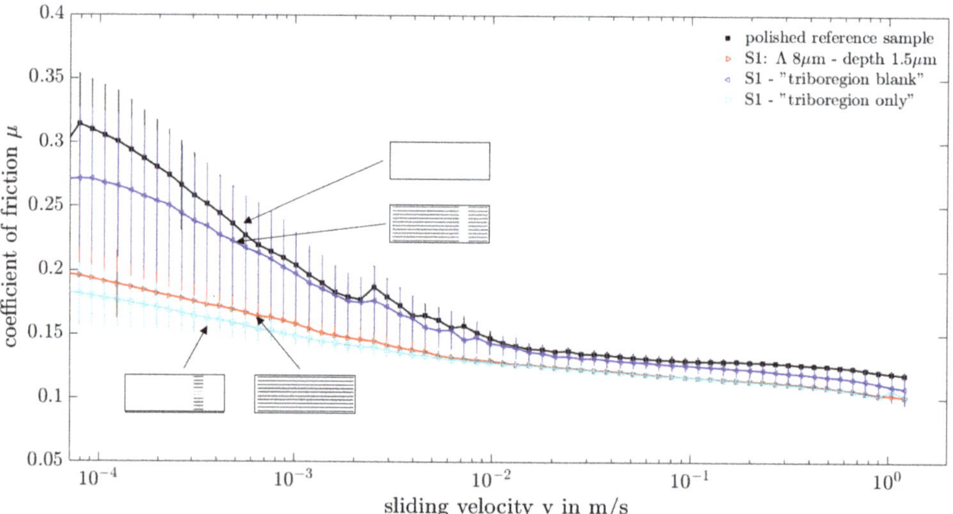

Figure 9. Experimental results from tribological evaluation of selectively structured samples. The Stribeck curves of three different samples with the same laser texture S1 ($\lambda = 8$ µm, depth $= 1.5$ µm) but with differently structured regions ("fully textured", "triboregion blank" or "triboregion only") are compared to the Stribeck curve of a polished reference sample.

3.2.4. Transition of Fluid out of Laser Textured Surfaces into Tribocontact

In the previous section it was shown that it makes a difference whether the triboregion is textured or not (meaning that after wear textures end directly at the tribocontact). In this section, we outline one possible effect that explains this behavior. The focus of this section lies on the transition region out of the laser textures into the tribocontact. However, the setup of the tribometer does not allow a precise observation of the transition region and the oil flow in it. Therefore, a different setup is applied to elucidate the phenomena occurring in this transition region. Nonetheless, the alternative setup has certain limitations:

1. Due to a necessary amount of light for microscopy a glass lens is used instead of the 100Cr6 ball. The lens has the same radius of curvature and similar wetting properties.
2. The phenomena in the tribometer occur under dynamic conditions. In the alternative setup it is only possible to observe effects under static conditions.
3. The test samples first run in the tribometer and then the transition effects with the glass lens are observed. Therefore, the tribocontact exists on the sample and the glass lens is put manually into this contact.

Figure 10a shows a side view of a sample with the lens on top. Figure 10b corresponds to the setup explained in Section 2.3.4 and shows a top view of the textures through the glass lens. The picture illustrates the tribocontact, the lubrication front with the indicated flow direction and the final meniscus at the lens.

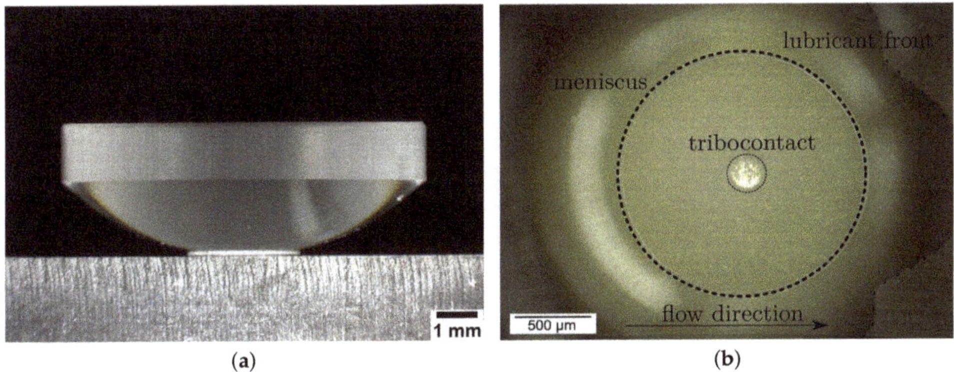

(a) (b)

Figure 10. In order to elucidate the flow behavior in the transition region between channels and tribocontact a transparent glass lense made of N-BK7 is used as a substitute for the counterbody. (a) A side view of the glass lens on top of a textured sample; in (b) the lens is manually put into the tribocontact and the flow around the lens is observed through the lens with a microscope.

The temporal course of the fluid flow is shown in Figure 11. In Figures 11a–c it is possible to see the fluid front moving from left to right and the tribocontact. In Figure 11a, the front is approaching the tribocontact. In Figure 11b, it is possible to see that a lee-region is formed behind the contact point of the lens where initially no oil arrives. In Figure 11c, this region is also filled with oil. In Figure 11d, the fluid front has passed the field of view and a meniscus "climbing" the glass lens has formed. The driving force for this meniscus is the surface tension. It stops when an equilibrium state has been reached. This phenomenon can explain how oil is "pulled out" of the surface textures. As long as a continuous oil supply through the textures is guaranteed, the meniscus can grow until equilibrium state. Hence, it marks one possible explanation for the transition of the oil out of the textures into the tribocontact. This phenomenon can only occur when the textures are present exactly where the triboregion begins which explains why the case "triboregion blank" is worse in terms of reduction of COF than the case "triboregion only".

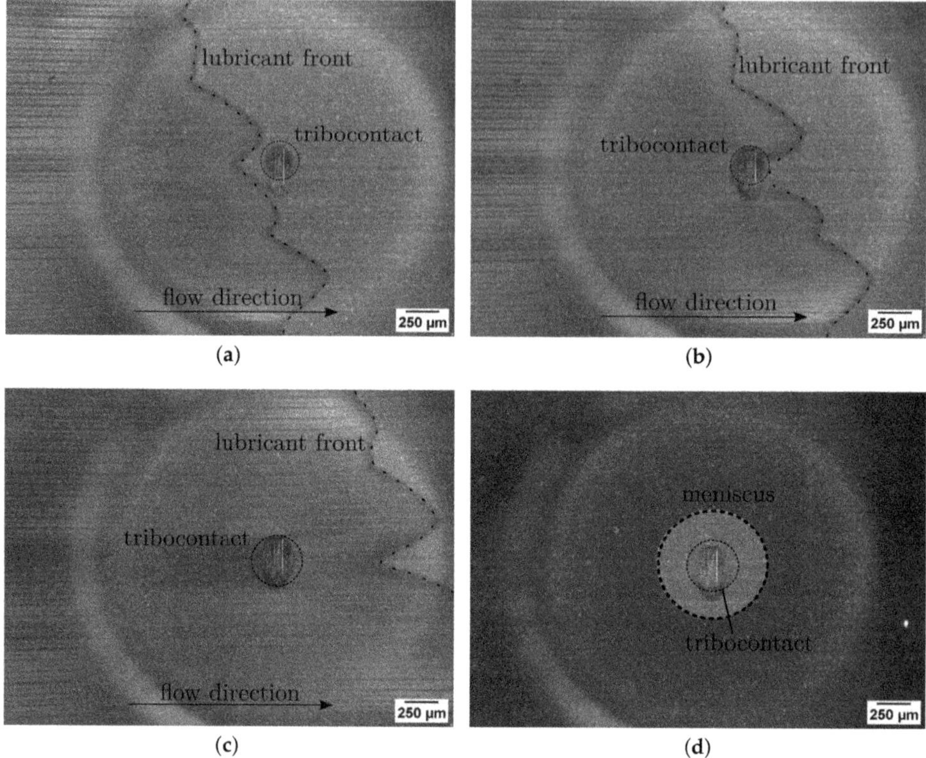

Figure 11. Temporal course of fluid flow around glass lens in tribocontact and building of meniscus. (**a**) Fluid front approaching the tribocontact; (**b**) fluid front passing by the tribocontact; (**c**) fluid filling the channels behind the tribocontact and (**d**) building meniscus started at the contact of the lens with the sample.

4. Conclusions

In this study, the fluid transport in DLIP textures was evaluated. From the numerical investigations it was shown that the capillary force plays an important role in small channels and the spreading factor is higher in channels with a width of 10 µm compared to channels with a width of 30 µm. DLIP textures were created with two different periods (4 µm and 8 µm) and three depths (1.5 µm, 1.0 µm and 0.7 µm). First, the fluid transport in these textures was evaluated. It was found that all textures show a Washburn-like behavior with the fastest transport in the texture with a period of 4 µm and a depth of 1 µm. In a second step, the tribological performance of four different DLIP textures was evaluated. It was shown that the level of COF could be reduced significantly compared to a polished reference but only negligible differences were found between the four textures. In a third step, the effect of differently structured zones was evaluated. For this test, one representative DLIP texture was chosen. In the evaluation, three samples with the same texture but with differently structured zones were compared to a polished reference sample. The samples differed as follows. One sample was fully textured (case "fully textured"), on one sample only the triboregion was textured (case "triboregion only") and on the third sample the triboregion was blank and the rest was textured (case"triboregion blank"). The resulting Stribeck curves showed similar behavior for the polished sample and case"triboregion blank" and a much lower COF but also similar behavior for case "fully textured" and case "triboregion only". The result shows the necessity of available textures directly at the position where the tribocontact ends. At this position the requirement for capillary forces breaks down if the channel is fully filled because then there is no meniscus at the gas–liquid interface. Therefore, in a last step, one possible explanation for this behavior was outlined. For this, a glass lens was chosen as a substitute for the tribological counterpart and was put manually into the tribocontact. Then, the fluid flow around the lens and the building meniscus could be observed. The prerequisite for the building meniscus is that the structures end in the tribocontact such that the fluid can flow until the contact point of the lens. For the abovementioned case "triboregion blank" this is not the case which might explain the higher COF. One limitation is that the lens experiments were conducted under static conditions but the effects in the tribometer are highly dynamic. However, the principle of this hypothesis also holds for the dynamic case.

Author Contributions: Conceptualization, all authors; Methodology, T.S. and H.M; Software, H.M. and T.S.; Validation, T.S. and H.M.; Formal Analysis, T.S. and H.M.; Investigation, T.S. and H.M.; Resources, all authors; Data Curation, T.S.; Writing—Original Draft Preparation, T.S.; Writing—Review and Editing, all authors; Visualization, T.S.; Supervision, T.K., H.M. and A.F.L; Project Administration, T.K. and A.F.L.; Funding Acquisition, T.K., H.M. and A.F.L.

Funding: This research was funded by the European Union's Horizon 2020 Research and Innovation Programme under the Marie Skłodowska-Curie grant agreement No. 675063. The work of A.F.L. is also supported by the German Research Foundation (DFG) under Excellence Initiative program by the German federal and state governments to promote top-level research at German universities.

Acknowledgments: The author H.M. thanks the Collaborative Research Center CRC 1194 "Interaction between Transport and Wetting Processes" by the German Research Foundation (DFG). The authors also thank Gerd Dornhöfer, Joachim Klima and Sarona Frank from the Robert Bosch GmbH for the fruitful discussions. Furthemore, they thank Marius Hofmann for his contribution in this study.

Conflicts of Interest: The authors declare no conflict of interest. The funders had no role in the design of the study; in the collection, analyses, or interpretation of data; in the writing of the manuscript, or in the decision to publish the results.

Abbreviations

The following abbreviations are used in this manuscript:

LST	laser surface texturing
COF	coefficient of friction
DLW	Direct Laser Writing
DLIP	Direct Laser Interference Patterning

Appendix A

Appendix A.1. Tribological comparison of all DLIP textures introduced in this study

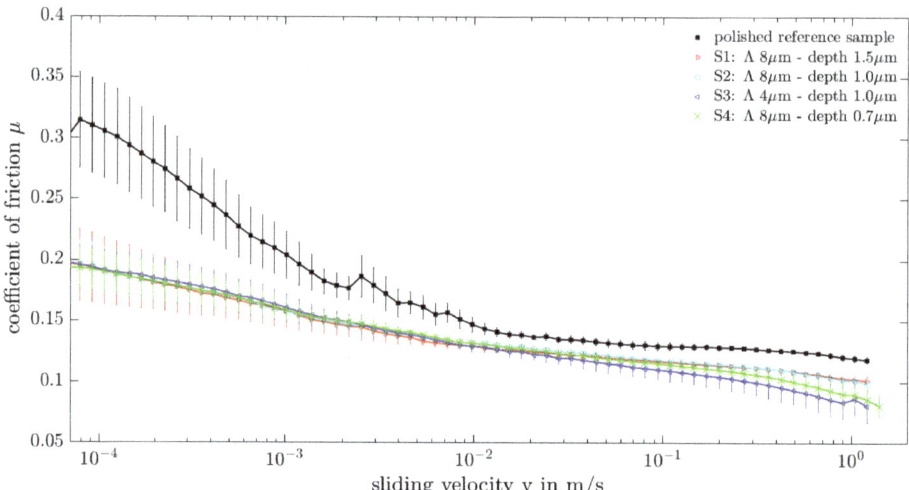

Figure A1. Tribological comparison of Stribeck curves for all DLIP textures introduced in this study and a polished reference sample.

Appendix A.2. Numerical Parameters and Solver Settings

In OpenFOAM further specification can be made with respect to the algorithms used to solve equations and matrices numerically. Tables A1 and A2 show the specifications in *fvSolution* and *fvSchemes*, respectively.

Table A1. Specifications *fvSolution*.

Field	Solver	Preconditioner	Smoother
pd	PCG	DIC	GaussSeidel
U	BiCGStab	DILU	none
C	PBiCG	DILU	none
Ccoupled	GMRES	Cholesky	none

Table A2. Specifications *fvSchemes*.

Operation	OpenFOAM	Scheme
Time derivative	ddt	Euler
Gradient	grad	Gauss linear
Divergence	div	Gauss Gamma
Laplacian	laplacian	Gauss linear uncorrected
Interpolation	interpolation	linear

References

1. Holmberg, K.; Andersson, P.; Erdemir, A. Global energy consumption due to friction in passenger cars. *Tribol. Int.* **2012**, *47*, 221–234. [CrossRef]
2. Baumgartner, W.; Saxe, F.; Weth, A.; Hajas, D.; Sigumonrong, D.; Emmerlich, J.; Singheiser, M.; Böhme, W.; Schneider, J.M. The sandfish's skin: Morphology, chemistry and reconstruction. *J. Bionic Eng.* **2007**, *4*, 1–9. [CrossRef]

3. Kovalchenko, A.; Ajayi, O.; Erdemir, A.; Fenske, G. Friction and wear behavior of laser textured surface under lubricated initial point contact. *Wear* **2011**, *271*, 1719–1725. [CrossRef]

4. Borghi, A.; Gualtieri, E.; Marchetto, D.; Moretti, L.; Valeri, S. Tribological effects of surface texturing on nitriding steel for high-performance engine applications. *Wear* **2008**, *265*, 1046–1051. [CrossRef]

5. Blatter, A.; Maillat, M.; Pimenov, S.; Shafeev, G.; Simakin, A.; Loubnin, E. Lubricated sliding performance of laser-patterned sapphire. *Wear* **1999**, *232*, 226–230. [CrossRef]

6. Andersson, P.; Koskinen, J.; Varjus, S.e.; Gerbig, Y.; Haefke, H.; Georgiou, S.; Zhmud, B.; Buss, W. Microlubrication effect by laser-textured steel surfaces. *Wear* **2007**, *262*, 369–379. [CrossRef]

7. Duarte, M.; Lasagni, A.; Giovanelli, R.; Narciso, J.; Louis, E.; Mücklich, F. Increasing lubricant film lifetime by grooving periodical patterns using laser interference metallurgy. *Adv. Eng. Mater.* **2008**, *10*, 554–558. [CrossRef]

8. Rosenkranz, A.; Martin, B.; Bettscheider, S.; Gachot, C.; Kliem, H.; Mücklich, F. Correlation between solid–solid contact ratios and lubrication regimes measured by a refined electrical resistivity circuit. *Wear* **2014**, *320*, 51–61. [CrossRef]

9. Etsion, I. State of the art in laser surface texturing. *J. Tribol.* **2005**, *127*, 248–253. [CrossRef]

10. Braun, D. Größeneffekte bei strukturierten tribologischen Wirkflächen, Ph.D. Thesis, KIT, Karlsruhe, Germany, 2015.

11. Stark, T.; Alamri, S.; Aguilar-Morales, A.I.; Kiedrowski, T.; Lasagni, A.F. Positive effect of laser structured surfaces on tribological performance. *J. Laser Micro Nanoeng.* **2019**, *14*, 13–18.

12. Lasagni, A.F. Laser interference patterning methods: Possibilities for high-throughput fabrication of periodic surface patterns. *Adv. Opt. Technol.* **2017**, *6*, 265–275. [CrossRef]

13. Klima, J. *Lubricant Transport Towards Tribocontact in Capillary Surface Structures*; KIT Scientific Publishing: Karlsruhe, Germany, 2018; ISBN 978-3-7315-0814-4.

14. Lucas, R. Ueber das Zeitgesetz des kapillaren Aufstiegs von Flüssigkeiten. *Kolloid Zeitschrift* **1918**, *23*, 15–22. [CrossRef]

15. Washburn, E.W. The dynamics of capillary flow. *Phys. Rev.* **1921**, *17*, 273. [CrossRef]

16. Yang, D.; Krasowska, M.; Priest, C.; Popescu, M.N.; Ralston, J. Dynamics of capillary-driven flow in open microchannels. *J. Phys. Chem.* **2011**, *115*, 18761–18769. [CrossRef]

17. Chen, T. Capillary Force-Driven Fluid Flow in Open Grooves with Different Sizes. *J. Thermophys. Heat Transf.* **2015**, *29*, 594–601. [CrossRef]

18. Grützmacher, P.G.; Rosenkranz, A.; Gachot, C. How to guide lubricants–Tailored laser surface patterns on stainless steel. *Appl. Surf. Sci.* **2016**, *370*, 59–66. [CrossRef]

19. Lambert, P. *Surface Tension in Microsystems*; Springer: Berlin/Heidelberg, Germany, 2013.

20. Jacqmin, D. Contact-line dynamics of a diffuse fluid interface. *J. Fluid Mech.* **2000**, *402*, 57–88. [CrossRef]

21. Cai, X.; Marschall, H.; Wörner, M.; Deutschmann, O. Numerical Simulation of Wetting Phenomena with a Phase-Field Method Using OpenFOAM®. *Chem. Eng. Technol.* **2015**, *38*, 1985–1992. [CrossRef]

22. Fink, V.; Cai, X.; Stroh, A.; Bernard, R.; Kriegseis, J.; Frohnapfel, B.; Marschall, H.; Wörner, M. Drop bouncing by micro-grooves. *Int. J. Heat Fluid Flow* **2018**, *70*, 271–278. [CrossRef]

23. Zolper, T.; Li, Z.; Chen, C.; Jungk, M.; Marks, T.; Chung, Y.W.; Wang, Q. Lubrication Properties of Polyalphaolefin and Polysiloxane Lubricants: Molecular Structure—Tribology Relationships. *Tribol. Lett.* **2012**, *48*, 355–365. [CrossRef]

24. Deubelbeiss, Y.; Kaus, B. Comparison of Eulerian and Lagrangian numerical techniques for the Stokes equations in the presence of strongly varying viscosity. *Phys. Earth Planet. Int.* **2008**, *171*, 92–111. [CrossRef]

25. Niethammer, M.; Marschall, H.; Kunkelmann, C.; Bothe, D. A numerical stabilization framework for viscoelastic fluid flow using the finite volume method on general unstructured meshes. *Int. J. Numer. Methods Fluids* **2018**, *86*, 131–166. [CrossRef]

26. Lasagni, A.; Roch, T.; Berger, J.; Kunze, T.; Lang, V.; Beyer, E. To use or not to use (direct laser interference patterning), that is the question. In Proceedings of the Laser-Based Micro-and Nanoprocessing IX, San Francisco, CA, USA, 10–12 February 2015; International Society for Optics and Photonics: Bellingham, WA, USA, 2015; Volume 9351, p. 935115.

© 2019 by the authors. Licensee MDPI, Basel, Switzerland. This article is an open access article distributed under the terms and conditions of the Creative Commons Attribution (CC BY) license (http://creativecommons.org/licenses/by/4.0/).

Article

The Influence of Surface Texturing on the Frictional Behaviour in Starved Lubricated Parallel Sliding Contacts

Dariush Bijani [1,*], Elena L. Deladi [2], Matthijn B. de Rooij [3] and Dirk J. Schipper [3]

1   Materials Innovation Institute (M2i), P.O. Box 5008, 2600 GA Delft, The Netherlands
2   Bosch Transmission Technology, Dr. Hub van Doorneweg 120, 5026 RA Tilburg, The Netherlands
3   Laboratory for Surface Technology and Tribology, Faculty of Engineering Technology, University of Twente, P.O. Box 217, 7500 AE Enschede, The Netherlands
*   Correspondence: d.bijani@utwente.nl; Tel.: +31-(0)68-113-3568

Received: 2 July 2019; Accepted: 8 August 2019; Published: 9 August 2019

Abstract: Starvation occurs when the lubricated contact uses up the lubricant supply, and there is not enough lubricant in the contact to support the separation between solid surfaces. On the other hand, the use of textures on surfaces in lubricated contacts can result in a higher film thickness. In addition, a modification of the surface's geometrical parameters can benefit the tribological behaviour of the contacts. In this article, for parallel sliding surfaces in starved lubricated conditions, the effect of surface texturing upon the coefficient of friction is investigated. It is shown that surface texturing may improve film formation under the conditions of starvation, and as a result, the frictional behaviour of the parallel sliding contact. Furthermore, the effect of starved lubrication on textured surfaces with different patterns in the presence of a cavitation effect, and its influence on frictional behaviour, is investigated. It is shown that surface texturing can reduce the coefficient of friction, and that under certain conditions, the texturing parameter could have an influence on the frictional behaviour of parallel sliding contacts in the starved lubrication regime.

Keywords: mixed lubrication; starvation; deterministic asperity model; surface texturing; film thickness; texturing patterns; numerical modelling

---

1. Introduction

Many experimental investigations on the frictional behaviour of textured surfaces in sliding contacts have been done, and recently the theoretical research in this field gets more attention. In machine components like gears and bearings, the influence of a mixed lubrication regime could be significant where asperities and the fluid contribute to carrying the load [1]. Moreover, in these applications, the existence of starvation can influence the effect of lubricant film in contact, and increase the friction in contact [2]. Investigations of high-speed bearings have shown in many cases that they operate under the starved lubrication regime [3–5]. In many cases, the lubricant cannot ensure a full separation of surfaces, which can cause higher friction and wear. Therefore, in starved lubricated conditions it is important to identify the influential surface parameters in order to understand the frictional behaviour of these lubricated contacts.

The effect of starvation on lubrication performance is analyzed experimentally in studies by Wedeven et al. [6], Pemberton et al. [7] and Kingsbury [8], and theoretically by Chiu [9], Damiens et al. [10] and Chevalier et al. [11]. It is worth to mention that in these theoretical studies, the concept of "fractional film", introduced by Jakobsson and Floberg [12] and Olsson [13], has an important role.

In the theoretical study of Brewe and Hamrock [14] on the effect of starvation in hydrodynamically-lubricated contacts, they chose the start of the pressure build-up at the inlet meniscus boundary, and by employing a systematic reduction of the fluid inlet level, they observed an increase in the contact pressure for a specified film thickness. By considering a wide range of geometry parameters, they solved the Reynolds equation to simulate the film thickness in the contact area; moreover, in their study the film thickness formula in the hydrodynamic lubrication regime is modified to incorporate the starvation effect into it. In the work of Boness [15] on the cage and roller slip, it was shown experimentally that the oil supply can have a significant effect upon the cage and roller motion, and that limiting the oil supply decreases the amount of slip. Chevalier et al. [16] employed an iso-viscous hydrodynamic model to analyse a non-deformable body; in this study the flow continuity equation is based upon the Elrod [17,18] theory. They conclude that the inlet film shape could affect the film thickness. In Cann and Lubrecht [19], a study of the relationship between the film thickness, velocity, load and viscosity, was the focus of investigation.

Investigations on starved lubrication show that the employment of surface modification methods could be a practical and efficient method for the reduction of friction. One of the first rational methods of decreasing the friction can be by a reduction of roughness, making the surfaces smooth; however, producing an extra smooth surface is expensive [20]. In this case, surface texturing proved to be a reliable method to influence the frictional behaviour in the contacts. A well-designed use of this technique can modify the hydrodynamic component of mixed lubrication, which results in the enhancement of several tribological parameters, such as load carrying capacity and friction coefficient [21–29]. In general, when the temperature increases, the shear stress of the boundary layer decreases except when the temperature is passing from the melting point of boundary layer, moreover an increase in temperature of the parts due to interaction between asperities can change the situation from effective lubrication to high wear [30]. In work of Kango et al. [31], based on the Reynolds equation and the JFO (Jakobsson and Floberg and Olsson) boundary conditions for non-Newtonian fluids, temperature effects on textured surfaces are theoretically studied. They show that in presence of surface texturing, the average temperature of the lubricating film reduces. In work of Guzek et al. [32], upon an optimization of the surface texturing parameters in parallel bearings, they numerically solved the Reynolds equation, considering mass-conserving cavitation and viscosity changes due to temperature change. They showed that the decrease in viscosity due to the temperature rise can reduce the load carrying capacity. Therefore, cavity height ratios should be higher in order to have a similar load carrying capacity to textured surfaces with a constant temperature assumption. In work of Gu et al. [33] on the performance of surface texturing under starved and mixed lubrication, they employed a thermal mixed lubrication model considering the oil supply. They found that the start-up conditions can affect the friction coefficient. Moreover, they showed that it is easier for the textured surface to form the hydrodynamic lubrication than it is for the smooth surface, which is helpful to separate the mixed lubricated contact surfaces, and thus less friction heat is generated at the start-up phase. In the work of Bijani et al. [34] the influence of surface texturing on mixed lubricated contacts, different texturing patterns and cavity shapes are studied, and a numerical model to predict the friction is proposed.

Although the starved lubrication influence on film thickness in different applications is studied extensively in more recent times, not much work has been done on mixed lubrication under starved lubrication conditions, and in the case of friction in starved lubricated textured surfaces, even fewer studies have been done. When the lubricant in contact is limited to a specific amount, a correction of the film thickness formula is necessary, so in this study, a corrected film thickness is presented for textured surfaces under starved lubrication conditions.

In order to develop the starved lubrication model, the modified film thickness relation for starved contacts is solved, by taking the limited input film thickness into account, then the corrected film thickness is combined with the deterministic contact model. In this article the consequences of the existence of starvation in lubricated contacts on friction is discussed.

During the past decades, several efforts were devoted to study this mixed lubrication frictional behaviour [35–38]. Based on the contact model, mixed lubrication models can be divided in two types: Statistical and deterministic contact models. In the statistic models, the parameters represent the random characteristics of surface roughness. A major shortcoming of this model is its inability to provide detailed information on local roughness, which has an influence on the mechanisms of lubrication and friction. Another approach to simulate the frictional behaviour of contacting asperities results in a deterministic model, which employs the deterministic information of surface roughness. In these models, for a given separation, by summing up the local components of load and contact area, it is possible to deterministically calculate the real contact area and the total force carried by the contact.

In 1972 Johnson, Greenwood and Poon [39] developed a model in which the load carried by a contact in the mixed lubrication regime is shared between the asperity contact and the fluid film. In their model, they combined the well-known Greenwood and Williamson [40] theory of random rough surfaces in contact with the Elasto-hydrodynamic lubrication theory. This model was extended in 1999 by Gelinck and Schipper [41] to calculate the Stribeck curve for line contacts. Shi and Salant [42] introduced a mixed lubrication model, considering the inter-asperity cavitation and surface shear deformation for soft materials, and showed the occurrence of local cavitation. For moderately-loaded lubricated systems, the Jakobsson-Floberg-Olsson [12,13] cavitation theory is used. In 1970, Greenwood and Tripp introduced a deterministic contact model between two identical rough surfaces.

The flow factor method was introduced by Patir and Cheng [43,44]. They solved the Reynolds equation on a small area of the rough lubricating gap. The calculated micro flow is related to the flow of a perfectly smooth lubricating gap with similar mean height, resulting in flow factors. Fluid flow assumed to have two sources, shear driven flow and pressure. The flow factors are calculated independently by solving the local deterministic flow problem for a specified roughness topography. The main drawback of this method is due to nature of the roughness asperities that are not identical to the coordinate axes; this method is not effective in modelling the cross-flow of anisotropic roughness. Hu et al. [45] present a numerical solution for the contact of elastic bodies with three-dimensional roughness. The elastic contact has been modelled as a linear complementarity problem, and was solved by the Conjugate Gradient Method. Yu et al. [46,47] developed a full numerical solution to mixed lubrication in point contacts. They viewed the asperity contact as a result of a continuous decrease in the film thickness. By employing this assumption, the transition from contact to non-contact is continuous, and as a result, the same mathematical model should work for both regions. To calculate the asperity contact problem a multi-level integration method is used. In the work of Faraon et al. [48], by developing a numerical model for a real distribution of the asperities, the Stribeck curves were calculated; based on this model they compared the Stribeck curves between the deterministic and statistic model. They showed that the Stribeck curve results obtained with the statistic and the deterministic contact models are significantly different when the distribution of the surface heights deviates from the Gaussian height distribution; then by performing experimental measurements they showed that the deterministic mixed lubrication model is in good agreement with the measurements.

Recent developments in texturing techniques made it possible to employ different geometrical micro- and meso-scale patterns on the surface. These surface modification techniques include machining, photoetching, etching techniques, ion beam texturing and laser texturing [49]. Laser surface texturing proves to be more efficient, accurate, convenient and controllable for many materials [50], and is used to study the effect of micro-scale cavities on the frictional behaviour of contacts [25,27,29,37,51–56]. Kovalchenko et al. [27] show the influence of texturing on the transitions between the different lubricating regimes. They show that LST is able to enhance the hydrodynamic lubrication regime and thus increase the load carrying capacity of the contact; moreover, they found that the lapping after laser texturing that is carried out to remove the bulges at the edges of dimples is essential for increasing the positive effect of LST. In another study, Ryk et al. [29] theoretically and experimentally investigated on the beneficial effects of applying LST on piston rings.

They observed that the benefits of LST in both full and starved lubrication conditions results in fuel consumption reduction in combustion.

In the work of Bijani et al. [34] on the influence of surface texturing on mixed lubricated sliding contacts, the deterministic asperity contact model is applied. By employing this contact model and solving the Reynolds equation, Stribeck-like curves for several cavity patterns with different geometry were plotted, and the behaviour of the coefficient of friction based on these parameters was investigated.

2. Materials and Methods

Here a deterministic mixed lubrication model is developed; in this model, the lubrication is based on a limited amount of lubricant supplied to include the effect of starvation on friction in parallel sliding contacts. This is realized by combining a model calculating the film thickness in a sliding contact under starved conditions [57], with a deterministic asperity contact model to calculate and study the friction. The results are presented in Stribeck-like curves. By using this model, the frictional performance of starved lubricated contacts as a function of velocity and texturing parameters like cavity depth, size and density will be analysed, in particular within hydrodynamic and mixed lubrication regimes under starved conditions.

2.1. Deterministic Asperity Model

The Greenwood and Williamson contact model [40] assumes that asperities are spheres with a similar radius, and that asperity heights can vary randomly with a Gaussian probability distribution. However, in reality, this is rarely the case, as all of the asperities have the same radius; also, representing them as spheres or ellipsoids is more accurate. Moreover, the Gaussian height distribution is not an accurate approximation for most of the surfaces. As a recent advancement in optical measurement tools, the interference microscope can provide more accurate digital data of the surface topography which could be applicable for different applications, such as the calculation of the load carried by the deformed asperities of a rough surface when it is in contact with a flat surface (see Figure 1).

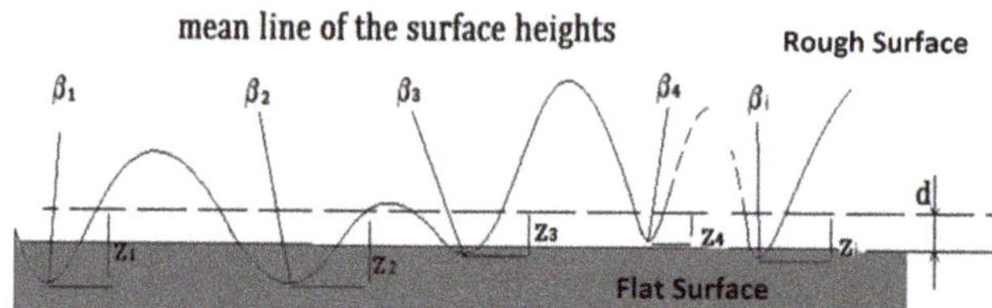

Figure 1. The schematic illustration of a flat surface and rough surface contact [34].

Figure 1 shows a flat on rough surface contact. For a given separation of two surfaces (d), the total force carried by the contact, the real contact area and the number of asperities s, are deterministically measurable by summing up the local contributions of the above-mentioned parameters. The asperities are assumed as ellipsoids with different radii (β_{xi} are the asperity radii in the sliding direction, and β_{yi} is in perpendicular direction), as well as the fact that they deform independently from each other. From Figure 2. the deformation of an asperity can be defined as Equation (1):

$$w_i = z_i - d \tag{1}$$

where z_i is the individual summit height, and w_i is the indentation of each deformed asperity.

For a given value of w_i, by adding the individual components of each asperity contact, the asperities' normal load (F_C) and the real contact area can be determined.

2.2. Load Sharing Concept

According to Johnson [39], in the case of a mixed lubrication regime (ML), the total normal load (F_T) is equal to the load carried by the boundary lubrication BL force component (F_C) plus the hydrodynamic lubrication HL force component (F_H), therefore:

$$F_T = F_C + F_H \tag{2}$$

Based on (Equation (2)), coefficients (γ_1) and (γ_2) are introduced:

$$\gamma_1 = \frac{F_T}{F_H}, \ \gamma_2 = \frac{F_T}{F_C} \tag{3}$$

The two coefficients (γ_1 and γ_2) are dependent of each other through the equation:

$$1 = \frac{1}{\gamma_1} + \frac{1}{\gamma_2} \tag{4}$$

In the deterministic asperity contact model, for the contact between a rigid flat surface against a rough surface, the Stribeck curve can be calculated by employing the two (γ_1) and (γ_2) parameters, where these two coefficients are defined in Equation (3). In the work of Gelinck [41], these two parameters are presented in terms of pressure (see Equation (5)). By combining the well-known Greenwood and Williamson [40] contact model under a classical hypothesis of the Reynolds isothermal equation, the entire Stribeck curve can be calculated.

$$\gamma_1 = \frac{p_T}{p_H}, \ \gamma_2 = \frac{p_T}{p_C} \tag{5}$$

where p_T, is the total pressure carried by the contact, p_C is the pressure on asperities, and p_H is the pressure carried by the hydrodynamic component of mixed lubrication [48].

In mixed lubricated contacts, in order to calculate the coefficient of friction, the asperity and hydrodynamic load components (F_C and F_H), as well as the related film thickness, must be determined. By solving the following three equations, it is possible to determine the above-mentioned three parameters:

I. The first equation is Equation (2) ($F_T = F_C + F_H$), in order to consider the load components in the BL component and the HL component, and their relation with the total load.

II. The second equation is the film thickness relation. In this study, the film thickness calculation is based on solving the Reynolds equation for textured surfaces [57].

This Reynolds equation is derived from the Navier-Stokes equation by taking the narrow gap assumption into account. In the Cartesian coordinate system, the Reynolds equation can be written as:

$$\frac{\partial}{\partial x}\left(\frac{\rho h^3}{\eta}\frac{\partial p}{\partial x}\right) + \frac{\partial}{\partial y}\left(\frac{\rho h^3}{\eta}\frac{\partial p}{\partial y}\right) = 6(u_0)\frac{\partial(\rho h)}{\partial x} + 6\rho h\frac{\partial(u_0)}{\partial x} + 12\frac{\partial(\rho h)}{\partial t} \tag{6}$$

where p is the lubricant pressure, η is the viscosity, h is the film thickness, and u_0 is the sum velocity. In absence of textures over the surface in parallel sliding contacts, the right side of the Reynolds equation equals zero; therefore there will be no pressure build up in contact, and no film can get formed.

At the outlet of the cavity, the lubricant is dragged through a converging region, and as a result, pressure is generated. The flow divergence at the entry of the cavity results in a negative pressure. This negative pressure is suppressed by cavitation, and as a cavitation product, vapour bubbles are appearing in the lubricant film. The Jakobsson-Floberg-Olsson model is dividing the lubrication film into two zones. The first zone is the lubricant film without the cavitation effect; therefore no vapour bubble exists in this zone, and the Reynolds equation is valid. In the second zone, where cavitation does take place, the lubricant occupies just a fraction of the film gap, and the vapour bubbles exist in the void fraction. In this zone the pressure is taken as a constant [58].

By suggesting the use of a switch function, Elrod introduced a universal solution for cavitated and full film zones (see Equation (7)). In Equation (7), φ represents a dimensionless dependent variable, and F is the aforementioned switch function, and these parameters are defined as in a liquid zone, where $F = 1$, $\varphi \geq 0$ and $p = \varphi$, and in the cavitated region, $F = 0$, $\varphi < 0$ and $p = 0$.

The steady-state mass-conservation Reynolds equation, taking the Elrod cavitation algorithm into account, can be written in a Cartesian coordinate system as (Equation (7)) [59]:

$$\frac{\partial}{\partial x}\left(\frac{h^3}{\eta}\frac{\partial(F\varphi)}{\partial x}\right) + \frac{\partial}{\partial x}\left(\frac{h^3}{\eta}\frac{\partial(F\varphi)}{\partial y}\right) = \frac{6u_0}{p_a - p_c}\frac{\partial((1+(1-F)\varphi)h)}{\partial x} \tag{7}$$

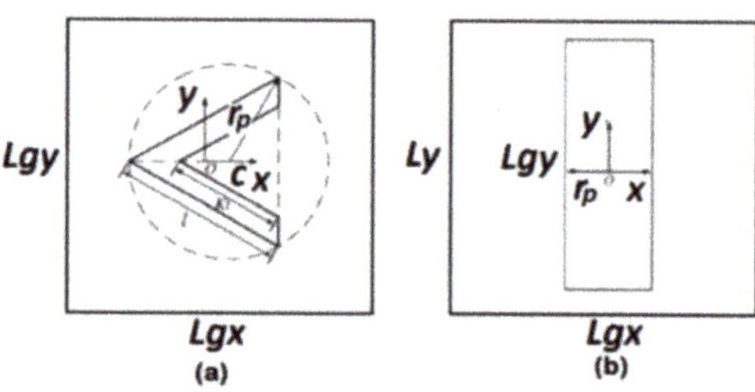

Figure 2. Geometrical scheme of patterns: (**a**) Chevron, (**b**) Groove (Reproduced with the permission of Mingfeng Qiu, Bret R. Minson, Bart Raeymaekers, *Tribology International*, published by Elsevier, 2013) [60].

In this study, chevron and groove patterns have been investigated. Figure 2, shows the different cavity shapes and the parameters characterizing their geometry. The chevron pattern is defined by two similar equilateral triangles of different sizes. When the inner edge length of the chevron approaches zero, the chevron pattern transforms to a triangular pocket. For these two cases, the centre of the unit cell coincides with the midpoint of the altitude line of the triangle or chevron shape; (see also [60]). All patterns have a rectangular cross-sectional profile; (see Figure 3). The general film thickness formula can be written as (Equation (8)):

$$h = h_0 + h_{macro} + h_{texture} \tag{8}$$

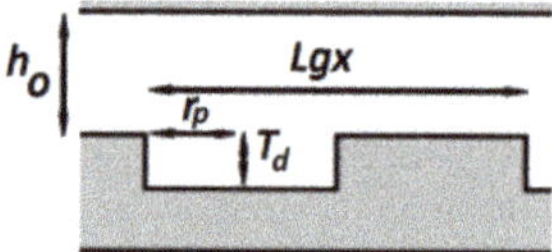

Figure 3. Schematic illustration of the cavity profile.

When both interacting surfaces of contact are flat, then h_{macro} is negligible. In case of a flat on flat sliding contact, the film thickness (Equation (8)) reduces to (Equation (9)) [60,61]:

$$\frac{h(x,y)}{h_0(x,y)} = 1 + H(x,y) \tag{9}$$

The film thickness formula for the chevron can be written as (Equation (10)):

$$
H(x,y) = \begin{cases} 0, & \{X, Y \notin \Omega \\ \frac{T_d}{h_0}, & \{X, Y \in \Omega \end{cases} \quad \Omega : -\frac{3}{4} \le X \le \frac{3}{4} \; and \begin{cases} \frac{1}{\sqrt{3}}X + \frac{\sqrt{3}}{2}\{K - \frac{1}{2} \le \\ Y \le \frac{1}{\sqrt{3}}X + \frac{\sqrt{3}}{4} \\ -\frac{1}{\sqrt{3}}X - \frac{\sqrt{3}}{4} \le \\ Y \le -\frac{1}{\sqrt{3}}X + \frac{\sqrt{3}}{2}\{\frac{1}{2} - K \end{cases} \tag{10}
$$

The film thickness formula for the grooves is given in (Equation (11)):

$$
\Omega : -\frac{1}{2} \le X \le \frac{1}{2} \; and \; \frac{1}{2} \le Y \le \frac{1}{2} \tag{11}
$$

In this simulation r_p is the characteristic radius for the chevron patterns and the half width of the grooves.

To solve the Elrod cavitation algorithm for Reynolds equation (Equation (7)), the tri-diagonal matrix algorithm (TMDA) is used, and in order to reduce the storage needed for calculation, the line-by-line TDMA solver (Patankar [62]) is employed. The TDMA is a direct method for a one-dimensional situation, but by solving it iteratively line-by-line, it is possible to apply it for two- and three-dimensional problems, as well. [63]. The algorithm for this numerical solution is presented in Appendix A.

III. The third equation is derived from the equilibrium of the modified relation for the central pressure and average contact pressure carried by asperities, represented as [48].

$$
F_C = \sum_{i=1}^{N} \frac{2}{3} E' R_i (z_i - d) \tag{12}
$$

In Equation (12), E' is the combined elasticity modulus and R_i is the reduced radius of the cylinder. The reduced elastic modulus is given by:

$$
\frac{2}{E'} = \frac{1 - v_1^2}{E_1} + \frac{1 - v_2^2}{E_2} \tag{13}
$$

where $E_1 = E_2 = E$ and $v_1 = v_2 = v$.

The total friction force (F_f) in the ML regime can be calculated as the sum of the shear force of the lubricant (F_{fH}) and friction force of the contacting asperities and (Equation (14)):

$$
F_f = \sum_{i=1}^{N} \iint_{A_{ci}} \tau_{ci} dA_{ci} + F_{fH} \tag{14}
$$

In Equation (14), τ_{ci} is the shear stress at the asperity contact, N is the number of contacting asperities and A_{ci} the area of contact of a single asperity contact. For the hydrodynamic component of shear force (F_{fH}), the friction force can be written as:

$$
F_{fH} = \tau_H A_H \tag{15}
$$

where τ_H is the shear stress of the lubricant and A_H is the contact area of the hydrodynamic component. Friction is assumed to be of the Coulomb type for the contacting asperities:

$$
f_{ci} = \frac{\tau_{ci}}{p_{ci}} \tag{16}
$$

with p_{Ci} the average contact pressure on the ith asperity, and f_{Ci} the coefficient of friction which is assumed constant for all asperities; then the double integral in Equation (14) can be written as:

$$\sum_{i=1}^{N} \iint_{A_{ci}} f_C p_{ci} dA_{ci} = f_C F_C \tag{17}$$

The value of f_C is measureable from experiments, and in these calculations this value is set to 0.1 based on the measurements presented in Appendix B. The coefficient of friction is given by:

$$f = \frac{F_f}{F_N} = \frac{f_C F_C + F_{fH}}{F_N} \tag{18}$$

It is worth mentioning that, in the absence of texturing features in parallel sliding, no lubricant film in contact can be formed, and the two sliding surfaces will stick to each other. Therefore, the coefficient of friction will be constant and equal to the coefficient of friction in the boundary lubrication regime (f_C).

In order to study the effect of starvation for textured surfaces with different texturing patterns, and to investigate the frictional behaviour based on the different texturing parameters, several simulations are carried out.

3. Problem Definition and Its Solution

In order to investigate the effect of lubricant supply on the coefficient of friction for different texturing parameters and patterns, several simulations were performed.

In Figure 4., when the input film thickness (h_{oil}) is smaller than the calculated film thickness for non-starved lubrication (h_s), the contact is operated under starved conditions. The limited amount of lubricant in the input region of contact (h_{oil}) can affect the coefficient of friction for the textured surfaces. To gain a better understanding of the starved lubrication phenomenon, the influence of input film thickness and texturing parameters, such as texture pitch (P_x), texture depth (T_d) and texture size (S), on the coefficient of friction, is investigated. These simulations are based on the linear groove and chevron patterns (see Figure 5).

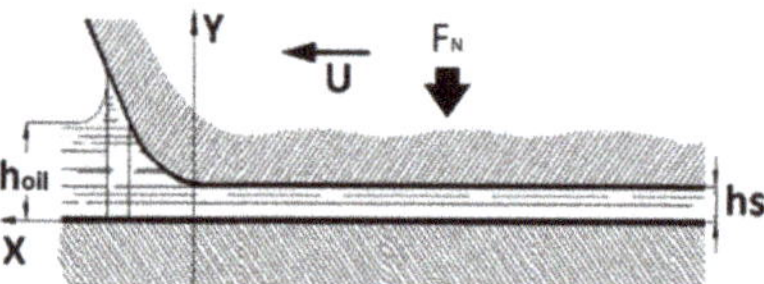

Figure 4. Schematic illustration of the input film thickness (h_{oil}) and the calculated film thickness (h_s) [64].

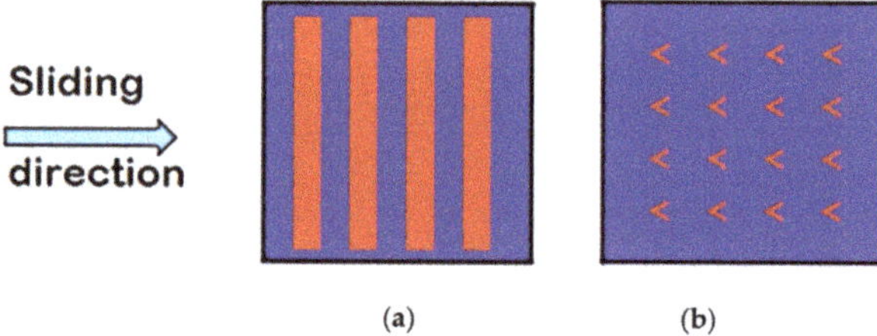

Figure 5. Schematic illustration of different patterns (a) Grooves, (b) Chevron pattern [34].

In order to investigate the effect of texturing, the geometrical parameters are studied. In this parametric study, the effect of the geometrical parameters determining the texture shape and the effect of the texture area fraction upon the frictional behaviour of the starved lubrication conditions are studied. By introducing texture pitch (P_x), it is possible to define the texture area fraction.

The pitch is calculated as:

Cavity size: $S = 2r_p$ and Pitch in x direction: $P_x = S/L_{gx}$.

Also, texture depth (T_d) can affect the film formation in friction in textured surfaces. It is possible to define the geometry of grooves by these three parameters. Another parameter that can help to define the geometry of chevrons is the cavity width ratio, which is represented by K in Figure 6.

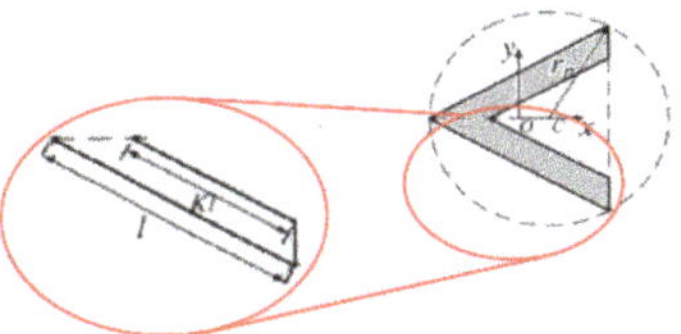

Figure 6. Chevron cavity width ratio.

Prior study on film thickness [57] showed that the rectangular cross section can build a thicker lubricant film in contact, and it is more efficient in comparison with the circular cross-sectional patterns; therefore in the present article, the rectangular cross-section patterns are employed (see Figure 3). Here, linear grooves and chevrons will be analysed. The operating conditions applied in this calculation are presented in Table 1. The analyses of the roughness measurement and surface topography were performed using a Keyence Color 3D LASER Scanning Microscope (Keyence, Osaka, Japan), which uses a violet LASER $\lambda = 388$ nm.

Table 1. Operating conditions.

Properties	Value
Normal load	5 N
Average contact pressure	0.05 MPa
Contact area	10^{-4} m^2
Lubricant viscosity	8 mPa·s

Where K = Inner wall length/outer wall length, and in these calculations (K) is constant, and equal to 0.5.

To validate the model and algorithm, a comparison was performed between the experimental measurements of Kovalchenko et al. [37] and the numerical results from the algorithm developed in this work. It is worth it to mention that, although the developed algorithm can calculate the friction in all three lubrication regimes within this section just to validate the code, the comparison takes place in the hydrodynamic lubrication regime. Therefore, the numerical and measurement results are just dealing with the hydrodynamic lubrication regime. Kovalachenko et al. [37], investigated the effect of size and the density of circular pockets on the coefficient of friction (see Figure 7). In this study the Disk 3 results are compared with results from the developed numerical algorithm.

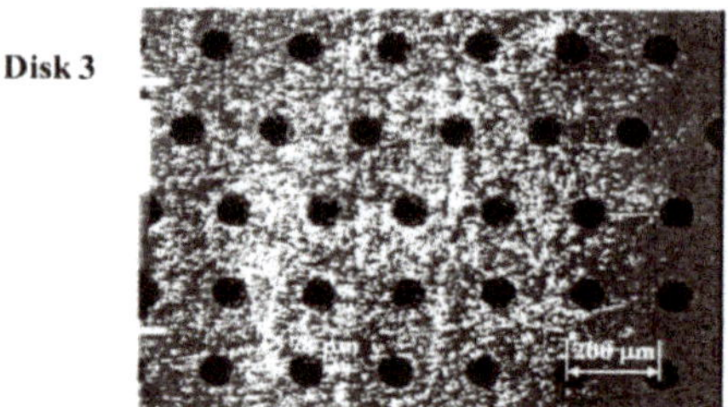

Figure 7. Disk 3 dimple array, (Reproduced with permission from Andriy Kovalchenko, Oyelayo Ajayi, Ali Erdemir, et al., Tribology Transactions, published by Taylor and Francis, 2004) [37].

In Disk 3 the cavity depth is 5.5 µm, the cavity size ($S = 2r_p$) is 78 µm and the texture density is 28%. The measurement results for this disk are presented in Figure 8.

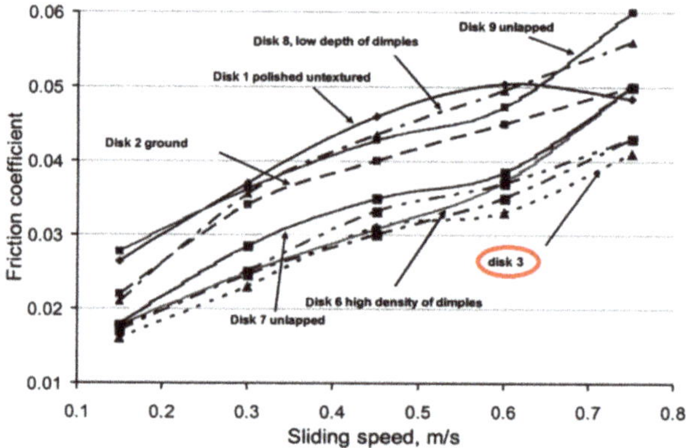

Figure 8. Measurement results (Reproduced with permission from Andriy Kovalchenko, Oyelayo Ajayi, Ali Erdemir, et al., Tribology Transactions, published by Taylor and Francis, 2004) [37].

Experimental data extracted from Figure 8 and the numerical results are based on the calculation of coefficient of friction for the full film condition, when the lubricant kinematic viscosity is 1247 cSt at 40 °C, and the normal load of 20 N is applied.

The texture array schematically illustrated in Figure 9. The comparison between the numerical results and the experimental measurements for circular pockets with $T_d = 5$ µm, $r_p = 15$ µm and $P_x = 0.3$, are presented in Figure 10.

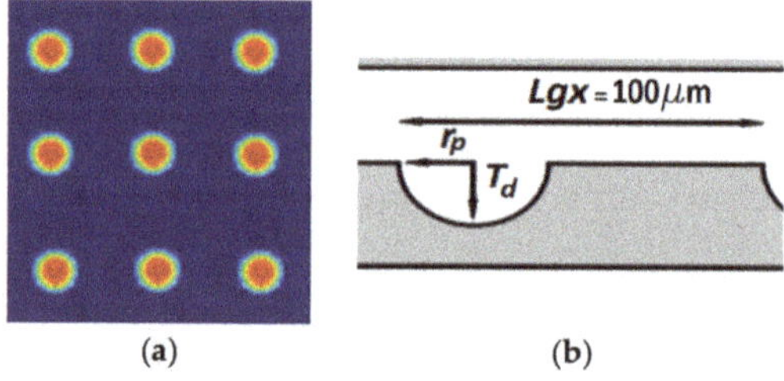

Figure 9. (a) Simulated texture array and (b) cavity profile.

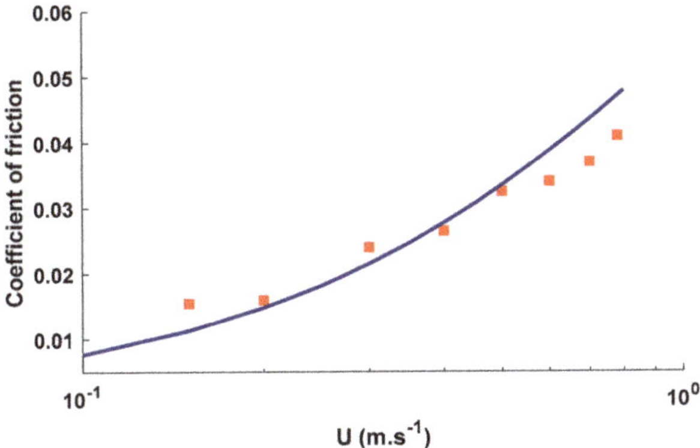

Figure 10. Comparison between numerical and experimental results.

From Figure 10 it is possible to see that there is a good agreement between the values of the coefficient of friction calculated by the numerical model (blue line) and the experimentally measured results (red diamonds). When the velocity is 0.5 m·s^{-1}, the calculated value is around 1% less than the measured value for Disk 3 that is in reasonable range. Moreover, results show that the calculated coefficient of friction has the same trend as the measured results for the coefficient of friction.

4. Results

4.1. The Effect of Texture Depth (T_d)

In order to study the influence of the effect of texturing in starved lubricated contact with respect to the texture depth, several simulations were carried out, based on groove and chevron texturing patterns. In these calculations, the texturing parameters other than the texture depth are constant. The coefficient of friction is calculated based on three different values of texture depth, and ranges from 5 to 10 µm. In Figures 11 and 12, the simulation results show the effect of different values for the texture depth based on starved and non-starved lubrication for grooves and chevrons. In Figure 11a, it can be seen from the results that a depth of 10 µm gives the lowest coefficient of friction. In Figure 11b, a limited input film thickness is applied in order to simulate the starvation. The results in Figure 11b show that the calculated coefficient of friction for different values of texture depth tend to the same values, therefore the Stribeck curves are merging together. In the calculations shown, the input film thickness (h_{oil}) is set to 10 µm. The texturing properties of following calculations are given in Table 2.

Table 2. Texturing properties.

Parameter	Value
Texture pitch (P_x)	0.4
Cavity size (S)	100 µm

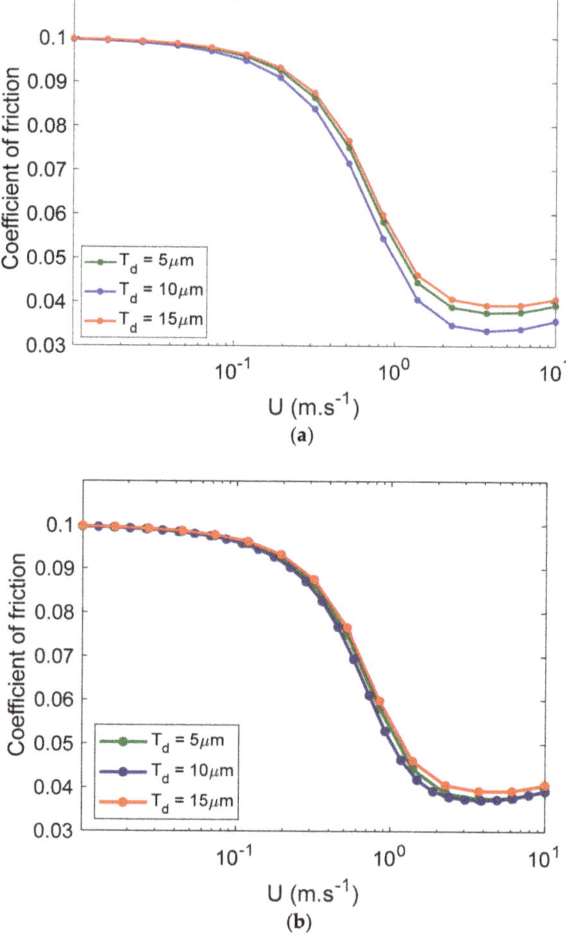

Figure 11. Stribeck curves as a function of texture depth for grooves when the lubrication is: (**a**) Non-starved, (**b**) starved, and $h_{oil} = 10$ μm.

In Figure 12a, in case of chevron patterns similar to the groove pattern, the lowest coefficient of friction in the case of non-starved lubrication is achievable when the depth is around to 10 μm. In Figure 12b in the case of the starved lubrication, results show that the calculated coefficient of friction for different values of texture depth are tending to the same values, therefore the Stribeck curves are merging. By comparing the curves in Figure 12, and based on the results from the study on film thickness [57], (which showed that a higher film thickness is achievable for grooves and chevrons when the depth is around 10 μm), it is possible to conclude that the lowest coefficient of friction is achievable when the film thickness has the highest value. This is because in the case of mixed lubrication an increasing film thickness separates the surfaces and reduces contact. Although the growth in the depth of cavities to the optimum values leads to the higher film thicknesses, in the case of starved lubrication, this growth in lubricant film thickness is limited due to the limit in oil supply; this will result in a limitation in effect of this parameter on the coefficient of friction as shown in Figure 12.

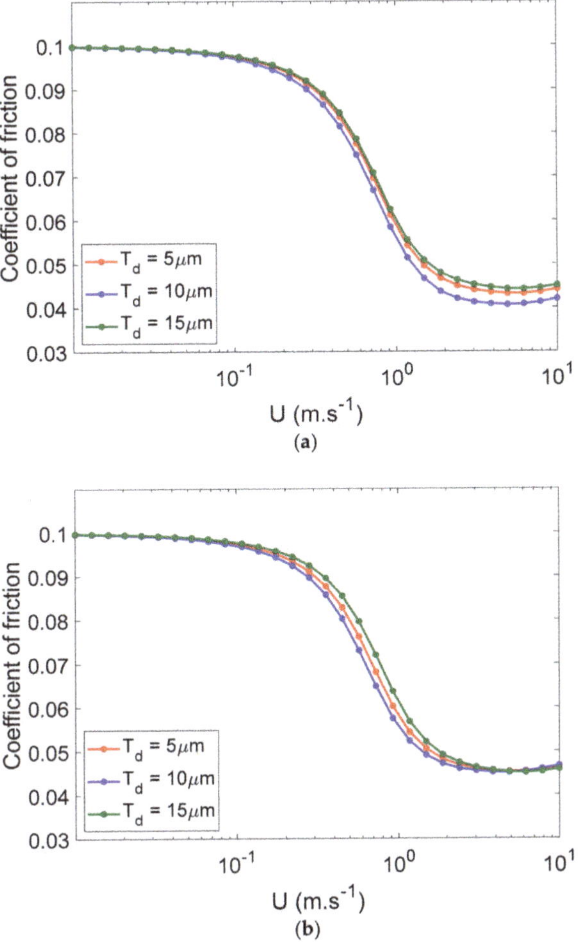

Figure 12. Stribeck curves as a function of texture depth for chevrons when the lubrication is: (**a**) Non-starved, (**b**) starved, and $h_{oil} = 10$ μm.

4.2. The Effect of Texture Size (S)

In this section, to study the effect of size on the coefficient of friction for starved lubricated contacts, the texture depth and pitch are set as constant, and the size is changed. For three different texture size parameters (S), the coefficient of friction is calculated. In Figures 13 and 14, the simulation results show the effect of different values for texture size-based-starved and non-starved lubrication for grooves and chevrons. In Figure 13a, in the case of groove patterns when the size is around 100 μm, the friction force has the lowest value. In Figure 13b, the limited input film thickness is applied in order to simulate the starvation. The results in Figure 13b show that the calculated Stribeck curves for different values of texture size are merging in the case of starved lubrication. Texturing properties for calculations of Figures 13 and 14 are shown in Table 3.

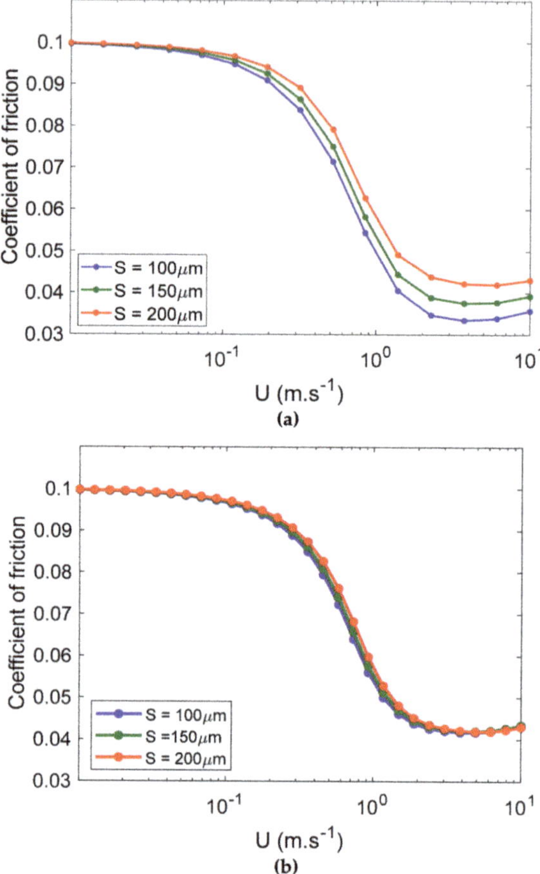

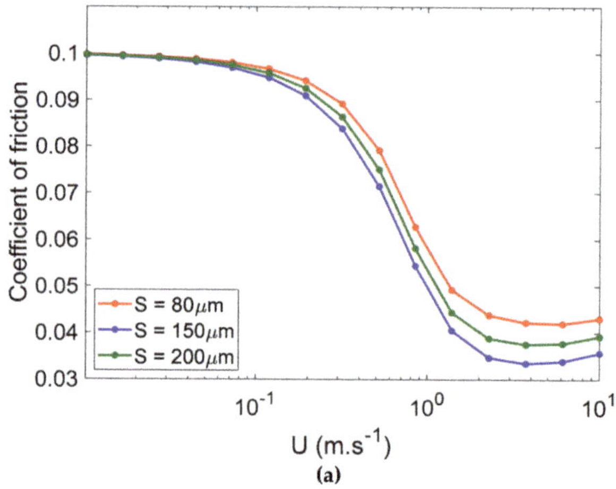

Figure 13. Stribeck curves as a function of cavity size (S) for grooves when the lubrication is: (**a**) Non-starved, (**b**) starved, and $h_{oil} = 10$ µm.

Figure 14. *Cont.*

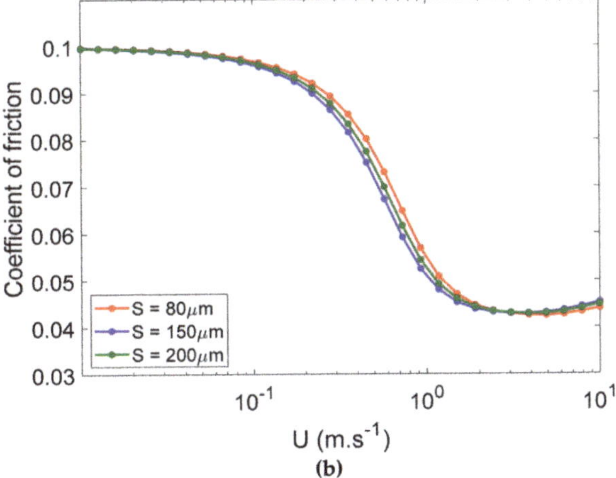

Figure 14. Stribeck curves as a function of cavity size for chevrons when the lubrication is: (a) Non-starved, (b) starved, and $h_{oil} = 10$ μm.

Table 3. Texturing properties.

Parameter	Value
Texture depth (T_d)	10 μm
Texture pitch (P_x)	0.4

In Figure 14a, in the case of a chevron pattern similar to the groove pattern, the lowest coefficient of friction in the case of non-starved lubrication is achievable when the size is around 150 μm. Also for this case, starved lubrication results in the merging of the curves as shown in Figure 14b. Consequently, as mentioned and achieved from Figure 14, changing the size parameters has an influence on the coefficient of friction, but this effect is limited with respect to values close to the optimum value for this parameter.

4.3. The Effect of Texture Pitch (P_x)

In Figures 15 and 16, the simulation results show the effect of different values for texture pitch based on starved and non-starved lubrication for grooves and chevrons. In Figure 15a, in the case of a groove pattern when the pitch is around 0.4, according to the results from the study on film thickness [57] the lubricant film is thicker; when the lubrication is in a mixed lubrication regime the thicker film results in a lower solid-solid contact. Therefore, the friction has the lowest value when $P_x = 0.4$. In Figure 15b, the limited input film thickness is applied in order to calculate the friction in the starved lubrication. The results in Figure 15b show that when the lubrication is in the starved regime the calculated coefficient of friction for different values of texture size is tending to the same values, therefore the Stribeck curves are merging together. Texturing properties for calculations of Figures 15 and 16 are shown in Table 4.

Table 4. Texturing properties.

Parameter	Value
Texture depth (T_d)	10 μm
Cavity size (S)	100 μm

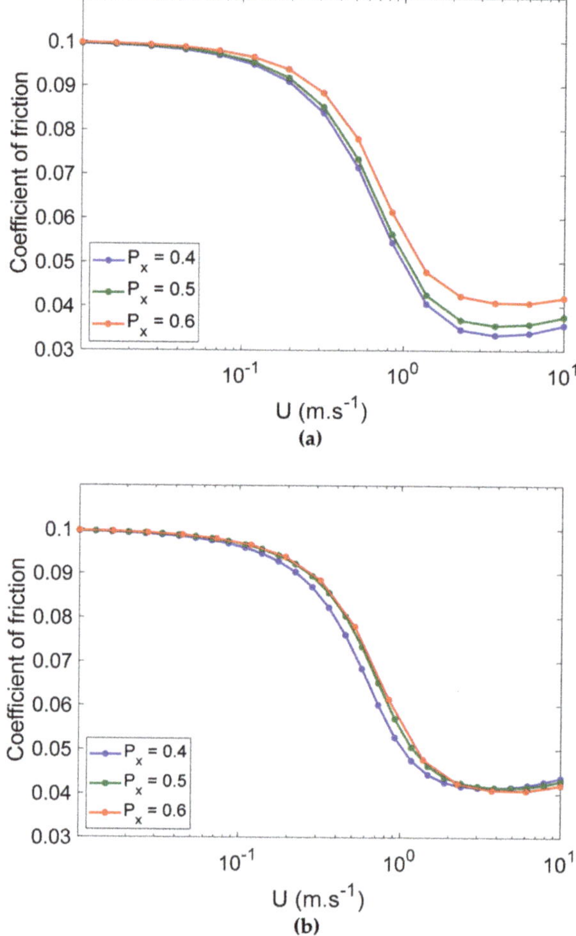

Figure 15. Stribeck curves as a function of texture pitch for grooves when the lubrication is: (**a**) Non-starved, (**b**) starved, and $h_{oil} = 10$ μm.

In Figure 16a, in case of a chevron pattern similar to the groove pattern, the lowest coefficient of friction in the case of non-starved lubrication is achievable when the pitch is around 0.5. In Figure 16b, it is possible to observe the same behaviour of the merging in Stribeck curves for chevrons in Figure 12b, for the groove pattern. Therefore, due to the limit in lubricant supply, which results in limited lubricant film growth, the same trend of behaviour is also predictable for other patterns when the starvation in lubrication happens, i.e. triangular pockets and circular pockets.

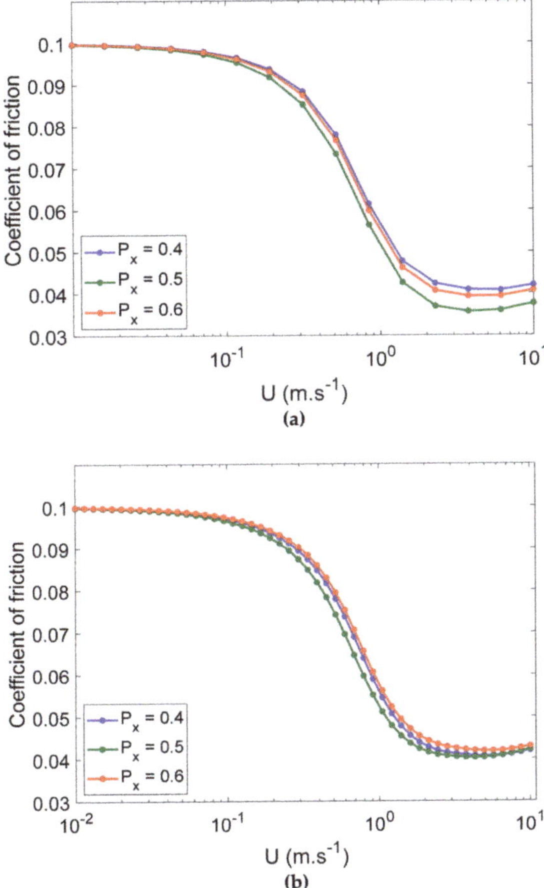

Figure 16. Stribeck curves as a function of texture pitch for chevrons when the lubrication is: (**a**) Non-starved, (**b**) starved, and $h_{oil} = 10$ µm.

4.4. Effect of Input Film Thickness on the Coefficient of Friction

In Figure 17, the calculated film thickness for grooves, when $T_d = 10$ µm, $S = 100$ µm and $P_x = 0.4$, is presented. In this calculation, the effect of different input film thicknesses is studied. This effect is studied for four different input film thicknesses, $h_{oil} = 2.5$, 4, 6 and 8 µm, the operational conditions and texturing properties are presented in Table 5.

Table 5. Texturing properties and operating conditions.

Parameter	Value
Texture depth (T_d)	10 µm
Texture pitch (P_x)	0.4
Cavity size (S)	100 µm
Normal load (F_T)	5 N
Average contact pressure	0.05 MPa
Lubricant viscosity	0.08 Pa·s

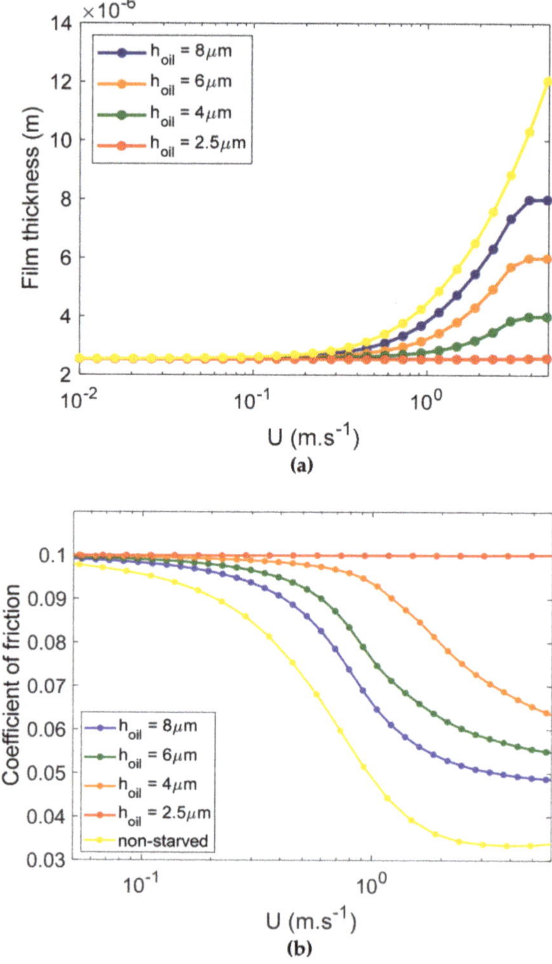

Figure 17. Effect of input film thickness on (**a**) film thickness, (**b**) coefficient of friction.

The results of the calculation of the Stribeck curve and the film thickness for different limited values of input film thickness are presented in Figure 17a,b. The tendency of the starved Stribeck curve and the corresponding film thickness as a function of input film thickness can be described as follows:

1. When the input film thickness (h_{oil}) is lower than 2.5 μm, the Stribeck curve transforms into a straight line, and the film thickness for different velocities stays constant in the same level as it is in the BL regime.

2. If the h_{oil} varies between 2.5 and 8 μm, the friction level in the HL and ML regimes starts to decrease, and the film thickness increases. Figure 13b shows that transitions from BL to ML and from ML to HL stay approximately at the same transition velocity for different values of h_{oil}.

It concludes that, although the starvation has no influence on the transition of lubrication regimes, the friction level changes due to starved lubrication.

5. Discussion

In this article, the influence of starvation and surface texturing on each other by performing a set of parametric studies has been investigated. The variation range of texturing parameters in these parametric studies was chosen based on the sizes which can give more fluctuation in the frictional behaviour of contacts; these ranges were chosen from the previous study on the effect of surface texturing on film thickness [57].

Based on the results in this section, texturing for flat-flat sliding contacts with a starved lubrication regime can be helpful for reducing the coefficient of friction. However, from results (Figures 7–13), it must be considered that increasing texturing parameters to the optimum values in the case of contacts with higher sliding velocities (lubrication regime in hydrodynamic lubrication) may not be as beneficial as at lower velocities. In other words, starvation can limit the beneficial influence of surface texturing at higher velocities. Therefore, surface texturing in the case of starved lubrication may not be advantageous to reducing the friction based on the operating conditions and any limit in lubricant input film.

From the parametric study in this article, it is possible to achieve that, although the growth in the depth of cavities to the optimum values leads to the higher film thicknesses in the case of starved lubrication, this growth in lubricant film thickness is limited due to the limit in oil supply, as is shown in Figure 9. The same frictional behaviour and merging in Stribeck curves is observable in the case of the study of the effect of size and pitch parameters. Changing the size and pitch parameters has an influence on the coefficient of friction, but for the values close to the optimum value for this parameter, this effect is limited. Furthermore, it is possible to obtain that, due to the limit in lubricant supply. The same trend of behaviour is also predictable for other patterns, i.e., triangular pockets and circular pockets when the starved lubrication regime is occurring in contact. Moreover, in the case of a chevron pattern similar to the groove pattern, the lowest coefficient of friction in the case of non-starved lubrication is achievable when the depth is around 10 μm and the pitch is around 0.5.

Based on the results for the section on the effect of the input film thickness on the coefficient of friction, this parameter can have a vital role on the efficiency of texturing. In particular, when the input film thickness (h_{oil}) is lower than 2.5 μm, the Stribeck curve transforms into a straight line and the film thickness does not change compared to the boundary lubrication regime film thickness.

If the h_{oil} varies between 2.5 and 8 μm, the friction level in the HL and ML regimes starts to decrease, and the film thickness increases.

The influence of roughness is more considerable in the boundary lubrication regime and mixed lubrication to the point at which the transition between ML and HL is happening. Therefore, employing surfaces with a higher roughness can shift the Stribeck curve; this shift in addition to the effect of starvation can reduce the influence of texturing and application of optimized texture properties.

6. Conclusions

The goal of this investigation was to study and predict the effect of surface texturing on frictional behaviour for parallel sliding contacts under starved lubrication condition. In addition, the efficiency of surface texturing as a method to reduce the friction in starved lubricated contact is also studied. This model is a numerical algorithm based on the Reynolds equation with the Elrod cavitation algorithm formulation. By applying the value of calculated film thickness in the deterministic asperity model, the coefficient of friction is calculated. In this article, Stribeck curves for different situations are plotted. The effect of several parameters on starved regime frictional behaviour, such as depth, size and texture pitch, has been studied.

In this study, the deterministic asperity contact model is employed efficiently, considering the effect of different scales of surface features (roughness and texture) on the coefficient of friction that is not dependent on the directions of the asperities.

1. This approach allows the effect of texture and roughness to influence the friction independently; this may be beneficial in optimizing the surface texture.

2. In order to reduce the friction in starved lubrication conditions, surface texturing has a beneficial effect, and this effect is also presented in numerical study of Gu et al. [33,65].

3. The positive effect is more sensible mostly when contacts are in lower sliding velocities. When the sliding velocity reaches higher values, the effect of texturing can be influenced by the input film thickness.

4. In the case of starved lubrication, when the value of this input film thickness (h_{oil}) decreases, the starvation effect gains a greater influence upon the film thickness. Therefore, the effect of variation of the texture parameters (pitch, depth and size) on the coefficient of friction is also decreasing.

5. Surface texturing in starved lubricated conditions, based on operating conditions and a limit in lubricant input film, may not be advantageous as a method to reduce the fiction.

6. In order to apply the texturing in starved lubricated contacts a simulation of the coefficient of friction and film thickness based on texturing and lubricant properties can help to avoid the unnecessary surface texturing.

It is worth mentioning that based on the operational conditions, thermal and atmosphere effects could play an important role in lubrication. These effects should be included in the numerical simulation for accurate performance predictions. However, since these effects were not considered texture-specific, which is the focus of this article, thus they have not been included in this article.

Author Contributions: Conceptualization, D.J.S., E.L.D. and D.B.; Methodology, D.J.S. and D.B.; Software, D.B.; Validation, D.B. and D.J.S.; Formal Analysis, D.B. and D.J.S.; Investigation, D.B. and D.J.S.; Writing—Original Draft Preparation, D.B.; Writing—Review & Editing, D.B., E.L.D., M.B.d.R. and D.J.S.; Supervision, D.J.S.

Funding: This research was funded by Materials Innovation Institute (M2i) grant number M21.1.11448.

Conflicts of Interest: The authors declare no conflict of interest.

Nomenclature

Parameters	Description	Unit
A_C	Real area of asperity contact	m^2
D	Separation	m
f	Coefficient of friction	–
f_C	Coefficient of friction in BL regime	–
F	Elrod cavitation algorithm switch function	–
F_C	Load carried by the asperities	N
F_H	Load carried by the hydrodynamic component	N
F_f	Friction force	N
F_{fH}	Hydrodynamic friction force	N
F_T	Normal load on the contact	N
h	Film thickness	m
h_{oil}	Limited value of input film thickness	m
h_0	Film thickness	m
H	Dimensionless local depth of textured surface	–
L_{gx}	Texture cell length in x-direction	m
L_x	Textured area in x-direction	m
L_y	Textured area in y-direction	m
p	Pressure	Pa
p_a	Ambient pressure	Pa
p_c	Cavitation pressure	Pa
P	Dimensionless pressure	–
P_a	Average contact pressure	Pa
P_x	Texture pitch	–
r_p	Cavity characteristic depth	m
S	Cavity size = $2r_p$	m
T_d	Texture depth	m

Parameters	Description	Unit
U_0	Sum velocity	–
X	Dimensionless Cartesian coordination	–
Y	Dimensionless Cartesian coordination	–
w_i	Compliance of an asperity	m
z_i	Asperity height	m
ρ	Density	$kg \cdot m^{-3}$
γ_1	Adaption parameter for hydrodynamic component in ML	–
γ_2	Adaption parameter for asperity contact component in ML	–
η	Dynamic viscosity	Pa·s
τ_C	Shear stress of asperity contact	Pa
φ	Cavitation dimensionless variable	–

Appendix A

Mixed lubrication Model Algorithm

For a starved lubrication condition, the coefficient of friction in the ML regime can be calculated as explained in the flowchart below (see Figure A1).

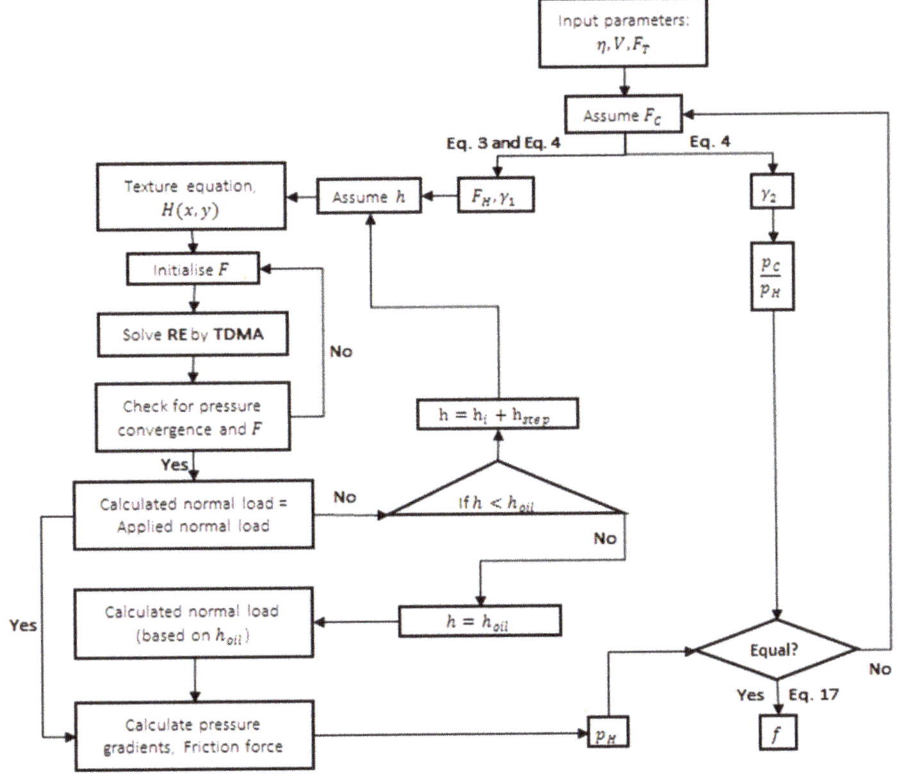

Figure A1. Flowchart of mixed lubrication under starved lubrication condition.

Appendix B

Determination of Roughness Parameters

In this study, calculations were performed using the roughness measured by images extracted from the laser microscope; these images are obtained from roughness height measurements. In this appendix, one of these images is shown as an example, and the equations for calculating the roughness parameters are presented.

Roughness Measurement

In order to calculate the boundary lubrication component, the roughness measurement is essential. As mentioned in the article, in the case of the deterministic approach, the real measured height of asperities is needed so as to calculate the separation between the opposing surfaces. In order to achieve the height of the asperities, the surface topography for textured surfaces is measured by using microscopic images. These analyses have been performed using a Keyence Color 3D Scanning Microscope (Keyence, Osaka, Japan), which uses a violet Laser $\lambda = 388$ nm. The result from a roughness measurement in the case of a chevron textured sample is presented in Figure A2, (standard lens 50× is employed).

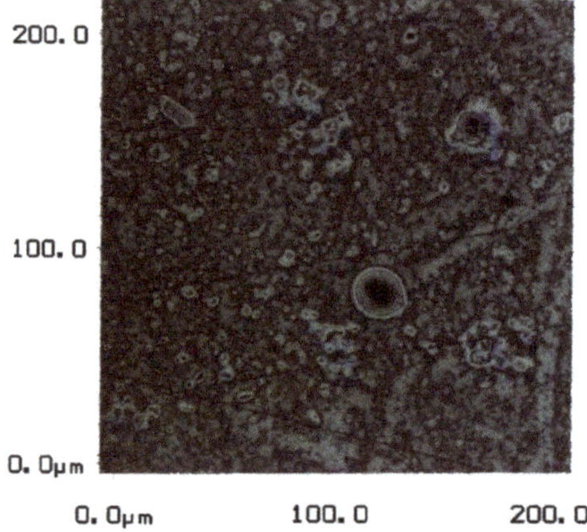

Figure A2. Surface image by Laser Scanning Microscope.

The roughness data measured by confocal microscope is illustrated in Figure A3.

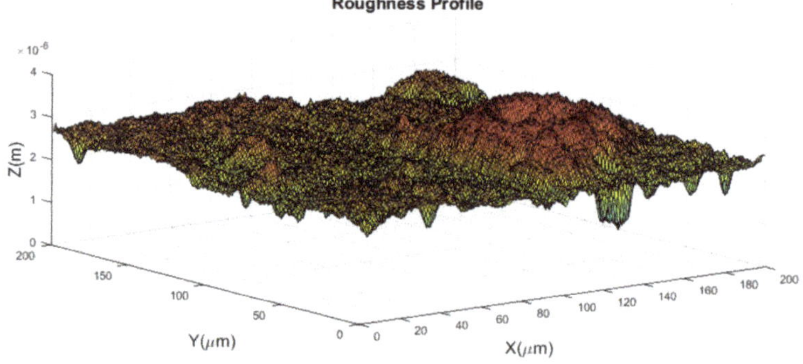

Figure A3. Surface roughness profile based on the Laser Scanning Microscope measurements ($R_a = 0.11$ μm).

Figure A4 shows the real measured probability density of asperities against the dimensionless asperity height, in order to compare the measured roughness with the Gaussian roughness distribution. The red line represents the Gaussian probability density distribution of one surface.

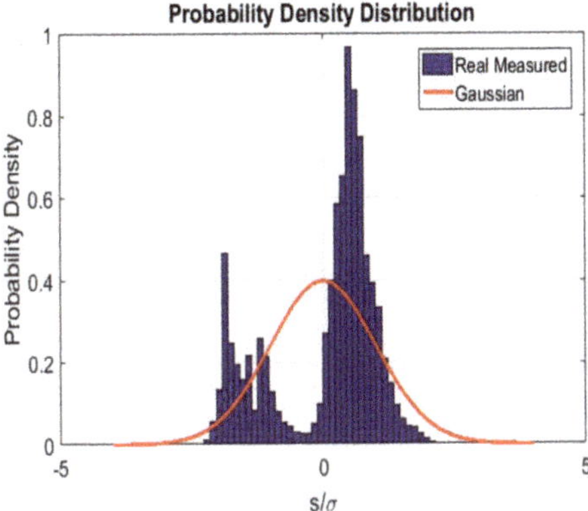

Figure A4. Real measured and Gaussian distribution of surface heights as a function of the dimensionless asperity height (s/σ).

The Summit is defined as a point that is higher than its eight neighbour points (see Figure A5).

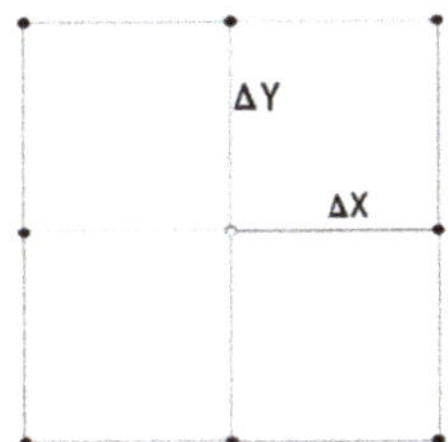

Figure A5. Definition of a Summit.

To determine the radius of an asperity the 3-point definition is employed. In Figure A5, Δx, Δy are the steps or pixel size, and the asperity radii in both the x and y directions can be calculated as:

$$\beta_x^{-1} = \frac{z_{x-\Delta x,y} - 2z_{x,y} + z_{x+\Delta x,y}}{\Delta x^2} \tag{A1}$$

$$\beta_y^{-1} = \frac{z_{x-\Delta x,y} - 2z_{x,y} + z_{x+\Delta x,y}}{\Delta y^2} \tag{A2}$$

In Equation (A1), β_x is the asperity radii in x direction respectively and in Equation (A2), β_y is the asperity radii in y direction and $z_{x,y}$ is the local surface height at location (x, y). The combined summit radius β_i of the radii in the two perpendicular directions β_x and β_y is obtained by:

$$\beta_i = \sqrt{\beta_{xi} \cdot \beta_{yi}} \tag{A3}$$

To calculate the average summit radius ($\bar{\beta}$) we have:

$$\bar{\beta} = \frac{1}{n} \sum_{i=1}^{n} \beta_i \tag{A4}$$

Therefore, the calculated average radius of asperity ($\bar{\beta}$), is equal to 4.8×10^{-8} m.

References

1. Evans, H.P.; Snidle, R.W.; Sharif, K.J. Deterministic mixed lubrication modelling using roughness measurements in gear applications. *Tribol. Int.* **2009**, *42*, 1406–1417. [CrossRef]
2. Faraon, I.; Schipper, D.J. Stribeck curve for starved line contacts. *J. Tribol.* **2007**, *129*, 181–187. [CrossRef]
3. Gethin, D.; Medwell, J. Paper III (iv) Starvation effects in two high speed bearing types. *Tribol. Ser.* **1987**, *11*, 81–92.
4. Eugen, B.T.; Sandi, B.C. Specific problems of high speed bearings. *J. Eng. Stud. Res.* **2012**, *18*, 31–38.
5. Olaru, D.; Gafitanu, M. A new methodology to estimate starvation in ball bearings. *Lubr. Sci.* **1997**, *4*, 93–106. [CrossRef]
6. Wedeven, L.D.; Evans, D.; Cameron, A. Optical analysis of ball bearing starvation. *J. Lubr. Technol. Trans. ASME* **1971**, *93*, 349–363. [CrossRef]
7. Pemberton, J.; Cameron, A. A mechanism of fluid replenishment in elastohydrodynamic contacts. *Wear* **1976**, *37*, 185–190. [CrossRef]
8. Kingsbury, E. Cross flow in a starved EHD contact. *ASLE Trans.* **1973**, *16*, 276–280. [CrossRef]
9. Chiu, Y.P. An analysis and prediction of lubricant film starvation in rolling contact systems. *ASLE Trans.* **1974**, *17*, 22–35. [CrossRef]
10. Damiens, B.; Venner, C.H.; Cann, P.M.E.; Lubrecht, A.A. Starved lubrication of elliptical EHD contacts. *J. Tribol.* **2004**, *126*, 105–111. [CrossRef]
11. Chevalier, F.; Lubrecht, A.A.; Cann, P.M.E.; Colin, F.; Dalmaz, G. Film thickness in starved EHL point contacts. *J. Tribol.* **1998**, *120*, 126–132. [CrossRef]
12. Jakobsson, B.; Floberg, L. The finite journal bearing, considering vaporization: (Das Gleitlager von endlicher Breite mit Verdampfung). In *Chalmers Tekniska Högskolas Handlingar*; Avd. Maskinteknik Chalmers University of Technology: Gothenburg, Sweden, 1957.
13. Olsson, K.O. Cavitation in Dynamically Loaded Bearings. In *Chalmers University of Technology, Institute of Machine Elements*; Chalmers University of Technology: Gothenburg, Sweden, 1965.
14. Brewe, D.E.; Hamrock, B.J. Analysis of starvation effects on hydrodynamic lubrication in nonconforming contacts. *ASME J. of Lubr. Tech.* **1982**, *104*, 410–417. [CrossRef]
15. Boness, R.J. Effect of oil supply on cage and roller motion in a lubricated roller bearing. *J. Lubr. Technol. Trans. ASME* **1970**, *92*, 39–53. [CrossRef]
16. Chevalier, F.; Lubrecht, A.A.; Cann, P.M.E.; Colin, F.; Dalmaz, G. Starved Film Thickness: A Qualitative Explanation. *Tribol. Ser.* **1995**, *30*, 249–257.
17. Elrod, H.G. A Cavitation algorithm. *J. Lubr. Technol.* **1981**, *103*, 350–354. [CrossRef]
18. Elrod, H.G.; Adams, M.L. A computer program for cavitation and starvation problems. *Cavitation Relat. Phenom. Lubr.* **1975**, *1*, 37–41.
19. Cann, P.M.E.; Lubrecht, A.A. Bearing performance limits with grease lubrication: The interaction of bearing design, operating conditions and grease properties. *J. Phys. D Appl. Phys.* **2007**, *40*, 5446–5451. [CrossRef]
20. Hirst, W. Scuffing and its prevention. *Chart. Mech. Eng.* **1974**, *21*, 88–92.
21. Glavatskih, S.B.; McCarthy, D.M.C.; Sherrington, I. Hydrodynamic Performance of a Thrust Bearing with Micropatterned Pads. *Tribol. Trans.* **2005**, *48*, 492–498. [CrossRef]
22. Brizmer, V.; Kligerman, Y.; Etsion, I. A Laser Surface Textured Parallel Thrust Bearing. *Tribol. Trans.* **2003**, *46*, 397–403. [CrossRef]
23. Etsion, I. State of the Art in Laser Surface Texturing. *J. Tribol.* **2005**, *127*, 248–253. [CrossRef]
24. Etsion, I.; Burstein, L. A Model for Mechanical Seals with Regular Microsurface Structure. *Tribol. Trans.* **1996**, *39*, 677–683. [CrossRef]
25. Etsion, I.; Halperin, G.; Brizmer, V.; Kligerman, Y. Experimental Investigation of Laser Surface Textured Parallel Thrust Bearings. *Tribol. Lett.* **2004**, *17*, 295–300. [CrossRef]
26. Etsion, I.; Kligerman, Y.; Halperin, G. Analytical and Experimental Investigation of Laser-Textured Mechanical Seal Faces. *Tribol. Trans.* **1999**, *42*, 511–516. [CrossRef]
27. Kovalchenko, A.; Ajayi, O.; Erdemir, A.; Fenske, G.; Etsion, I. The effect of laser surface texturing on transitions in lubrication regimes during unidirectional sliding contact. *Tribol. Int.* **2005**, *38*, 219–225. [CrossRef]
28. Ronen, A.; Etsion, I.; Kligerman, Y. Friction-Reducing Surface-Texturing in Reciprocating Automotive Components. *Tribol. Trans.* **2001**, *44*, 359–366. [CrossRef]

29. Ryk, G.; Kligerman, Y.; Etsion, I. Experimental Investigation of Laser Surface Texturing for Reciprocating Automotive Components. *Tribol. Trans.* **2002**, *45*, 444–449.
30. Briscoe, B.; Cruton, B.; Willis, F. The shear strength of thin lubricant films. *Proc. R. Soc. Lond. A Math. Phys. Sci.* **1973**, *333*, 99–114. [CrossRef]
31. Kango, S.; Sharma, R.; Pandey, R. Thermal analysis of microtextured journal bearing using non-Newtonian rheology of lubricant and JFO boundary conditions. *Tribol. Int.* **2014**, *69*, 19–29. [CrossRef]
32. Guzek, A.; Podsiadlo, P.; Stachowiak, G.W. Optimization of textured surface in 2D parallel bearings governed by the Reynolds equation including cavitation and temperature. *Tribol. Online* **2013**, *8*, 7–21. [CrossRef]
33. Gu, C.; Meng, X.; Xie, Y.; Kong, X. Performance of surface texturing during start-up under starved and mixed lubrication. *J. Tribol.* **2017**, *139*, 011702. [CrossRef]
34. Bijani, D.; Deladi, E.; Akchurin, A.; de Rooij, M.; Schipper, D. The influence of surface texturing on the frictional behaviour of parallel sliding lubricated surfaces under conditions of mixed lubrication. *Lubricants* **2018**, *6*, 91. [CrossRef]
35. Hamilton, D.B.; Walowit, J.A.; Allen, C.M. A Theory of Lubrication by Microirregularities. *J. Basic Eng.* **1966**, *1966*, 177–185. [CrossRef]
36. Qiu, Y.; Khonsari, M.M. Experimental investigation of tribological performance of laser textured stainless steel rings. *Tribol. Int.* **2011**, *44*, 635–644. [CrossRef]
37. Kovalchenko, A.; Ajayi, O.; Erdemir, A.; Fenske, G.; Etsion, I. The effect of laser texturing of steel surfaces and speed-load parameters on the transition of lubrication regime from boundary to hydrodynamic. *Tribol. Trans.* **2004**, *47*, 299–307. [CrossRef]
38. Raeymaekers, B.; Etsion, I.; Talke, F.E. Enhancing tribological performance of the magnetic tape/guide interface by laser surface texturing. *Tribol. Lett.* **2007**, *27*, 89–95. [CrossRef]
39. Johnson, K.L.; Greenwood, J.A.; Poon, S.Y. A simple theory of asperity contact in elastohydro-dynamic lubrication. *Wear* **1972**, *19*, 91–108. [CrossRef]
40. Greenwood, J.A.; Williamson, J.B.P. Contact of Nominally Flat Surfaces. *Proc. R. Soc. Lond. A Math. Phys. Sci.* **1966**, *295*, 300–319.
41. Gelinck, E.R.M. *Mixed Lubrication of Line Contacts*; University of Twente: Enschede, The Netherlands, 1999.
42. Shi, F.; Salant, R.F. A mixed soft elastohydrodynamic lubrication model with interasperity cavitation and surface shear deformation. *J. Tribol.* **2000**, *122*, 308–316. [CrossRef]
43. Patir, N.; Cheng, H.S. Application of Average Flow Model to Lubrication Between Rough Sliding Surfaces. *J. Lubr. Technol.* **1979**, *101*, 220–229. [CrossRef]
44. Patir, N.; Cheng, H.S. An Average Flow Model for Determining Effects of Three-Dimensional Roughness on Partial Hydrodynamic Lubrication. *J. Lubr. Technol.* **1978**, *100*, 12–17. [CrossRef]
45. Hu, Y.-Z.; Barber, G.C.; Zhu, D. Numerical analysis for the elastic contact of real rough surfaces. *Tribol. Trans.* **1999**, *42*, 443–452. [CrossRef]
46. Hu, Y.-Z.; Zhu, D. A full numerical solution to the mixed lubrication in point contacts. *J. Tribol.* **2000**, *122*, 1–9. [CrossRef]
47. Hu, Y.-Z.; Wang, H.; Wang, W.; Zhu, D. A computer model of mixed lubrication in point contacts. *Tribol. Int.* **2001**, *34*, 65–73. [CrossRef]
48. Faraon, I.C. Mixed Lubricated Line Contacts. Ph.D. Thesis, University of Twente, Enschede, The Netherlands, 2005.
49. Wakuda, M.; Yamauchi, Y.; Kanzaki, S.; Yasuda, Y. Effect of surface texturing on friction reduction between ceramic and steel materials under lubricated sliding contact. *Wear* **2003**, *254*, 356–363. [CrossRef]
50. Wang, W.-Z.; Huang, Z.; Shen, D.; Kong, L.; Li, S. The effect of triangle-shaped surface textures on the performance of the lubricated point-contacts. *J. Tribol.* **2013**, *135*, 021503. [CrossRef]
51. Wang, X.; Kato, K.; Adachi, K.; Aizawa, K. The effect of laser texturing of SiC surface on the critical load for the transition of water lubrication mode from hydrodynamic to mixed. *Tribol. Int.* **2001**, *34*, 703–711. [CrossRef]
52. Etsion, I. Improving tribological performance of mechanical components by laser surface texturing. *Tribol. Lett.* **2004**, *17*, 733–737. [CrossRef]
53. Ryk, G.; Etsion, I. Testing piston rings with partial laser surface texturing for friction reduction. *Wear* **2006**, *261*, 792–796. [CrossRef]

54. Vilhena, L.M.; Sedlaček, M.; Podgornik, B.; Vižintin, J.; Babnik, A.; Možina, J. Surface texturing by pulsed Nd:YAG laser. *Tribol. Int.* **2009**, *42*, 1496–1504. [CrossRef]

55. Dunn, A.; Wlodarczyk, K.L.; Carstensen, J.V.; Hansen, E.B.; Gabzdyl, J.; Harrison, P.M.; Shephard, J.D.; Hand, D.P. Laser surface texturing for high friction contacts. *Appl. Surf. Sci.* **2015**, *357*, 2313–2319. [CrossRef]

56. Saeidi, F.; Parlinska-Wojtan, M.; Hoffmann, P.; Wasmer, K. Effects of laser surface texturing on the wear and failure mechanism of grey cast iron reciprocating against steel under starved lubrication conditions. *Wear* **2017**, *386*, 29–38. [CrossRef]

57. Bijani, D.; Deladi, L.E.; Schipper, D.J. The Influence of Surface Texturing on the Film Thickness in Parallel Sliding Surfaces. In Proceedings of the 20th International Colloquium Tribology: Technische Akademie Esslingen, Stuttgart/Ostfildern, Germany, 12–14 January 2016; p. 12.

58. Coyne, J.C.; Elrod, H.G. *Conditions for the Rupture of a Lubrication Film*; The ASME Digital Collection: New York, NY, USA, 1965.

59. Xiong, S.; Wang, Q.J. Steady-state hydrodynamic lubrication modeled with the payvar-salant mass conservation model. *J. Tribol.* **2012**, *134*, 031703. [CrossRef]

60. Qiu, M.; Minson, B.R.; Raeymaekers, B. The effect of texture shape on the friction coefficient and stiffness of gas-lubricated parallel slider bearings. *Tribol. Int.* **2013**, *67*, 278–288. [CrossRef]

61. Qiu, M.; Delic, A.; Raeymaekers, B. The effect of texture shape on the load-carrying capacity of gas-lubricated parallel slider bearings. *Tribol. Lett.* **2012**, *48*, 315–327. [CrossRef]

62. Patankar, S.V. Series in computational methods in mechanics and thermal sciences. In *Numerical Heat Transfer and Fluid Flow/Suhas V. Patankar*; Hemisphere Pub. Corp.: New York, NY, USA, 1980.

63. Versteeg, H.K.; Malalasekera, W. *An Introduction to Computational Fluid Dynamics: The Finite Volume Method*; Longman Group Ltd.: London, UK, 1995.

64. Bijani, D.; Deladi, E.; de Rooij, M.; Schipper, D. The influence of surface texturing on the film thickness in starved lubricated parallel sliding contacts. *Lubricants* **2018**, *6*, 61. [CrossRef]

65. Gu, C.; Meng, X.; Xie, Y.; Yang, Y. Effects of surface texturing on ring/liner friction under starved lubrication. *Tribol. Int.* **2016**, *94*, 591–605. [CrossRef]

© 2019 by the authors. Licensee MDPI, Basel, Switzerland. This article is an open access article distributed under the terms and conditions of the Creative Commons Attribution (CC BY) license (http://creativecommons.org/licenses/by/4.0/).

Article

Production and Tribological Characterization of Tailored Laser-Induced Surface 3D Microtextures

Joel Voyer [1,*], Johann Zehetner [2,*], Stefan Klien [1], Florian Ausserer [1] and Igor Velkavrh [1]

[1]   V-Research GmbH, Stadtstrasse 33, 6850 Dornbirn, Austria
[2]   Research Center for Microtechnology, Vorarlberg University of Applied Sciences, Hochschulstrasse 1, 6850 Dornbirn, Austria
*   Correspondence: joel.voyer@v-research.at (J.V.); johann.zehetner@fhv.at (J.Z.); Tel.: +43-5572-394-15936 (J.V.); +43-5572-792-7203 (J.Z.)

Received: 28 June 2019; Accepted: 1 August 2019; Published: 6 August 2019

Abstract: The aim of the present study was firstly to determine the manufacturing feasibility of a specific surface 3D-microtexturing on steel through an ultra-short pulsed laser, and secondly to investigate the tribological properties under 2 different lubrication conditions: oil-lubricated and antifriction coated. The selected 3D-microtexture consisted of 2 different levels of quadratic micropillars having side dimensions of approximately 45 μm, heights of about 35 μm and periods of 80 μm. It was shown that the production of specific 3D-microtextures on steel substrates using an ultra-short pulsed laser was feasible, and that the reproducibility of the texture dimensions over the entire textured region was extremely good. Frictional investigations have shown that, in comparison to the benchmark (untextured samples), the 3D-microtextured samples do not induce any significant improvements in the coefficient of friction (COF) under oil-lubricated conditions, but that under antifriction coated conditions, significant improvements in the friction coefficients may be achieved. Wear-based tribological tests have shown that the antifriction coating on benchmark samples was completely depleted, which greatly influenced their friction and wear behavior, since steel-steel contact occurred during testing. For the 3D-microtexture, the antifriction coating was also partially depleted; however, it accumulated itself in the microtexture which acted as a potential lubricant reservoir.

Keywords: laser microtexturing; ultra-short pulsed laser; surface characterization; tribological characterization; friction; wear

---

1. Introduction

Achieving greater productivity at lower costs represents a significant challenge for the modern manufacturing industry. In this endeavor, new advanced materials and surface technologies are needed in order to enhance the efficiency of mechanical systems and reduce their energy consumption, which may be achieved through a reduction of the friction and wear of tribological contacts. This reduction of the tribological properties may be obtained through several different approaches, such as modification of the component's geometry, application of hard protective coatings, optimizing surface roughness and/or topography [1–5] or by the introduction of specific surface textures, which play a major role in lubrication, level of friction and the wear rates of tribological systems. Surface microtexturing has established itself over the last few years as a very promising approach to reduce friction in different materials in various lubricated applications [6–18].

Over recent years, a diverse array of manufacturing processes has been developed for the production of surface microtextures, such as mechanical micromachining, chemical etching or laser ablation, to name a few. All these industrial surface microtexturing processes show advantages or disadvantages. The latter process (laser ablation) was selected and used to produce the desired specific

3D microtexture to be investigated in the present study. The investigations consisted of producing, characterizing and tribologically testing the specific 3D microtexture.

Contour shaping, surface patterning and surface functionalization by femtosecond laser ablation (FLA) are becoming fundamental technologies in fields like tribology, microfluidic transport, fuel cells and medicine. Self-ordered hierarchical micro- and nano- structures can be generated on the majority of metals, semiconductors and dielectrics. Laser-induced periodic surface structures (LIPSSs) are significantly involved in the formation of such self-ordered microstructures and pinholes. LIPSSs have been attracting research interest for decades, but their formation is still not fully understood. The latest on the LIPSS formation mechanism and its potential deployment in tribological applications is outlined in [19]. However, the main interests of the authors are focused on a usable technology for contour shaping and surface patterning for specific tribological applications. In order to obtain the desired surface structure or surface quality, the understanding and handling of any feedback mechanisms occurring during the production of LIPSS, self-ordered microstructures or contour shaping is essential. The present authors recently focused themselves on the fabrication of membranes (based on AlGaN/GaN hetero-structure layers grown on 4H–SiC) for sensor fabrication. Feedback effects, which occur between small surface distortions and the slot waveguide function of LIPSS in SiC, were found to promote the formation of micropores in the membranes. To cope with this unwanted and challenging effect, the present authors developed a polarization steering procedure which interrupted the feedback loop and terminated the growth cycle of pinholes [20,21]. In the present study on steel samples, a similar pinhole formation phenomenon at locations where LIPSSs were interrupted by distortions was also observed. Such distortions can be triggered by the scanning laser beam itself, side wall irregularities at the laser generated cube structures, inhomogeneities inside the metal grain, laser generated debris particles or self-ordered bumps. In the current experiments, it is demonstrated that, among all the laser parameters contributing to the formation of such distortions, the most effective parameter is the orientation of the laser polarization with respect to the scan direction, which may suppress or promote the formation of self-ordered bumps and pinholes.

Tribological tests under lubricated conditions and also with an antifriction coating using the aforementioned 3D microtexture along with untextured (benchmark) samples were undertaken to evaluate the effectiveness of reducing the overall friction by decreasing the nominal contact areas through surface microtexturing.

2. Materials and Methods

2.1. Samples Geometry and Dimensions

For the present tribological investigations, the upper specimen consisted of a convex cylinder made of hardened 31CrMoV9 (around 700HV), while the lower specimen consisted of a disc made of hardened 42CrMo4 (around 650HV; hardness values were measured with a hardness tester from EMCO-TEST Prüfmaschinen GmbH, Kuchl, Austria) whose dimensions are shown schematically in Figure 1. For all tribological tests performed, cylinders were used as received, but disc specimens were further treated in order to evaluate the effect of a 3D surface texturing. Furthermore, the influence of an antifriction coating was also studied for the present manuscript. The four different disc specimen surface states investigated in the present work are shown schematically in Figure 2. The as-received convex cylinders had surface roughness values of $R_a = 0.12 \pm 0.01$ µm and $R_z = 1.68 \pm 0.38$ µm, and the benchmark disc samples had surface roughness values of $R_a = 0.06 \pm 0.01$ µm and $R_z = 0.65 \pm 0.19$ µm.

The antifriction coating used was applied on one side of the previously cleaned discs (blank or surface microtextured) through the coating manufacturer (Carl Bechem GmbH, Hagen, Germany). The thickness of the antifriction coating ranged between 15 and 20 µm. The choice of this coating was based on the fact that it should be especially designed for a reduction of the friction coefficient (AF320E). Some details on the antifriction coating used in the present study are listed in Table 1.

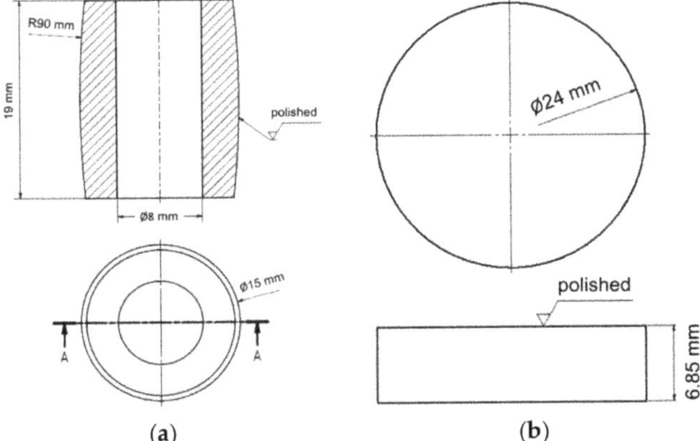

Figure 1. Geometries of the samples used in the present study: (**a**) cylinder (upper specimen) and (**b**) disc (lower specimen) (Note: drawings are not to scale).

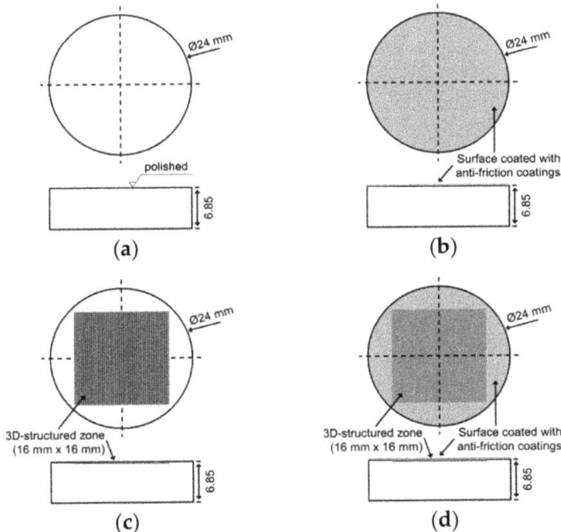

Figure 2. Different processed disc surfaces investigated in the present study: (**a**) benchmark disc, (**b**) benchmark disc with an antifriction-coating, (**c**) 3D microtextured disc and (**d**) 3D microtextured disc with an antifriction coating.

Table 1. Excerpt of important properties of the antifriction coating investigated in the present study.

Coating	Density (g/cm^3 @ 20 °C)	Solid Lubricant	Solid Lubricant Amount	Binder	Solvent
AF320E	0.97	PTFE	33%	organic	organic

Disc specimens were 3D surface microtextured using a femtosecond laser; the procedure used will be explained in detail later. It is worth noting that only a small central quadratic region (16 mm × 16 mm) of one side of the disc specimens was actually microtextured as shown in Figure 2c. This 3D microtexture was chosen based on previously performed tribological investigations, in which the influence of a reduction of the effective nominal contact area through the production of microridges on elastomer pads was analyzed and reported [22]. The very promising results obtained from the aforementioned studies led the authors to produce, characterize and evaluate a further optimized surface microtexture in the present study, which consisted of a combination of 2 microridged textures perpendicular to each other, as schematically shown by the red arrows in Figure 3. Again, based on the results of previous studies [7,22], the desired dimensions of the plateaus of the 3D microtexture should range between 40 and 50 μm for the sides, between 30 and 40 μm for the height, and about 80 μm for the period in both horizontal and vertical directions. With these desired dimensions for the uppermost plateaus (which define the contact area), a calculated ratio of approximately 20% for the nominal contact area of the 3D-textures samples in comparison to the untextured (benchmark) specimens may be obtained (benchmark: nominal contact area of 100%, without taking into account its surface roughness).

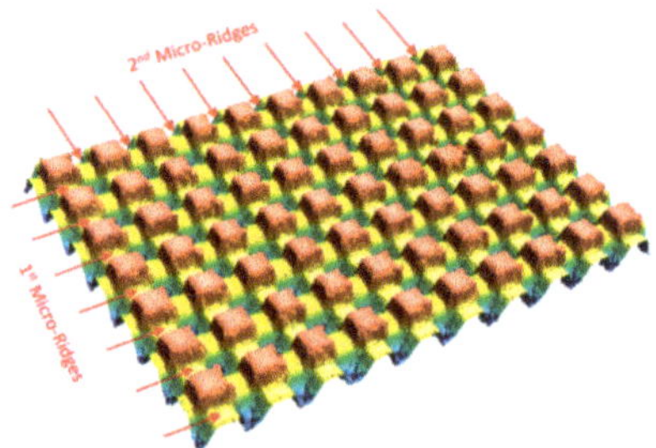

Figure 3. Example of the desired 3D surface microtexture to be produced using an ultra-short pulsed laser on one side of disc specimens.

All specimen surfaces were characterized using a non-contact laser optical surface roughness measuring apparatus (VK-X250/260, Keyence International NV/SA, Mechelen, Belgium). The surface roughness values of the convex cylinders and benchmark disc specimens (without and with coating) were evaluated. Furthermore, the characteristics of the 3D microtexture (without and with coating), such as side dimensions, heights and periods of the different plateaus, to name a few, were measured and reported as average values.

2.2. Production of 3D-Microtexture through Laser Ablation

For the production of the desired cube-shaped 3D microtexture, a laser work station (microSTRUCTvario, 3D-MICROMAC, Chemnitz, Germany) in combination with a femtosecond laser (SPIRIT, Spectra Physics, Rankweil, Austria) were used. The laser delivered 350 fs pulses at 200 kHz with an average output power of 4 W at a wavelength of 1040 nm and 1.6 W at 520 nm. The laser beam was linearly polarized and the polarization direction could be flipped by 90° during the ablation process. To focus the laser beam, a telecentric scanner optic (Linos F-Theta Ronar, QIOPTIQ Photonics GmbH & Co. KG, Feldkirchen, Germany) with a focal length of 100 mm providing a focus spot radius of 6 μm for the 520 nm wavelength was used. For all test samples, a scan speed of 1000 mm/sec and a hatch distance of 5 μm were used.

The disc specimens for the tribological tests were produced in two steps using an average laser power of 385 mW. Firstly, a pattern along the x direction consisting of stripes with a width of 40 μm was scanned using the aforementioned hatch distance of 5 μm. After 15 consecutive scans, the polarization was flipped by 90° (symbolized by the yellow cross in Figure 4a), and this procedure was repeated 11 times. The laser polarization of the final scan cycle was in the x direction and parallel to the scan direction (the red arrow in Figure 4a represents the scan direction and the yellow arrow represents the polarization of the final scan cycle). The bottom of level 3 of the first trench, as shown in Figure 4a, was formed by this first ablation cycle, and the depth with respect to level 1 was approximately 35 μm. Secondly, the same procedure was performed with scans along the y direction. The polarization was flipped similarly as in the first step, and consequently, the final scan ended up with a polarization along the x direction (perpendicular to the scan in the y direction). This second step formed the 35 μm deep bottom level of Section 2 (shown in Figure 4a) and a deeper bottom level (level 4) at the intersection with level 3. Surprisingly, the surface quality of Section 2 was not equal to Section 3. The fact that the laser beam is astigmatic could be one of the reasons for such an outcome. To rule out the laser beam astigmatism in our experiments, it was decided to do only scans along the x direction but to keep the polarization flip procedure after 15 consecutive scans. By using this procedure, the trench width was increased from 40 to 200 μm and the interfering effects from the generated side walls of a too narrow trench were therefore limited. In addition, the number of repetitions was reduced from 11 to 7 and, as a consequence, the depth of the obtained trench was reduced from 35 to 25 μm respectively, as shown in Figure 4b. One can clearly see traces from the 5 μm hatch pattern (small lines) along which pinholes with diameters of approximately 2 to 3 μm were formed. Because the polarization of the final scan cycle was along the x direction, the generated LIPSSs are orientated in the y direction [23,24]. As a consequence, the hatch lines intersect the LIPSS, or at least modulate the depth of the LIPSS, which leads to the growth of sub-micrometer pinholes of approximately 2 to 4 μm in diameter. Such a growth mechanism in SiC was previously simulated and investigated, and showed that LIPSSs correlate well with the slot waveguide characteristics in high refractive index material [20]. It is believed that such a model may also be applied to metals because, according to theoretical and experimental research, deep grooves in LIPSSs behave as plasmonic slot waveguides [25]. Moreover, previously published models predict a field distribution for deep slots which can generate nanometer-sized pinholes or cross periodic structures respectively [25,26]. The model previously developed by the present authors describes the growth of such nanometer pinholes towards micrometer sizes at the locations of interruptions or distortions of LIPSSs [20]. In earlier work on SiC, a certain threshold for pinhole growth was identified and the laser power was consequently reduced from 385 to 238 mW, while all other conditions were kept constant [20]. There was a significant reduction in the pinhole formation rate but at the cost of a reduction of the trench depth from 25 to 18 μm, as shown in Figure 5a.

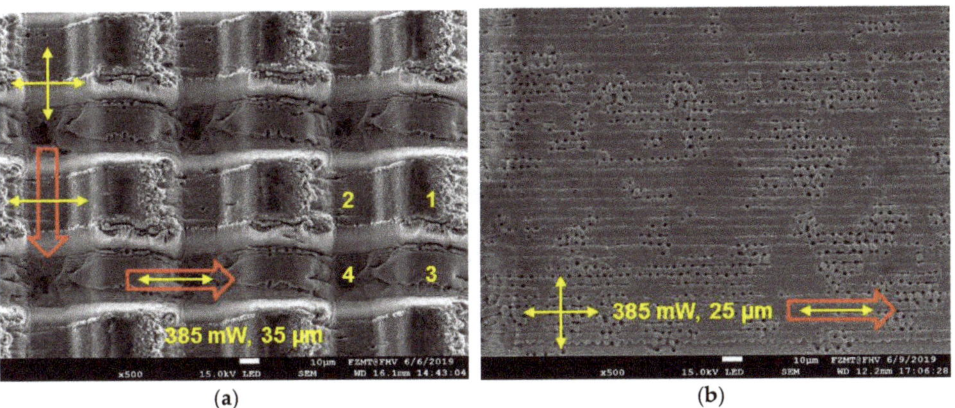

Figure 4. Micrographs: (**a**) 3D microtextured disc; (**b**) test trench (hatch intersect the LIPSS).

An increase of the number of scans in order to reach a depth of 25 μm would have required an unwanted longer production time; therefore, it was decided to test the approach for which the final scan cycle was a combination of a scan in the x direction with a polarization in the y direction. By using such a procedure, it is expected that the generated LIPSS would be nearly parallel to the hatch lines, the number of intersections among hatch and LIPSS would be lower and the formation and growth rate of the pinholes may be reduced.

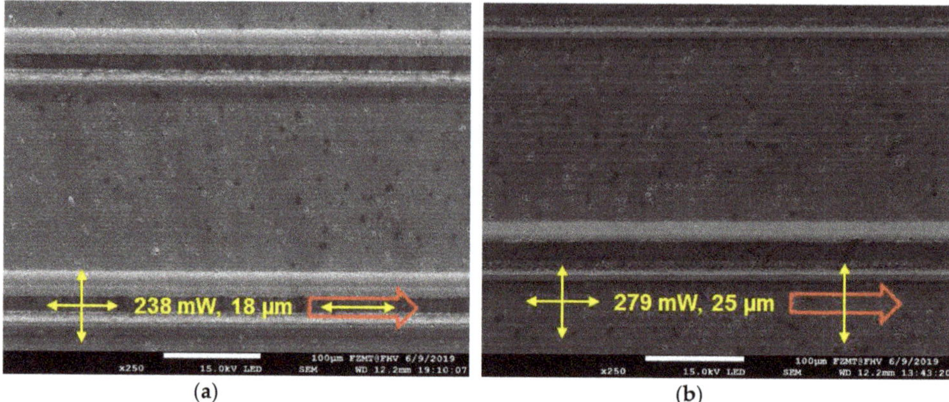

(a)   (b)

Figure 5. Surface quality after laser ablation: (**a**) reduced laser power and polarization of final scan cycle parallel to scan direction; (**b**) increased laser power and number of scans, but polarization of the last scan cycle perpendicular to the scan direction.

Furthermore, the laser power was increased from 238 to 279 mW, while the number of scans per constant polarization cycle was reduced from 15 to 10; in return, the number of cycles increased from 7 to 14. The obtained results verified our hypothesis and expectations: the obtained surface quality was close to that of the previous test performed with a laser power of 238 mW along with the fact that a trench depth of 25 µm could be again obtained as previously, as shown in Figure 5b. As more than one parameter for this test was simultaneously changed, two supplementary tests were conducted in order to confirm that our concept of polarization flipping after a certain number of scans, combined with a final scan cycle having a laser polarization perpendicular to the scan direction, significantly contributes to a smooth surface quality after laser ablation. Using the same parameters and procedure as previously, but skipping the polarization flipping, it is possible to compare the results obtained from a laser polarization and a scan both parallel to the x direction to the results obtained from a laser polarization along the y direction and perpendicular to a scan in the x direction, as shown in Figure 6a,b. The obtained results obviously support the hypothesis that in metals (as in SiC), pinhole growth is linked to the number of interruptions of the LIPSS. After these experiments, it came to the authors' attention that the simple rule of orientating the laser polarization perpendicular to the scan direction is the most effective measure to suppress pinhole growth, from negligible nanometer size pores up to micrometer dimensions. However, additional polarization flipping still has its benefits, i.e., contributing to pinhole suppression, because practically, it is impossible to obtain undisturbed gratings like LIPSS patterns on a larger ablation area; eventually, some pinholes will grow over time. LIPSS formation and subsequent pinhole growth are favored either by long durations at low scan speeds or a high number of scans at higher scan speeds. If the polarization is flipped within such a time or scan interval, the existing LIPSSs are reorganized along a new direction, the already generated pinholes are then removed (providing that their size is still not too big) and the growth cycle has to start again. Therefore, it was decided to use a growth cycle reset after every 10 consecutive scans and, as depicted in Figure 7b where the final cycle was performed with a laser polarization perpendicular to the scan direction, it may be observed that a further reduction in pinhole density with respect to the results shown in Figure 6b (same laser parameter but at constant polarization direction perpendicular to the scan direction) may be obtained.

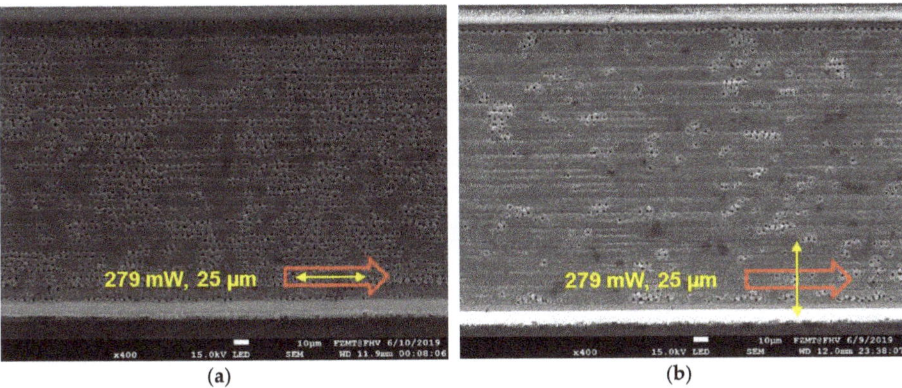

Figure 6. Surface quality after laser ablation without any polarization flipping: (**a**) polarization permanently parallel to scan direction; (**b**) polarization permanently perpendicular to scan direction.

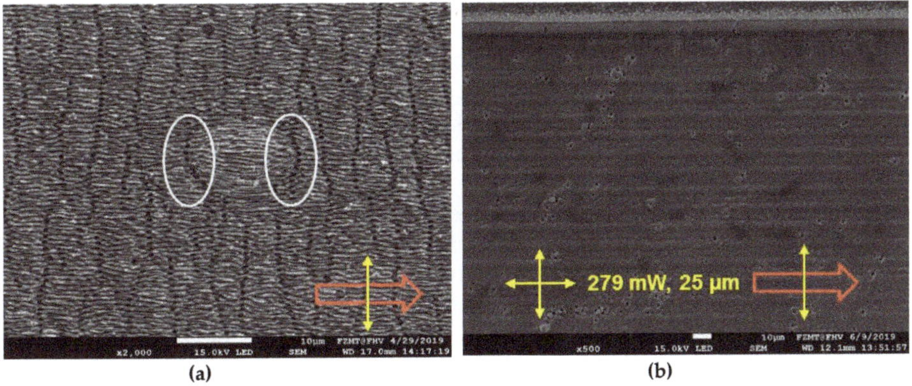

Figure 7. Micrographs: (**a**) LIPSS intersections perpendicular to their orientation where small pinhole formation starts; (**b**) same laser procedure as in Figure 5b but with enlarged scale (polarization of the last scan cycle was perpendicular to the scan direction).

However, the aforementioned simple rule of orientating the laser polarization perpendicular to the scan direction has one main disadvantage which becomes more important as a trench gets deeper after several scans. Due to Brewster angle effects and a resulting higher ablation rate in the direction of the laser polarization, small scratches in the trench wall grow faster and end up in a distinct score pattern [21]. After several consecutive scans, the pattern in the trench wall initiates a corresponding structure formation on the LIPSS-covered trench bottom, as shown in Figure 7a. Small pores are arranged in strings which are parallel to the polarization and perpendicular to the scan direction. The LIPSS are intersected perpendicular to their orientation, and after a number of several consecutive scans, the pores grow together to form bigger identities [21]. Any interruption of the LIPSS (for example a small bump) can trigger the formation and growth of pinholes, as depicted in Figure 7a. The front and end sides of the surface bump represent an interruption of the LIPSS (encircled areas in Figure 7a), and the growth cycle of pinholes predominately starts at this location. LIPSS striking tangentially the surface bump are not interrupted and a lower degree of pinhole formation was observed. The observations mentioned above (formation and growth of pinholes due to LIPSS intersections) are also interrupted by polarization flipping and have to start again from scratch, thus contributing to smoother surface quality and reduced pinhole formation. It is worth noting that for a better visualization of the aforementioned observations shown in Figure 7a, a laser with a wavelength of 1040 nm was used in order to obtain larger LIPSSs. Moreover, larger and deeper LIPSSs may provide further advantages when they are tribologically tested under oil-lubricated conditions. In the current work, 3D textures without any LIPSSs were used as the top contact area. It is believed that a further improvement of the current 3D microtexture may be achieved through a two-step procedure: firstly, the laser ablation of the trenches along both the x and y directions should be performed using a laser with a wavelength of 520 nm (which showed a lower tendency for pinhole formation in comparison to the laser with a wavelength of 1040 nm), and further applying polarization flipping under the condition that the final scan cycle should be made with a perpendicular polarization with respect to the scan direction; secondly, after producing the 3D microtexture, the whole surface area should be exposed to the laser with a wavelength of 1040 nm in order to cover the microtexture with a LIPPS-based nanotexture, such as that shown in Figure 7a. It is believed that the current tribological results under oil-lubricated conditions could be improved by such a hierarchical micro/nanotextured surface in future tribological investigations. Furthermore, it is believed that the adhesive strength and the wear resistance of the antifriction coating could be improved through this novel manufacturing procedure.

2.3. Evaluation of Dynamic Friction Coefficients

Stepwise load-varying tribological tests were performed using a SRV-tribometer (SRV®4, Optimol Instruments Prüftechnik GmbH, Munich, Germany). In these tests, friction coefficients were measured for short-time (5 min) oscillating tests at several different normal load values ranging from 25 to 200 N with steps of 25 N (as shown in Figure 8) under different lubrication conditions for as-received cylinders against different discs surface states: benchmark blank, antifriction coated benchmark, 3D microtextured or 3D microtextured/antifriction coated, as shown by the test matrix listed in Table 2. Furthermore, a supplementary test with a coated microtextured specimen, with the 3D microtexture oriented at an angle of 45° (instead of 90°) to the direction of motion of the cylinder, was also performed in order to investigate any possible influences of the microtexture orientation on its tribological behavior. Table 3 lists some of the important test parameters used for the determination of dynamic friction coefficients. For the determination of dynamic friction coefficients, the tribometer was programmed to perform high-frequency friction signal acquisition (FSA) at the beginning and end of each load level for duration of 0.2 s at a sampling rate of 1 kHz. This FSA signal analysis enables a highly precise time-resolved analysis of the friction force and coefficients of friction, which are otherwise not possible using the normal signal acquisition parameters of the SRV-tribometer. The obtained FSA raw signals were then postprocessed in order to calculate averages of dynamic coefficients of friction. This postanalysis consisted of eliminating 10% of the raw signal at the beginning and at the end of a half-cycle, i.e., keeping only 80% of the raw signals for each half-cycle and averaging these values over approximately 10 half-cycles, as shown by the light green zones in Figure 9. From Figure 9, it may also be observed that the first peak present at the onset of movement (in either direction) is due to the adhesive part of the friction, also known as static friction or stiction.

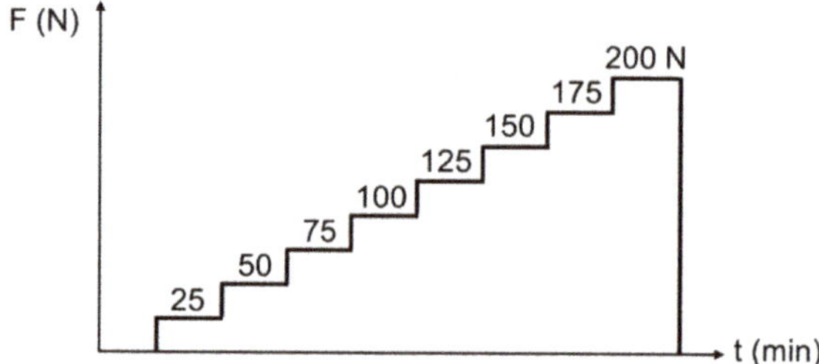

Figure 8. Schematic representation of the load-varying tribological tests used for the determination of the dynamic friction coefficients.

Table 2. Test matrix of all tribological investigations performed in the present study.

Test-ID	Texture Angle	Cylinder (Upper Specimen)		Disc (Lower Specimen)		Lubricant
		Material	Surface	Material	Coating	
Blank Benchmark	-	31CrMoV9	as-received	42CrMo4	-	Oil*
Coated Benchmark	-	31CrMoV9	as-received	42CrMo4	AF320E	-
Microtexture (90°)	90°	31CrMoV9	as-received	42CrMo4	-	Oil*
Coated Microtexture (90°)	90°	31CrMoV9	as-received	42CrMo4	AF320E	-
Coated Microtexture (45°)	45°	31CrMoV9	as-received	42CrMo4	AF320E	-

* Commercially available high-quality multi-grade hydraulic oil (ISO VG37).

Table 3. Main parameters used for the tribological investigations.

Normal Load (N)	Duration for Each Load (min)	Total Duration (min)	Stroke (mm)	Frequency (Hz)	Temperature (°C)
25, 50, 75, 100, 125, 150, 175, 200	5	40	4	25	22

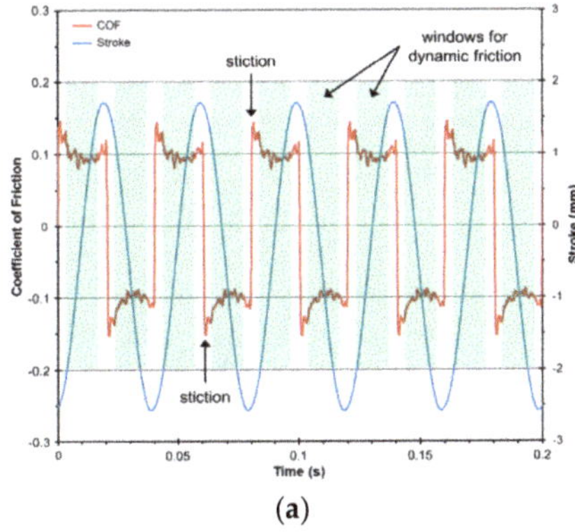

(a)

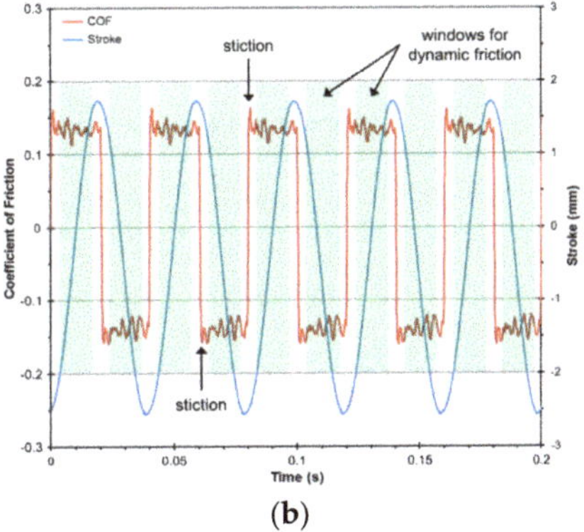

(b)

Figure 9. Example of raw signals obtained using FSA (high frequency signal acquisition) used for the calculation of both static and dynamic coefficients of friction (example shown: (**a**) benchmark (untextured samples) with oil at F_N = 175 N and (**b**) 3D microtexturing with oil at F_N = 175 N).

2.4. Evaluation of Wear Resistance and Long-Term Friction Behaviour

The wear resistance of the coated specimens (benchmark, 3D microtextured and 3D microtextured at 45°) also represents a major interest prior to their deployment in real industrial applications. The reason that only the coated specimens were selected for the investigations of wear resistance was based on the fact that:

1. oil-lubricated conditions may possess certain drawbacks for some specific highly technical industrial applications; thus, antifriction coated specimens were selected, since they are widely used in industrial applications.

2. it was considered useless to investigate the wear of uncoated specimens under unlubricated conditions, since these conditions are usually never used in the industry.

The wear resistance of the previously specified antifriction coated samples was determined through long duration tribological tests (120 min). Except for the total test duration (120 min at 1 load level instead of 10 load steps of 5 min each), all other test parameters were kept identical to those used for the determination of friction coefficients, as listed in Table 3. Due to the restricted number of available samples, these long-term tests were performed at a normal load value of 125 N only, i.e., at a load value high enough to produce a measurable wear on the investigated samples.

The overall wear of the tested samples (cylinders and discs) was determined using the aforementioned laser optical surface analysis apparatus (Keyence VK-X250/260). For both cylinders and discs, a direct measurement of the wear scar volume was performed using a height threshold-based volume measurement. It is worth noting here that for both antifriction coated 3D microtextured disc samples (90° and 45°), the measured wear volume values also account for the small proportion due to the valleys of the 3D-texture present at the bottom of the wear scar, and thus, the measured and presented values are slightly higher than the real wear volume values.

3. Results and Discussions

3.1. Characterization of Benchmark, Coated, 3D Microtextured and 3D Microtextured/Coated Samples

For the blank uncoated benchmark samples (cylinders and discs), the only characterization performed was the measurements of their surface roughness, which were already mentioned in Section 2.1 and are additionally listed in Table 4. For the untextured antifriction coated specimens (coated benchmark), the application of the aforementioned antifriction coating to their surfaces increased their surface roughness significantly, as shown in Table 4.

Table 4. Surface roughness values of untextured uncoated (blank benchmark) and untextured antifriction coated (coated benchmark) disc specimens.

Sample Denomination	Texture	Coating	R_a (µm)	R_z (µm)
Blank Benchmark	None	None	0.06 ± 0.01	0.65 ± 0.19
Coated Benchmark	None	AF320E	0.94 ± 0.10	5.74 ± 0.78

A detailed measurement and analysis of the microtextured disc specimens was performed in order to ascertain the accordance of the dimensions of the produced microtextures with their initial desired dimensions. Figure 10 shows a detailed typical topographical analysis of two selected 3D microtextured disc samples; their measured specific dimensions are listed in Table 5. These dimensions correspond quite exactly to the initial desired dimensions presented in Section 2.1, thus showing that the production of 3D-microtextures using an ultra-short pulsed laser may be performed with a good level of accuracy. The ratio of the nominal microtextured area of the top plateau in comparison to the benchmark (benchmark = 100%) was calculated using the measured side dimensions and periods listed in Table 5, resulting in a value of 20.9%, which corresponds quite adequately to the desired ratio (20%) calculated from the desired microtexture dimensions presented in Section 2.1. These results show clearly that the production of 3D microtextures on steel may be achieved with a very high level of precision.

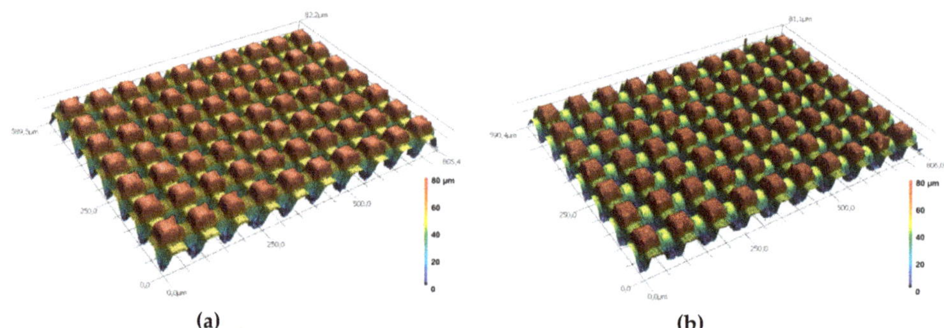

(a)　　　　(b)

Figure 10. Typical topographies of two selected 3D surface microtextures produced on disc specimens: (a) first sample; (b) second sample.

Table 5. Typical dimensions of produced 3D microtextures on disc specimens.

Plateau 1			Plateau 2		
Sides (b_x, b_y) (μm)	Height (Vs. Underlying Plateau 2) (μm)	Period (P_x, P_y) (μm)	Sides (b_x, b_y) (μm)	Height (Vs. Underlying Plateau 3) (μm)	Period (P_x, P_y) (μm)
37.2 ± 2.1	27.9 ± 0.6	82.3 ± 1.8	36.4 ± 1.5	41.2 ± 3.5	82.8 ± 0.8

As mentioned earlier, some of the 3D microtextured samples were then coated with an antifriction coating (properties shortly listed in Table 1). Typical topographies of such 3D microtextured and antifriction coated samples are shown in Figure 11. The first obvious observation that could be made from a comparison between Figures 10 and 11 is that the produced surface microtextures are no longer quadratic, and are not as sharply defined as previously: actually, the microtextures are in some way rounded by the presence of the antifriction coating. Furthermore, it may be easily observed that the valleys were partially filled with the antifriction coating, resulting in a reduction in the total height difference between the different plateaus of the 3D-texture. From the typical topographies presented in Figures 10 and 11, it may be easily concluded that the coating process used in the present study to apply the antifriction coating strongly modified the surface topography of the microtextured samples. The height of each plateau of the microtextured and coated samples was measured using the same procedure as for the microtextured samples. However, the lateral dimensions (b_x, b_y) and the periods (P_x, P_y) of the different plateaus were not determined due to their undefined rounded profiles, which are difficult to measure. The average height differences from 3 different measurements between each plateau are listed in Table 6, and show that the height differences between the plateaus had decreased due to a partial filling of the microtexture valleys by the antifriction coating, as mentioned previously.

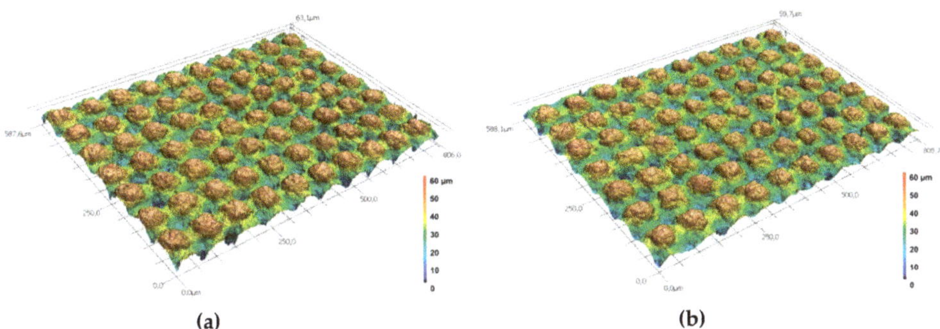

(a)   (b)

Figure 11. Typical topographies of two selected 3D microtextured and antifriction coated disc specimens: (a) first sample; (b) second sample.

Table 6. Typical dimensions of produced 3D microtextured and antifriction coated disc specimens.

Plateau 1 Height (Vs. Underlying Plateau 2) (μm)	Plateau 2 Height (Vs. Underlying Plateau 3) (μm)
18.1 ± 2.3	20.9 ± 2.5

3.2. Evaluation of Dynamic Friction Coefficients

The results obtained from the load-varying tests performed on all sample configurations under study are presented in Figure 12, in which each point represents the calculated average of the friction coefficient (calculated from the sliding section of the obtained curve as shown in Figure 9) for the specified applied normal force. For the benchmark sample under oil-lubricated conditions (Figure 12a, black curve), it may be observed that the sliding coefficient of friction is very stable throughout the load range under study (which is shown by the relatively small error bars), and that its value remains at around 0.10. For the coated benchmark sample (Figure 12a, red curve) under dry conditions, the coefficient of friction becomes very unstable (large error bars) and is higher than the value of the benchmark under oil-lubricated conditions. Furthermore, its value stays at around 0.22 up to a load of 100 N and then increases abruptly to a higher value of approximately 0.30 for higher loads. This increase of the friction coefficient has been visually correlated to the coating failure during the test.

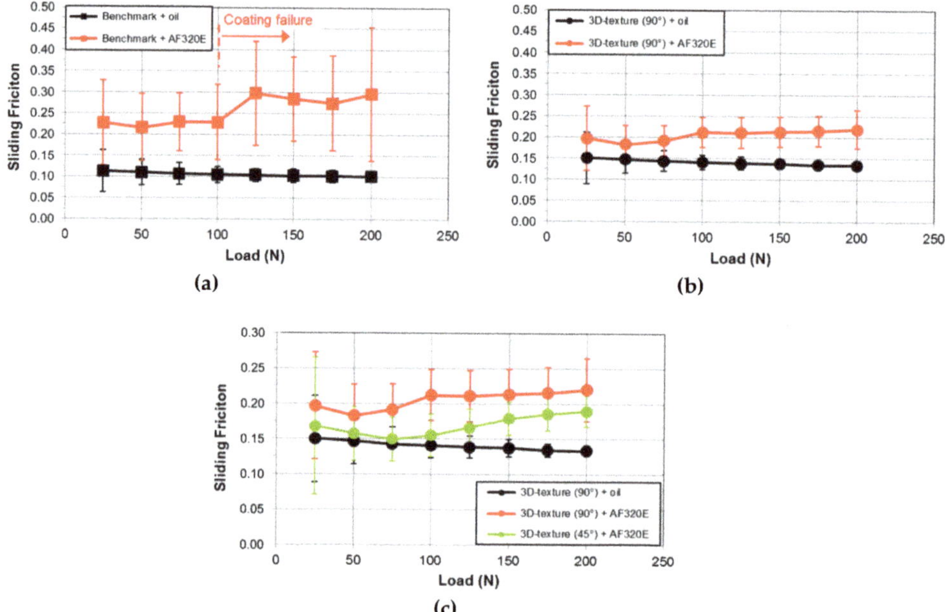

Figure 12. Sliding coefficients of friction as a function of applied load: (**a**) benchmark with oil or antifriction coating AF320E, (**b**) 3D microtexture (90°) with oil or antifriction coating AF320E, and (**c**) comparison of all 3D microtexture results (note: different scales for y-axis).

For the 3D microtextured samples (Figure 12b), both the sample under oil-lubricated conditions (black curve) and that with an antifriction coating (red curve) show relatively stable coefficients of friction (small error bars) throughout the load range under study. For the 3D microtexture under oil-lubricated conditions, the coefficient of friction remains around 0.14 and for the 3D microtexture with an antifriction coating, friction coefficient stays around 0.20. By comparing the results of Figure 12a,b, it may be observed that under oil-lubricated conditions (black curves), the 3D microtextured samples exhibit slightly higher coefficients of friction then the benchmark (COF$_{benchmark}$ ~0.10; COF$_{microtexture}$ ~0.14), and that the friction stability is similar for both samples, showing that the use of 3D microtextures does not bring any significant benefits under oil-lubricated conditions. However, with the use of an antifriction coating (red curves of Figure 12a,b), one may observe that the stability of the coefficients of friction is drastically increased (comparison of error bars), and that the coefficients of friction are significantly smaller for the 3D microtextured samples than for the benchmark samples, showing that the use of 3D microtextures along with an antifriction coating under unlubricated conditions represents an interesting alternative to oil-lubricated, untextured surfaces (benchmark).

It is believed that for the benchmark samples, a depletion of the antifriction coating occurs during the tribological tests, and that the coating is actually pushed out of the contact zone, thus exposing the underlying blank steel of the disc specimen to the cylinder, resulting in high and unstable coefficients of friction, especially for high load values. For the tests with the 3D microtextured and coated samples, a depletion of the antifriction coating also occurs, but only on the first microtexture level; the underlying levels of the 3D microtexture itself serve as lubricant reservoirs in which particles of the depleted antifriction coating from the top microtexture level are stored and continuously smeared back on the top level of the microtexture, thus enabling a kind of continuous lubrication regime which results in low and stable friction coefficients for all load values under study.

The results from tests performed on 3D microtextured samples with an antifriction coating but with the microtexture aligned at an angle of 45° from the sliding direction (instead of 90° for the previously presented results) are presented in Figure 12c (results of Figure 12b are incorporated in Figure 12c in order to compare easily the effect of the microtexture angle). From Figure 12c, it may be easily observed that the modification of the microtexture angle (from 90° to 45°) with respect to the sliding direction tends to decrease the coefficient of friction towards the value of the microtextured samples at an angle of 90° tested under oil-lubricated conditions for low load values only (<125 N), and for higher loads, the coefficient of friction of the 45°-oriented microtexture increases back to approximately the same values as for the 90°-aligned microtextured sample with an antifriction coating. However, these observations should be taken with caution, since error bars of the results obtained may indicate that this behavior is not as obvious as observed, and that these observations may be solely due to the intrinsic scattering nature of the measured coefficients of friction. More detailed investigations are necessary in order to assess the veracity of the previous observations. Nevertheless, by considering the obtained error bars, one may conclude without any doubt that the use of a 45°-oriented 3D microtexture (instead of a 90°-aligned 3D microtexture) with an antifriction coating tends to slightly lower the coefficients of friction, but that the coefficients of friction never fall below the values obtained for the 3D microtextured samples tested under oil-lubricated conditions.

3.3. Evaluation of Wear Resistance and Long-Term Friction Behaviour

As mentioned in Section 2.4, the determination of the wear resistance was performed only for the 3 samples having an antifriction-coating: coated benchmark, 3D microtextured (90°) coated and 3D microtextured (45°) coated samples. The wear resistance was determined as the wear rate (mm^3/h), which was calculated from wear volume values, measured using laser optical microscopy after the tribological tests, divided by the test duration. It is worth noting that since disc specimens possess an extremely high number of micropillars (and microvalleys) on their surfaces, it is almost impossible to precisely quantify their wear volume after the wear-based tribological tests, because wear measurements were performed using a normal height threshold technique using a defined level, which also encompasses intact microvalleys that did not experience any wear during the tribological tests. Therefore, reported wear values for the disc specimens should not be interpreted as absolute, but as relative, values, which still enable a valuable comparison to be made between the investigated tribological pairings.

But first of all, before presenting the measured wear rates of the 3 aforementioned sample variations, coefficients of friction during these wear resistance tests were also monitored and are presented in Figure 13a for a direct comparison to the friction coefficients measured during the load-varying tests (short-term tests; Figure 12). For this comparison, one should bear in mind that the wear tests were performed at a normal load of 125 N; therefore, on the corresponding diagrams of Figure 12, the friction coefficient value at a load of 125 N should be used as a comparison point. Furthermore, the first values of the coefficients of friction measured and presented in Figure 13a are after a test duration of 15 min, which is about 3 times longer than the short-term load-varying tests. In order to facilitate the direct comparison of the coefficients of friction related to both tribological tests, the results of Figure 12a,c were partially combined and are presented in Figure 13b, and the corresponding coefficients of friction are listed in Table 7.

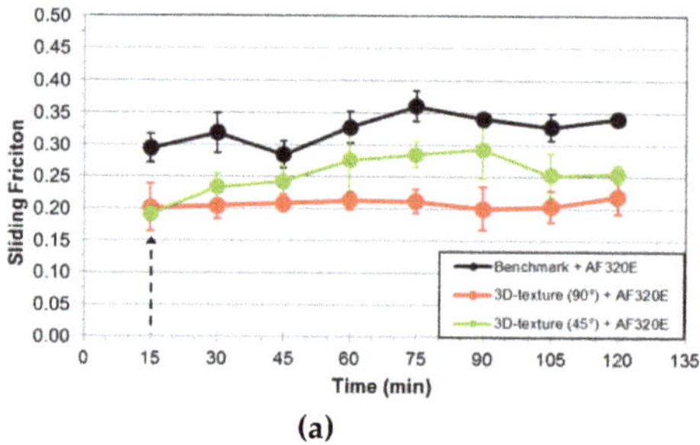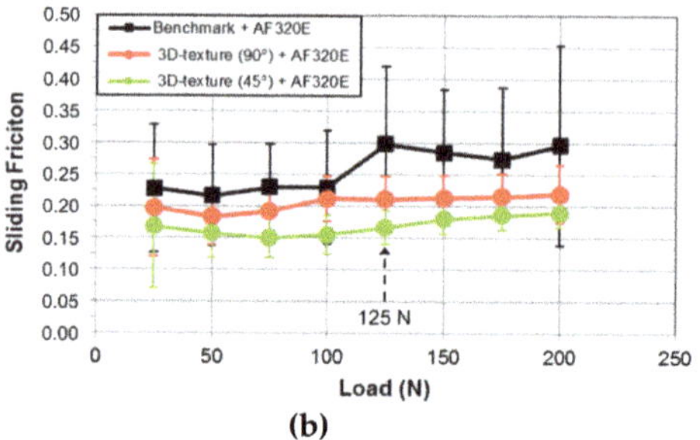

(a)   **(b)**

Figure 13. Average coefficients of friction measured: (**a**) during wear resistance tests (long-term tribological tests: 125 N, 2 h) and (**b**) during load-varying tests (short-term tribological tests: 25–200 N, 5 min. per load level) (Figure 13 = partial combination of Figure 12a,c).

Table 7. Comparison of coefficients of friction measured during wear resistance (long-term: 125 N, 2 h) and load-varying (short-term: 25-200 N, 5 min. per load level) tribological tests.

	Coefficient of Friction	
Sample Denomination	**Wear Resistance Test** (Figure 13a @ 15 min)	**Load-Varying Test** (Figure 13b @ 125 N)
Benchmark + AF320E	0.294	0.298
3D Microtexture (90°) + AF320E	0.201	0.211
3D Microtexture (45°) + AF320E	0.192	0.155

From Figure 13 and Table 7, it is obvious that the friction coefficients obtained either from wear resistance tests (@ 15 min) or from load-varying tests (@ 125 N) are quite similar, which gives a certain degree of validation for the friction coefficients obtained from both different tribological tests. The small discrepancy observed for the 3D microtexture with a 45°-angle may be due to the fact that the time-range of the compared values is different (15 min for wear tests, 5 min for load-varying tests).

It is worth noting that the error bars shown in Figure 13b are larger than those in Figure 13a due to different procedures used for friction measurements. The friction values shown in Figure 13b (short-term tribological tests with varying loads) were measured and averaged for a time duration of 0.2 s at the beginning and end of each load level ($t_{begin} = 0$ s; $t_{end} = 5$ min.); therefore, for each load level, and due to the short time period at a new constant load value, the friction values were partially measured when the system was still in the running-in phase (in which the friction may still be relatively unstable and may relatively differ from test to test), thus resulting in larger error bars when averaging values measured from 3 short-term tribological tests. On the other hand, the friction values shown in Figure 13a (long-term tribological tests at a constant load) were measured and averaged for a time duration of 0.2 s every 15 min ($t_1 = 15$ min.; $t_2 = 30$ min., etc.); therefore, for the first friction measurement at $t_1 = 15$ min., the running-in phase of the system was already finished and the system was already in a friction regime where the friction was relatively stable and relatively reproducible between different tests, resulting in smaller error bars when averaging 3 long-term tribological tests.

Typical top view pictures of wear scars after the performed wear resistance tests obtained using laser microscopy are shown in Figure 14. At first sight, one may observe that the scars of the benchmark samples are somewhat larger than those of the 3D-textured samples, with the smallest scars for the 3D-textured samples with a 90°-angle.

For the coated benchmark samples, it is obvious that the antifriction coating was completely depleted in the center of the wear scars on the disc specimen, and thus, that the cylinder came into contact with the base material of the disc specimen, which greatly influenced the friction and wear behavior of the coated benchmark samples. At the periphery of the wear scars on the cylinder, small areas may be observed in which material transfer of the antifriction coating from the disc specimens occurred.

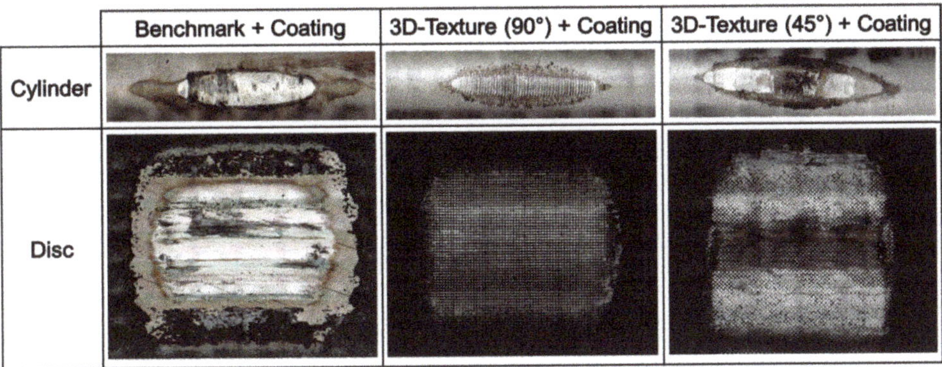

Figure 14. Typical wear scars on cylinder and disc specimens after wear resistance tests obtained using laser microscopy.

For the 3D-textured samples, the cylinder surface shows linear grooves produced by the lower 3D-texture for the samples aligned with an angle of 90°, but no linear grooves were visually observed for the 3D-textured sample aligned at an angle of 45°, which is simply due to the orientation angle of the 3D-texture with respect to the motion direction of the tribometer during the wear tests, as shown in Figure 15. Actually, linear grooves were also produced on the cylinders for the 3D-texture with an angle of 45°, but the extremely small distance between each groove renders their visual observation difficult: grooves produced on the cylinders when the angle was 90° had a width equal to the plateau side dimensions (~ 37 µm as reported in Table 5), while grooves produced on the cylinders when the angle was 45° had a width equal to the plateau diagonals (~52 µm = $\sqrt{2} \times 37$ µm).

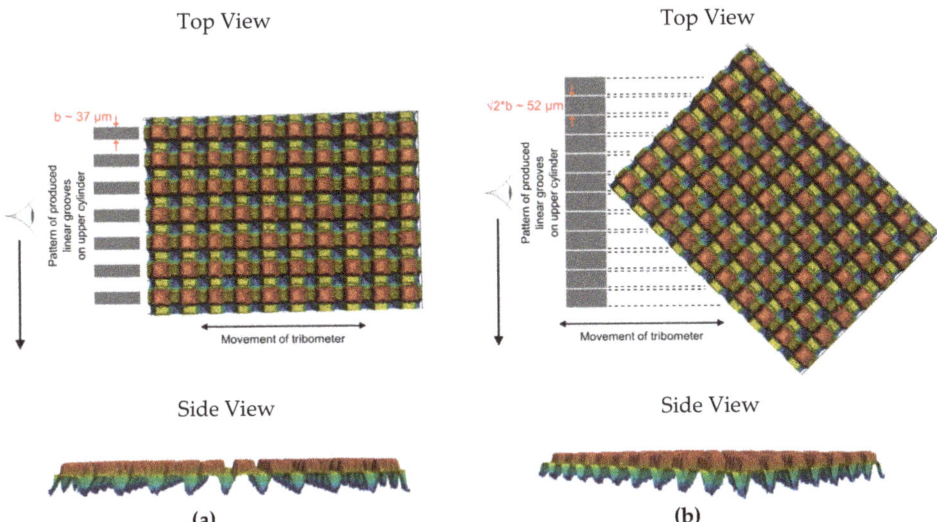

Figure 15. Explanation of (**a**) presence of visible linear grooves (width of grooves = sides of plateaus ~37 µm) on cylinders for the 3D-texture with an angle of 90° to the direction of motion and (**b**) near absence of visible linear grooves (width of grooves = diagonals of plateaus = ~52 µm) on cylinders for the 3D-texture with an angle of 45° to the direction of motion.

The averages of wear rates (mm^3/h) of discs and cylinders measured using laser optical microscopy are shown in Figure 16a and 16b for all 3 sample variations respectively. From both diagrams shown in Figure 16, it may be seen that the use of a 3D microtexture with an orientation of 90° to the direction of motion in combination with an antifriction coating may reduce the overall wear rate of the tribological system (upper and lower specimens) from an average of 0.5 mm^3/h to approximately 0.25 mm^3/h for the discs and from an average of 0.0125 mm^3/h to approximately 0.0075 mm^3/h for the cylinders. By orienting the 3D microtexture at an angle of 45° relative to the direction of motion, the wear rate of the discs stays at a value similar to the benchmark, while the wear rate of the cylinders increased slightly above 0.02 mm^3/h.

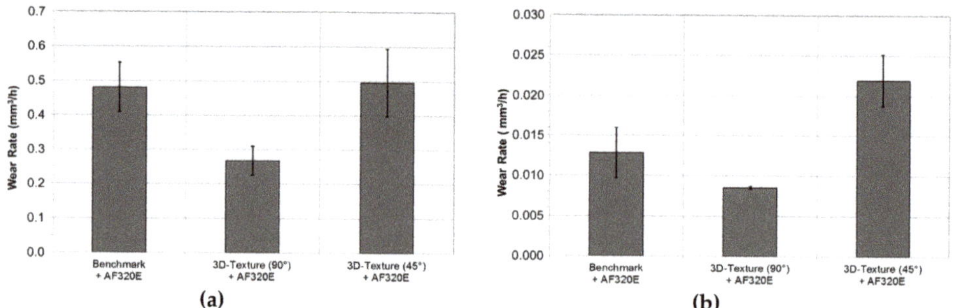

Figure 16. Average wear rate (mm^3/h) calculated from laser microscopy measurements after long-term tribological tests (2 h) of: (**a**) discs and (**b**) cylinders.

The increase in wear rates for the 3D-textured and coated samples with an angle of 45° in comparison with the similar samples with an angle of 90° may be explained by the fact that the contact width of the plateaus having a 45°-angle is significantly higher (~52 μm) than for a 90°-angle (~37 μm), as shown schematically in Figure 15, and therefore, the amount of worn material through ploughing is higher for both disc and cylinder specimens. Furthermore, the lower wear resistance of the microtexture with a 45°-angle (in comparison to the samples with a 90°-angle) is also believed to be due to the difference in real contact pressures between these two textures. For a defined unit surface area, the number of micropillars encompassed by this area is always about 20–25% lower for the samples with a 45°-angle than for the samples having a 90°-angle; thus, the real contact pressures experienced by the 45°-angle samples is 20–25% higher than the contact pressures encountered by the samples having a 90°-angle ($p = F_N/A$, where p = contact pressure in N/mm^2, F_N = load in N and A = contact surface in mm^2). Table 8 gives a short overview of the formulas used for the calculation of the ratio of the contact pressures for both microtextures having different orientation angles to the direction of motion during long-term tribological tests.

Table 8. Calculations of ratio of real contact pressures for both microtextures with different angles to the direction of motion during long-term tribological tests.

Angle	Number of Pillars/ Unit Surface Area	Normalized Contact Area	Calculated Contact Pressure		Ratio Contact Pressures
90°	100	$A_{90°} = A$	$p_{90°} = F_N/A_{90°}$	$p_{90°} = F_N/A$	$p_{45°}/p_{90°} = 1/0.8$
45°	~80	$A_{45°} \sim 0.8{*}A_{90°}$	$p_{45°} = F_N/A_{45°}$	$p_{45°} = F_N/0.8{*}A$	~1.25

For the textured samples with a 45°-angle, the fact that the wear rate of the disc specimens is similar to the coated benchmark samples (Figure 16a) while the wear rate of the cylinders is significantly higher than the coated benchmark samples (Figure 16b) needs to be analyzed separately in more detail; this will be the focus of a future study and publication.

4. Conclusions

The present study has shown that the production of specific 3D-microtextures on steel substrates using an ultra-short pulsed laser was feasible and the reproducibility of the texture dimensions over the entire textured regions was extremely good.

LIPSSs are the initial features produced during the formation of self-ordered hierarchical micro- and nano-textures. By nature, self-ordered structures are not predictable in any minute aspect of their shape. It was demonstrated that knowledge of optical and structural feedback mechanisms can become a tool to steer the metamorphosis from LIPSS towards desired surface structures for specific applications, or to suppress their formation (for example pinholes) when a smooth ablation area is required. A very simple approach for controlling pinhole formation was found: to suppress pinhole formation, the rule is to orientate the laser polarization perpendicular to the scan direction; to promote their formation, the rule is to orientate the laser polarization parallel to the scan direction. Additional polarization flipping by 90° after some consecutive laser ablation scans leads to further improvements in terms of surface texture quality (smoothness). Pinholes are the intermediate state towards the formation of self-ordered microbumps. Applying these simple ablation rules provides a tool to generate tailored self-ordered, hierarchical micro- and nano-textures for implementation in tribological applications.

Unfortunately, the application of an antifriction coating onto the 3D-textures did, however, greatly modify their topography: 3D-textures became more rounded and the height differences between each plateau were drastically reduced.

Tribological investigations under oil-lubricated conditions have shown that the 3D-microtexture does not show any significant COF improvements when compared to the benchmark over the entire load range studied: COF for the oil-lubricated benchmark was between 0.10 and 0.12 and for the oil-lubricated 3D-textured samples, between 0.13 and 0.15.

However, tribological investigations with an antifriction coating showed that the 3D-microtexture does show significant COF improvements when compared to the unlubricated coated benchmark over the entire load range studied: COF for the coated benchmark was between 0.25 and 0.30, and for the coated 3D-textured samples, between 0.20 and 0.22. However, the friction measured for the coated 3D-texture was still slightly higher than for the oil-lubricated benchmark; therefore, from the point of view of comparing only the friction coefficients, the 3D-texture under study may not be easily seen as an interesting alternative to oil-lubricated blank steel.

Wear-based tribological tests (125 N; 2 h) have shown that the antifriction coating on the benchmark samples was completely depleted, which greatly influenced their friction and wear behavior. Wear-based tribological tests with 3D-textured samples have shown that linear grooves were produced on the cylinder surface by the 3D-texture, but only for the samples aligned with an angle of 90° to the direction of motion. This phenomenon is due to the geometrical aspects caused by the orientation of the 3D-texture with respect to the direction of motion during the wear tests.

A comparison of wear rate results showed that 3D microtextures with an orientation of 90° in combination with an antifriction coating may drastically reduce wear of both the upper and lower samples. For 3D microtextures with a 45°-angle, disc wear remained similar to the benchmark, while cylinder wear slightly increased. The latter behavior for the 45°-angled 3D-texture was explained through a different nominal contact area and resulting real contact pressures.

Author Contributions: Conceptualization, J.V., J.Z., S.K.; methodology, J.V., J.Z., S.K.; formal analysis, J.V., J.Z., S.K., F.A., I.V.; investigation, J.V. and J.Z.; writing-original draft preparation, J.V. and J.Z.; writing-review and editing, J.V., J.Z., S.K., F.A., I.V.; project administration, J.V. and J.Z.

Funding: The work presented herein was funded by the Austrian COMET Programme (Project XTribology, no. 849109) and carried out at the "Excellence Centre of Tribology" (AC2T research GmbH) in cooperation with V-Research GmbH.

Acknowledgments: Carl Bechem GmbH (Hagen, Germany) is gratefully acknowledged for the suggestion and the free of charge application of an appropriate antifriction coating onto the disc specimens for the present tribological investigations. The authors would also like to thank Stephan Kasemann for tireless support in proofreading the manuscript, Heinz Duelli for valuable technical advises to obtain proper REM pictures, Thomas Auer for test sample preparation, Giovanni Piredda for many inspiring scientific discussions and Johannes Edlinger (head of Research Centre for Microtechnology) to cultivate a scope for interdisciplinary research. Open Access Funding by V-Research GmbH.

Conflicts of Interest: The authors declare no conflict of interest.

References

1. Holmberg, K.; Kivikytö-Reponen, P.; Härkisaari, P.; Valtonen, K.; Erdemir, A. Global Energy Consumption due to Friction and Wear in the Mining Industry. *Tribol. Int.* **2017**, *115*, 116–139. [CrossRef]
2. Sedlacek, M.; Podgornik, B.; Ramalho, A.; Cesnik, D. Influence of Geometry and the Sequence of Surface Texturing Process on Tribological Properties. *Tribol. Int.* **2017**, *115*, 268–273. [CrossRef]
3. Sergejev, F.; Peetsalu, P.; Sivitski, A.; Saarna, M.; Adoberg, E. Surface Fatigue and Wear of PVD Coated Punches during Fine Blanking Operation. *Eng. Fail. Anal.* **2011**, *18*, 1689–1697. [CrossRef]
4. Podgornik, B.; Hogmark, S.; Sandberg, O. Proper Coating Selection for Improved Galling Performance of Forming Tool Steel. *Wear* **2006**, *261*, 15–21. [CrossRef]
5. Leskovsek, V.; Podgornik, B. Vacuum Heat Treatment, Deep Cryogenic Treatment and Simultaneous Pulse Plasma Nitriding and Tempering of P/M S390MC Steel. *Mater. Sci. Eng. A* **2012**, *531*, 119–129. [CrossRef]
6. Voevodin, A.A.; Zabinski, J.S. Laser Surface Texturing for Adaptive Solid Lubrication. *Wear* **2006**, *261*, 11–12, 1285–1292. [CrossRef]
7. Voyer, J.; Klien, S.; Ausserer, F.; Velkavrh, I.; Ristow, A.; Diem, A. Friction Reduction Through Sub-Micro Laser Surface Modifications. *Tribologie und Schmierungstechnik* **2015**, *62*, 13–18.
8. Gachot, C.; Rosenkranz, A.; Hsu, S.M.; Costa, H.L. A Critical Assessment of Surface Texturing for Friction and Wear Improvement. *Wear* **2017**, *372–373*, 21–41. [CrossRef]
9. Podgornik, B.; Sedlacek, M. Performance, Characterization and Design of Textured Surfaces. *J. Tribol.* **2012**, *134*, 1–7. [CrossRef]
10. Vilhena, L.M.; Podgornik, B.; Vizintin, J.; Mozina, J. Influence of Texturing Parameters and Contact Conditions on Tribological Behaviour of Laser Textured Surfaces. *Meccanica* **2011**, *46*, 567–575. [CrossRef]
11. Voyer, J.; Ausserer, F.; Klien, S.; Ristow, A.; Velkavrh, I.; Diem, A.; Zehetner, J.; Stroj, S.; Heidegger, S.; Bertschler, C.; et al. Sub-Micro Laser Modifications of Tribological Surfaces. *Mater. Perform. Charact.* **2017**, *6*, 42–67. [CrossRef]
12. Vladescu, S.C.; Olver, A.V.; Pegg, I.G.; Reddyhoff, T. The Effects of Surface Texture in Reciprocating Contacts - An Experimental Study. *Tribol. Int.* **2015**, *82A*, 28–42. [CrossRef]
13. Andersson, P.; Koskinen, J.; Varjus, S.; Gerbig, Y.; Haefke, H.; Georgiou, S.; Zhmud, B.; Buss, W. Microlubrication Effect by Laser-Textured Steel Surfaces. *Wear* **2007**, *262*, 369–379. [CrossRef]
14. Cho, M.H.; Park, S. Micro CNC Surface Texturing on Polyoxymethylene (POM) and its Tribological Performance in Lubricated Sliding. *Tribol. Int.* **2011**, *44*, 859–867. [CrossRef]
15. Biboulet, N.; Bouassida, H.; Lubrecht, A.A. Cross Hatched Texture Influence on the Load Carrying Capacity of Oil Control Rings. *Tribol. Int.* **2015**, *82A*, 12–19. [CrossRef]
16. Huang, W.; Jiang, L.; Zhou, C.; Wang, X. The Lubricant Retaining Effect of Micro-Dimples on the Sliding Surface of PDMS. *Tribol. Int.* **2012**, *52*, 87–93. [CrossRef]
17. Suh, M.S.; Chae, Y.H.; Kim, S.S.; Hinoki, T.; Kohyama, A. Effect of Geometrical Parameters in Micro-Grooved Crosshatch Pattern under Lubricated Sliding Friction. *Tribol. Int.* **2010**, *43*, 1508–1517. [CrossRef]
18. Wakuda, M.; Yamauchi, Y.; Kanzaki, S.; Yasuda, Y. Effect of Surface Texturing on Friction Reduction Between Ceramic and Steel Materials under Lubricated Sliding Contact. *Wear* **2003**, *254*, 356–363. [CrossRef]
19. Bonse, J.; Höhm, S.; Kirner, S.V.; Rosenfeld, A.; Krüger, J. Laser-Induced Periodic Surface Structures—A Scientific Evergreen. *IEEE J. Sel. Top. Quantum Electron.* **2017**, *23*, 9000615. [CrossRef]
20. Zehetner, J.; Kraus, S.; Lucki, M.; Vanko, G.; Dzuba, J.; Lalinsky, T. Manufacturing of Membranes by Laser Ablation in SiC, Sapphire, Glass and Ceramic for GaN/Ferroelectric Thin Film MEMS and Pressure Sensors. *Microsyst. Technol.* **2016**, *22*, 1883–1892. [CrossRef]
21. Zehetner, J.; Vanko, G.; Dzuba, J.; Lalinsky, T. Femtosecond Laser Processing of Membranes for Sensor Devices on Different Bulk Materials. *Appl. Phys.* **2017**, *15*, 561–568. [CrossRef]

22. Voyer, J.; Ausserer, F.; Klien, S.; Velkavrh, I.; Diem, A. Reduction of the Adhesive Friction of Elastomers through Laser Texturing of Injection Molds. *Lubricants* **2017**, *5*, 45. [CrossRef]

23. Bonse, J.; Krüger, J.; Höhm, S.; Rosenfeld, A. Femtosecond Laser Induced Periodic Surface Structures. *J. Laser Appl.* **2012**, *24*. [CrossRef]

24. Varlamova, O.; Reif, J.; Varlamov, S.; Bestehorn, M. The Laser Polarization as Control Parameter in the Formation of Laser-Induced Periodic Surface Structures: Comparison of Numerical and Experimental Results. *Appl. Surf. Sci.* **2011**, *257*, 5465–5469. [CrossRef]

25. Yao, J.W.; Zhang, C.Y.; Liu, H.Y.; Dai, Q.F.; Wu, L.J.; Lan, S.; Gopal, A.V.; Trofimov, V.A.; Lysak, T.M. High Spatial Frequency Periodic Structures on Metal Surface by Femtosecond Laser Pulses. *Opt. Express* **2012**, *20*, 905–911. [CrossRef] [PubMed]

26. Ji, X.; Jiang, L.; Li, X.; Han, W.; Liu, Y.; Wang, A.; Lu, Y. Femtosecond Laser-Induced Cross-Periodic Structures on a Crystalline Silicon Surface under Low Pulse Number Irradiation. *Appl. Surf. Sci.* **2015**, *326*, 216–221. [CrossRef]

© 2019 by the authors. Licensee MDPI, Basel, Switzerland. This article is an open access article distributed under the terms and conditions of the Creative Commons Attribution (CC BY) license (http://creativecommons.org/licenses/by/4.0/).

 *lubricants* 

Article

Fabricating Laser-Induced Periodic Surface Structures on Medical Grade Cobalt–Chrome–Molybdenum: Tribological, Wetting and Leaching Properties

Sanne H. van der Poel [†], Marek Mezera [†], Gert-willem R. B. E. Römer , Erik G. de Vries and Dave T. A. Matthews *

Department of Mechanics of Solids, Surfaces and Systems (MS3), Faculty of Engineering Technology, University of Twente, Drienerlolaan 5, 7522 NB Enschede, The Netherlands
* Correspondence: d.t.a.matthews@utwente.nl
† These authors contributed equally to this work.

Received: 12 July 2019; Accepted: 8 August 2019; Published: 13 August 2019

Abstract: Hip-implants structured with anti-bacterial textures should show a low-friction coefficient and should not leach hazardous substances into the human body. The surface of a typical material used for hip-implants, namely Cobalt–Chrome–Molybdenum (CoCrMo) was textured with different types of laser-induced periodic surface structures (LIPSS)—i.e., low spatial frequency LIPSS (LSFL), hierarchical structures consisting of grooves superimposed with high spatial frequency LIPSS (HSFL) and Triangular shaped Nanopillars (TNP)—using a picosecond pulsed laser source. The effect of LIPSS on the wettability, friction, as well as wear of the structures, when slid against a polyethylene (PE) counter surface and biocompatibility was analyzed. Surfaces covered with LSFL show superhydrophobicity and grooves with superimposed HSFL, as well as TNP, show hydrophobic behavior. The coefficient of friction (CoF) of LIPSS against a polyethylene (PE) counter surface was found to be higher (ranging from 0.40 to 0.66) than the CoF of (polished) CoCrMo, which was found to equal 0.22. It was found that the samples release cobalt within biocompatible limits. Compared to polished reference surfaces, LIPSS cause higher friction of CoCrMo against PE contact. However, the wear of the PE counter surface only increased significantly for the LSFL textures. For these reasons, it is concluded that LIPSS are not suitable for a heavily loaded metal-on-plastic bearing contact.

Keywords: CoCrMo; LIPSS; LSFL; HSFL; grooves; triangular LIPSS

1. Introduction

At least 3% of the patients that require primary total hip arthroplasty surgery need a revision due to severe prosthetic joint infections [1]. This leads to additional hospitalizations, costs and compromises the patient's health. Surface textures in the nano- and micrometer scale are observed in nature, that have an antibacterial effect, such as butterfly wings and shark skin [2]. Anti-bacterial surface features must be of the same order or slightly smaller than the bacteria size, in order to influence the adherence behavior of the bacteria [3]. This effect is based on the reduction of the amount of available surface for the bacteria cell to adhere to. Bacteria that most often cause prostheses related infections are *Staphylococcus aureus* and *Escherichia coli* [4]. The characteristic dimensions of the bacteria are one to three μm in diameter [5,6].

Cobalt–Chrome–Molybdenum (CoCrMo) is an alloy that is most often used for the bearing surface of a hip implant, due to its high fatigue, wear and corrosion resistance [7]. In a metal-on-plastic artificial hip joint, the CoCrMo femur head articulates against an polyethylene (PE) acetabulum cup. The surface of the CoCrMo component is traditionally mirror polished [8].

An established method to alter surface properties on the nano- to micrometer scale is laser surface texturing using (ultra-) short laser pulses. Under specific conditions, this can lead to so called laser-induced periodic surface structures (LIPSS). LIPSS are nanometer sized, regular patterned surface textures and can improve tribological performance [9,10], wettability properties [10–15], anti-bacterial properties [15–18] and cell-tissue growth [11].

In order to reduce infections, LIPSS having dimensions (periodicity and amplitude) about the size of the bacterias on the CoCrMo material could improve the antibacterial performance of the CoCrMo femur. However, hip-implants also should show a low-friction coefficient and should not leach hazardous materials into the human body. Therefore, the aim of this study is to not only study the formation of LIPSS on CoCrMo, but also study the tribological properties of CoCrMo on PE as well as the wettability and the leaching properties of the textures. In this study, a picosecond pulsed laser is used to create different types of LIPSS on CoCrMo surfaces. To the best of the authors knowledge, it is the first report of hexagonally packed triangular nanopillar LIPSS produced with a picosecond laser on CoCrMo. The tribological, wetting and leaching properties of different types of LIPSS on CoCrMo are investigated and compared.

2. Materials and Methods

2.1. Laser Setup and Material

Figure 1a shows the experimental laser setup schematically. It consists of a pulsed Yb:YAG disk laser source (TruMicro 5050 of Trumpf GmbH, Ditzingen, Germany) emitting a laser beam with a wavelength of 1030 nm, maximum pulse frequency of 400 kHz, pulse energies up to 125 µJ and a fixed pulse duration of 6.7 ps. The fluence profile of the focal laser spot is nearly Gaussian ($M^2 < 1.3$). The polarization of the laser beam exiting the laser head is linear. Besides exposing the material to linear laser polarization, also a quarter wave plate was included in the setup to achieve circular polarization, which may lead to triangular shaped LIPSS textures. The beam was focused on the surface of the samples, using a telecentric Fθ lens (Ronar of Linos GmbH, Göttingen, Germany) with a focal length of 80 mm, resulting in a focal spot with an e^{-2}-diameter of $d = 33.6 \pm 1.6$ µm (see Section 2.2.1).

The samples consists of polished CoCrMo discs with a diameter of 25 mm and a thickness of 3 mm. The surface roughness ($R_a = 0.003 \pm 0.0003$ µm, $R_q = 0.004 \pm 0.0004$ µm) of the discs was measured with an atomic force microscope (NX10, Park Systems Corp., Suwon, South Korea). manufacturer's headquarters. The beam was scanned over the substrate using a galvoscanner (intelliSCAN14 of ScanLab GmbH, Puchheim, Germany) at normal incidence in air, see Figure 1b. Different shapes and sizes of LIPSS were produced by adjusting the type of polarization (linear or circular) and by adjusting the laser peak fluence (F_0) and the number of overscans of the laser spot over the surface (N_{OS}). The scan velocity of the laser spot (v), the laser pulse frequency (f_F) and the spatial pitch between laser pulses on the surface ($\Delta x, \Delta y$) were kept constant in this study at $v = 2$ m/s, $f_F = 1000$ Hz and $\Delta x = \Delta y = 5$ µm, respectively, see Figure 1b. This yields a geometrical pulse-to-pulse overlap (OL) in both, x- and y-direction, of $OL = 1 - v/(d \cdot f_F) \approx 0.85$. All samples were cleaned in an ultrasonic bath with ethanol for 20 min and dried in ambient air after laser treatment. Table 1 shows the chemical composition of the samples.

Table 1. Cobalt–Chrome–Molybdenum (CoCrMo) alloy composition in weight percent, the composition is balanced (Bal.) with Cobalt.

Element	Co	Cr	Mo	Ni	Si; Mn; Fe
wt %	Bal.	$27 - 30$	$5 - 7$	≤ 2	≤ 1

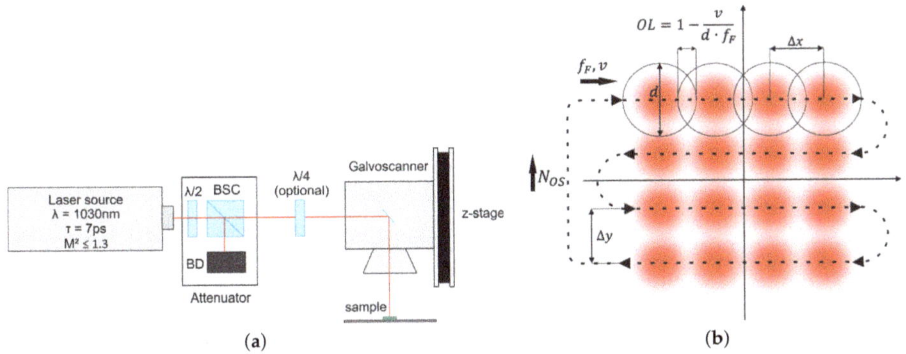

Figure 1. Schematic representations of the laser setup and the scanning trajectory of the laser spot. (**a**) Schematic representation of the laser setup; $\lambda/2$: half-wave plate; BSC: beam splitter cube; BD: beam dump; $\lambda/4$: quarter-wave plate (optional). (**b**) Scanning trajectory of the laser spot; f_F: pulse frequency; v: scan velocity; d: beam diameter; OL: geometrical pulse-to-pulse overlap; N_{OS}: number of overscans; $\Delta x, \Delta y$: geometrical pitch between subsequent laser pulses in x- and y-direction.

2.2. Analysis Tools

2.2.1. Laser Beam Characterization

The laser power was measured using an thermopile power sensor (PM30 of Coherent, Santa Clara, CA, USA) with a measurement uncertainty of $\pm 1\%$, connected to a readout unit (FieldMaxII-TOP of Coherent, Santa Clara, CA, USA). The focal spot diameter 33.6 ± 1.6 µm (e^{-2}) was measured using a laser beam characterization device (MicroSpotMonitor of Primes GmbH, Pfungstadt, Germany).

2.2.2. Surface Morphology Dimensions

Laser-induced surface structures were analyzed using a scanning electron microscope (JSM-7200F, Jeol, Tokyo, Japan). From SEM micrographs, the periodicity of LIPSS areas were analyzed with the help of a 2D fast Fourier transform (FFT) algorithm using a MATLAB [19] script. Details of the script are reported in our earlier work [20].

The roughness of the surface textures was analyzed by means of an atomic force microscopy (NX10, Park Systems Corp., Suwon, South Korea) in true non-contact™ mode using a non-contact cantilever (PPP-NCHR, $125 \times 30 \times 4$ µm, Tip < 10 nm, Park Systems Corp., Suwon, South Korea). The roughness parameters that were extracted from these measurements are used to characterize the surface by means of root mean square surface area roughness (R_q), average surface area roughness (R_a), maximum peak height (R_p), maximum valley depth (R_v), skewness (R_{sk}), kurtosis (R_{ku}) and the ratio between the real surface area and the projected area (σ).

2.2.3. Contact Mechanism and Friction Parameters

The wear of the samples was characterized in a wear test in which the textured sample was exposed to a polyehtylene (PE) ball (diameter of 9.5 mm) sliding over the surface in BCS as lubricant, in order to simulate the human body environment. The sample discs are clamped in a Universal Mechanical Tester (UMT, Bruker, Billerica, MA, United States). The PE ball was pressed against the sample surface and moved in a reciprocate fashion for 104 min with 11 mm/s and a normal load of 0.5 N (18 MPa contact pressure) [21]. The normal force and the shear force are measured and the coefficient of friction during steady state of the wear test was calculated. The wear of the PE ball surface is analyzed by observing the wear diameter with a confocal microscope (S Neox,

Sensofar, Terassa, Spain). The wear of the textures is investigated with a SEM (see Section 2.2.2) and energy-dispersive X-ray spectroscopy (EDS) analysis in the same SEM system.

2.2.4. Contact Angle Measurement

The wettability of the textures was characterized by means of a contact angle measurement device (SCA20, DataPhysics Instruments GmbH, Filderstadt, Germany). The sample surfaces were wiped with isopropanol and dried under ambient conditions prior to analysis. Next, a droplet of water of 5 μL was deposited on the surface and the angle between the surface and the droplet was measured. Three measurements were conducted for each sample.

2.2.5. Biocompatability

The substrates were sterilized by immersing them in ethanol for 15 min. A simulated body fluid (SBF) solution was made according to the instructions of Kokubo et al. [22]. That is, salts were dissolved in de-ionized water such that a solution was created with ion concentrations similar to that of blood plasma. The pH value of SBF (7.40) is comparable to the pH of human blood plasma, which ranges from 7.2 to 7.4. The substrates were immersed in 40 mL of SBF solution at 37 °C in a shaking incubator (160 rpm). The ion release was analyzed after 1, 7, 14, 21 and 26 days respectively. Inductively Coupled Plasma Atomic Emission Spectroscopy (ICP-AES) (Optima 5300 dual view, PerkinElmer Inc., Waltham, MA, USA) was used to analyze the leaching properties.

3. Results and Discussion

3.1. Surface Structures Processed with Linear Polarization

Two types of surface textures were processed using linear polarized laser irradiation, by applying increasing peak fluence levels and various number of overscans during the laser processing. These two types are low spatial frequency LIPSS (LSFL, see Figure 2) and hierarchical structures composed of micro-grooves and superimposed LSFL (see Figure 3).

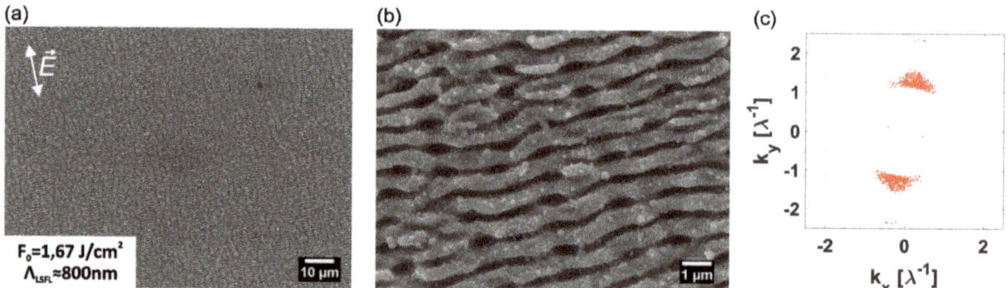

Figure 2. Scanning electron microscope (SEM) micrographs of low spatial frequency laser-induced periodic surface structures (LSFL) processed on CoCrMo with a laser peak fluence of $F_0 = 1.67\ \text{J/cm}^2$ and number of overscans $N_{OS} = 1$ ((**a**) 1000× magnification; and (**b**) 10,000× magnification). (**c**) 2D-fast Fourier transform (FFT) map of the SEM micrograph (**a**). The spatial periodicity of the LSFL is $\Lambda \approx 800$ nm. The arrow in micrograph (**a**) indicates the direction of the E-field of the laser polarization.

The periodicity of the LSFL in Figure 2 was found to equal $\Lambda \approx 800$ nm and are perpendicular to the E-field of the laser polarization direction, which is typical for LSFL on metals [23]. LSFL on the sample are spread homogeneously over the processed surface of $5 \times 5\ \text{mm}^2$.

Figure 3 shows SEM micrographs of hierarchical structures processed on CoCrMo with $N_{OS} = 5$ and various peak fluence levels. It can be observed in this figure, that with increasing peak fluence levels, the periodicity of the micro grooves increases from $\Lambda_{\text{Grooves}} \approx 3.55$ μm at a peak fluence level of $F_0 = 1.82\ \text{J/cm}^2$ to $\Lambda_{\text{Grooves}} \approx 7.9$ μm at $F_0 = 7.07\ \text{J/cm}^2$. The formation of micro-grooves and micro-bumps is attributed to an increased heat accumulation during processing [24–26].

The periodicity $\Lambda_{LSFL} \approx 920$ nm of the LSFL in Figure 3 was constant for all fluence levels within the fluence range studied. It is known that the LSFL periodicity increases with increasing fluence levels up to a certain fluence level, after which the periodicity does not vary with the fluence [27,28].

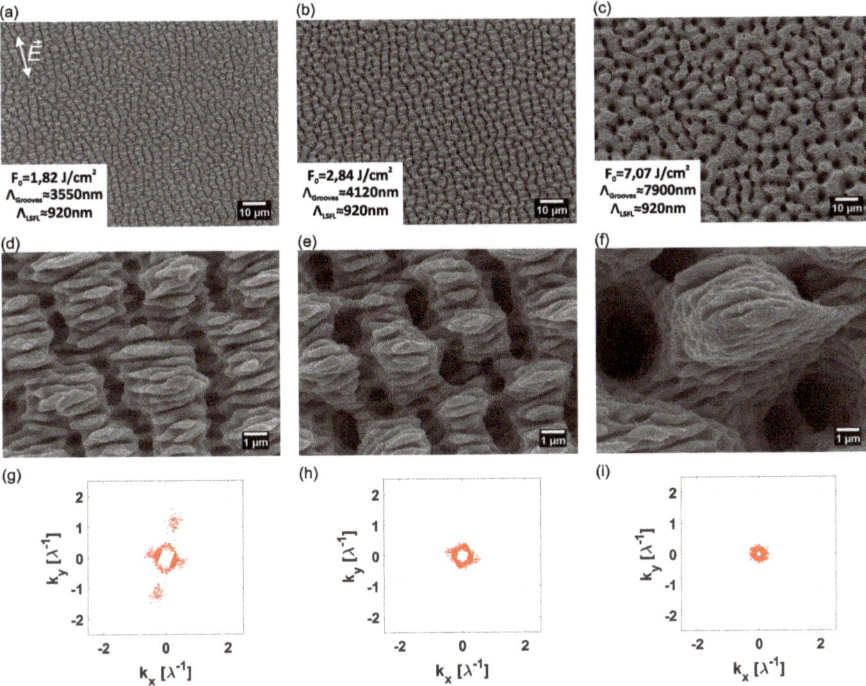

Figure 3. SEM micrographs of hierarchical structures processed on CoCrMo with $N_{OS} = 5$ and various peak fluence levels at two different magnifications ((**a–c**) 1000×; and (**d–f**) 10,000×). (**g–i**) 2D-FFT maps of the micrographs of the processed areas (**a–c**). The periodicity of the micro-grooves $\Lambda_{Grooves}$ increases with increasing peak fluence levels. The periodicity of the LSFL features are constant at $\Lambda_{LSFL} \approx 920$ nm for all micrographs. The arrow in micrograph (**a**) indicates the direction of the E-field of the laser polarization.

3.2. Surface Structures Processed with Circular Polarization

Triangular nanopillars (TNP), hexagonally packed, can be produced by exposing the surface to either circular polarized ultra-short laser pulses [13], or to double-pulsed (bursts of pulses), linear cross-polarized, ultra-short laser pulses [29,30]. These types of structures might be preferred over LSFL for the aimed application, since TNP are symmetric in three directions, whereas LSFL are symmetric in only one direction. Because hip joints rotate with respect to the x-, y- and z-axis, the tribological characteristics of the bearing should ideally be equal in any direction.

The physical phenomena behind the formation of triangular LIPSS are still under debate [13,29,30]. e.g., Fraggelakis et al. [30] proposes that the convection flow of the molten material layer as a cause for this type of LIPSS, whereas Liu et al. [29] claims the 2D nanotriangle structures develop due to the interference of surface plasmon polaritons (SPP's) with the incoming laser light. Since Liu et al. applied cross-polarized, time delayed double-pulses, these authors argue that the first pulse induces SPP's and the interference with the laser light leads to transient, spatially periodic meta-gratings of a modified refractive index on the surface with a wave vector parallel to the laser polarization. Further, they claim that the second cross-polarized pulse also induces SPP's at the surface due to surface roughness with a wave vector parallel to the laser polarization. The latter SPP then interferes with the transient refractive index meta grating of the first pulse and could diffract into two SPP's with different wave vectors. The interference of the laser light with these three SPP's in different directions leads to ablation of a hexagonal pattern, resulting in triangular shaped nanostructures.

Figure 4 shows SEM micrographs of surface structures processed on CoCrMo with circular polarization with $N_{OS} = 1$ and increasing peak fluence levels. It can be observed from Figure 4a, that LSFL with a periodicity of about $\Lambda_{LSFL} \approx 860$ nm form at a peak fluence level of $F_0 = 2.87$ J/cm^2. This structure may be an indication that the polarization is not perfectly circular, but actually elliptically polarized with the main axis perpendicular to the processed LSFL. It can also be observed in Figure 4a, that "interruptions" of the LSFL features start to appear in the direction and the periodicity of the hexagonal shapes, see Figure 4a and the indicated frequencies on the 2D-FFT map of Figure 4a. Further it can be recognized when comparing Figure 4a,b, that these "interruptions" are indeed a surface morphology "proceeding" the formation of grooves in two different directions, which then form the triangular nanopillars if the fluence is increased. At a fluence level of $F_0 = 5.23$ J/cm^2, regular TNP are formed, which become less regular and less pronounced for higher fluence levels, see Figure 4c. When comparing the laser processing conditions and groove periodicities between the hexagonal nanopillars with earlier studies (see Table 2), it becomes evident, that the hexagonal pattern processed either with single pulses of circular polarization or with cross polarized pulses with linear polarization origin from the same physical phenomena. Hence, the physical explanation of the origin for those patterns has to apply for each case of laser processing condition listed in Table 2. The physical explanation of hexagonal nanopillars exceeds the scope of this paper.

Table 2. Comparison of laser processing parameters and groove periodicity of hexagonal nanostructures with earlier studies.

Wavelength [nm]	Pulse Duration	Pulse Frequency [kHz]	Polarization	Period between Grooves	Reference
1030	7 ps	400	Circular, non-burst pulses	0.83 λ	This study
1032	310 fs	250	Circular, non-burst pulses	0.84–0.98 λ	[13]
710	50fs	1	Linear, cross-polarized bursts of pulses	0.89 λ	[29]
1030	350 fs	100	Circular, opposite direction of rotation & Linear, cross-polarized bursts of pulses	0.85 λ	[30]

High spatial frequency LIPSS (HSFL) were found between the formed LSFL in Figure 4a and the triangular nanopillars in Figure 4b,c, with a periodicity of $\Lambda_{HSFL} \approx 80$ nm. Liu et al. [29] processed triangular nanopillars with two consecutive, cross-polarized pulses with a pulse duration of 50 fs and with a time delay of 1.2 ps on tungsten in air and in vacuum. In the latter study, HSFL were not observed when processing tungsten in air, but have been observed when processing tungsten at low pressures of 10^{-3} Pa. It was claimed, that the formation of HSFL is attributed to a slower cooling rate of the molten, liquid material layer at lower pressures. In the latter case, less air exists in the experimental environment, to transfer the heat from the molten layer to. Therefore, heat remains in the molten layer for a longer period of time and the cooling rate decreases. Thus, when the liquid material cools down, there is more time for shrinking and film fragmentation of the melt into HSFL then when processing in air. Compared to the latter study, the pulse duration of the laser used in this work, is in the order of two magnitudes larger. Hence, more heat is introduced into the lattice, which might explain the occurrence of HSFL between LSFL and triangular nanopillars when processed in air.

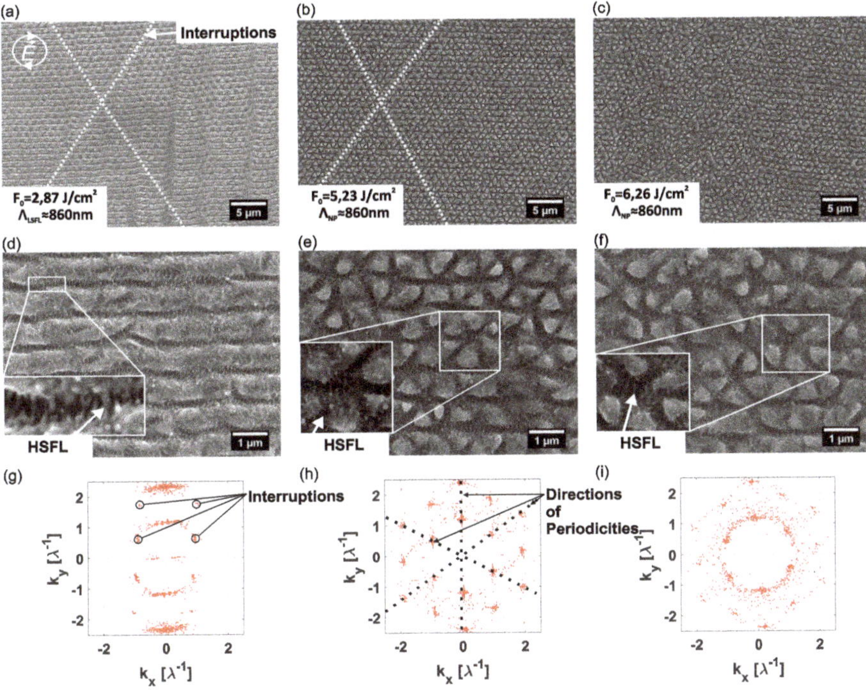

Figure 4. SEM micrographs of surface structures processed on CoCrMo using circular polarized laser radiation at $N_{OS} = 1$ and various peak fluence levels at two different magnifications (($\mathbf{a}$–$\mathbf{c}$) 3000×; and ($\mathbf{d}$–$\mathbf{f}$) 15,000×). ($\mathbf{g}$–$\mathbf{i}$) 2D-FFT maps if the micrographs of the processed areas ($\mathbf{a}$–$\mathbf{c}$). The periodicity of the LSFL in ($\mathbf{a}$) is about $\Lambda_{LSFL} \approx 860$ nm. The periodicity of the nanopillars is constant in the same range $\Lambda_{TNP} \approx 860$ nm. The arrow in micrograph ($\mathbf{a}$) indicates the direction of the E-field of the laser polarization.

3.3. Surface Morphology Dimensions

Three laser-textured surface patterns were chosen, based on their morphologies and uniformity, for the study of wettability properties, tribological performance and anti-bacterial behavior, see Figure 5. The laser parameters used to create these structures are listed in Table 3.

Table 3. Surface structures to be functionally evaluated processed with a pulse frequency of $f = 400$ kHz and a laser scanning velocity of 2 m/s.

Texture	Structure Types	Peak Fluence F_0 (J/cm^2)	N_{OS}	Laser Polarization	Figure
LSFL	LSFL	1.67 ± 0.01	1	linear	2a,b, 5b
Grooves	Grooves + LSFL	2.84 ± 0.01	5	linear	3a,d, 5c
TNP	hexagonal Nanopillars + HSFL	5.23 ± 0.01	1	circular	4b,e, 5d

The roughness parameters of these surface textures are listed in Table 4. As can be concluded from this table, the roughness parameters of these surfaces vary. Hence, significant differences in the functional properties (wetting, wear, biocompatibility) of these textures are expected. The higher value of R_a indicate that Grooves are more rough than a surface covered with only LSFL. Compared to the polished CoCrMo surface, the square root surface roughness R_q increases significantly due to laser-texturing. Quantitavely, 23 times $R_q^{polished}$ in the case of nanopillars and up to 130 times $R_q^{polished}$ in the case of grooves.

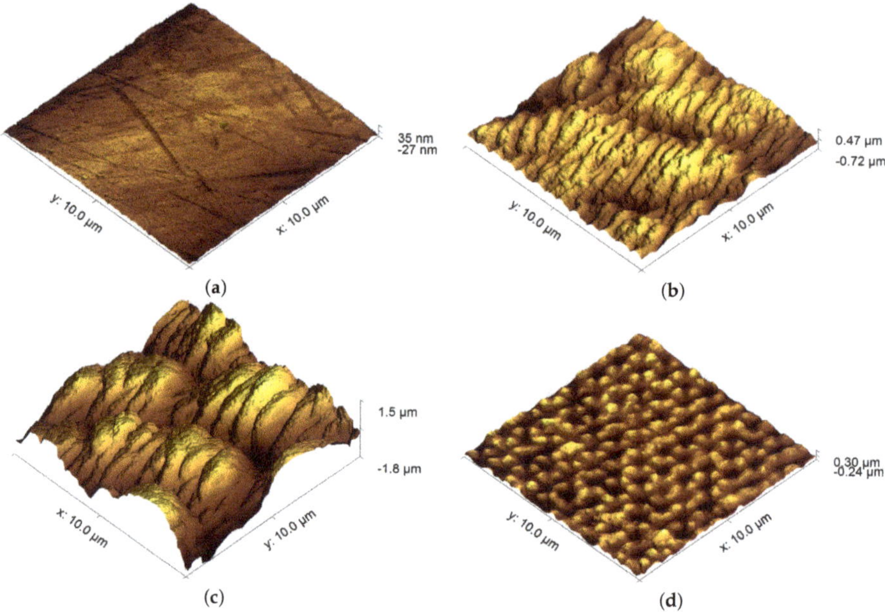

Figure 5. AFM micrographs of a CrCoMo surface: unprocessed (**a**), LSFL (**b**), grooves with superimposed LSFL (**c**) and triangular nanopillars (**d**). (**a**) AFM micrograph of polished CrCoMo sample. (**b**) AFM micrograph of LSFL. (**c**) AFM micrograph of Grooves + LSFL. (**d**) AFM micrograph of Triangular Nanopillars.

Table 4. Geometrical properties of the of the surface structures to be functionally evaluated.

Parameter	Polished	LSFL	Grooves	TNP
Periodicity (nm)	-	800	3550; 920 (LSFL)	860
R_q (µm)	0.004 ± 0.0004	0.163 ± 0.029	0.519 ± 0.130	0.092 ± 0.008
R_a (µm)	0.003 ± 0.0003	0.132 ± 0.026	0.423 ± 0.111	0.077 ± 0.007
R_p (µm)	0.010 ± 0.007	0.293 ± 0.082	1.038 ± 0.198	0.192 ± 0.030
R_v (µm)	-0.011 ± 0.007	-0.387 ± 0.113	-0.825 ± 0.321	-0.177 ± 0.023
R_{sk} (-)	-0.118 ± 0.038	-0.319 ± 0.488	-0.013 ± 0.455	0.038 ± 0.216
R_{ku} (-)	0.356 ± 0.812	-0.351 ± 0.859	-0.789 ± 0.306	-0.847 ± 0.308
σ (-)	1.005	1.328	1.752	1.307

As can be observed in Table 4, the dimensions of the chosen surface structures are indeed in the range of the sizes of the bacterias *S. aureus* and *E. coli*, which potentially gives these structures anti-bacterial properties [3].

3.4. Wetting Properties

When anti-bacterial properties of surfaces found in nature are studied, a correlation between hydrophobicity and anti-bacterial behavior is found [2,31,32]. Since LIPSS have been found to be hydrophobic [10–15] and also anti-bacterial [15–18], hydrophobicity is used in this paper as an indication of anti-bacterial behavior.

The three surface textures (see Figure 5) show hydrophobic behavior compared to the untextured, mirror polished surface, which shows a water contact angle of ($82.7° \pm 0.7°$, see Table 5).

Table 5. Contact angles of chosen surface structures.

Surface Structure	Polished	LSFL	Grooves	TNP
Contact angle (°)	82.7 ± 0.7	151 ± 2	141.1 ± 0.2	133 ± 2

The LSFL surface is superhydrophobic for water, whereas the contact angle of the mirror polished CoCrMo substrate is hydrophilic. The surfaces denoted Grooves and TNP are both hydrophobic, but less so than LSFL, as is shown in Table 5. The hydrophobicity of the textured surfaces is subject to variability due to the formation of oxide layers after laser micromachining over time. The polished sample will oxidize rapidly to a protective layer of CoO, Cr_2O_3 and MoO_3 [33–36]. For example, it was shown by Huerta–Murillo et al. [37], that the contact angles of laser textured titanium alloys increase over a time period of five weeks from about 90° to 130°. The contact angles in this study were measured after seven weeks for LSFL and Grooves and after 12 days for TNP. However, a positive effect of surface texturing (irrespective of morphological class) was seen on the hydrophobic behavior of the surfaces, in line with results found in literature [12,13,15].

The influence of the surface roughness on the contact angle can be described by either Wenzel [38], where it is assumed that the total surface will be in contact with the liquid, or by Cassie [39], where different materials or a combination of trapped air and a solid will be in contact with the liquid. In case of Wenzel the relation between the apparent contact angle (CA) θ_a and the intrinsic CA θ_i is given by

$$\cos(\theta_a) = \sigma(\cos\theta_i), \tag{1}$$

where σ is the ratio between the true surface area and the projected area. In case of hydrophilic surfaces an increase of the roughness will result in a decrease of the CA and in the case of hydrophobic surfaces an increase of the roughness will result in an increase of the CA.

In case air might be trapped due to surface morphology the contact angle according to Cassie–Baxter [39] is defined as

$$\cos(\theta_{CB}) = \sigma_{CB} f(\cos\theta_i) + f - 1. \tag{2}$$

In this equation, θ_{CB} is the apparent contact angle, f is the fraction of the projected area of the surface that is wet by the liquid and σ_{CB} is the roughness ratio of the wet area. This shows, that an increasing amount of trapped air, which means a smaller ratio f, will lead to an increase of the apparent contact angle.

This indicates that the measured contact angles on the CrCoMo samples due to laser processing can be explained by the increase of the ratio between the real surface area and the projected area σ and a reduction of the wetted area due to LIPSS [40,41]. Nonetheless, the contact angles are highly dependent on the formation of additional oxide layers. Interpretation of the origin of the hydrophobic properties would require a more thorough study of this surfaces.

3.5. Tribological Properties

The measured coefficient of friction (CoF) of the textured CoCrMo surfaces are listed in Table 6. The CoF of TNP with the hexagonal TNP is significantly lower than those of LSFL and Grooves. The friction coefficient of polished CoCrMo with 0.5 N (18 MPa), 11 mm/s and BCS lubricant was 0.22 ± 0.07. The friction coefficients of the textured surfaces are thus significantly higher than the CoF of the polished surface, due to the surface topograhphy changes.

Table 6. Coefficient of friction and polyethylene (PE) wear diameter of chosen surface structures.

Surface Structure	Polished	LSFL	Grooves	TNP
Coefficient of friction [-]	0.22 ± 0.07	0.66 ± 0.05	0.62 ± 0.09	0.40 ± 0.07
PE wear diameter [mm]	0.8 ± 0.1	3.7 ± 0.26	1.3 ± 0.03	0.8 ± 0.04

Figure 6 shows SEM micrographs of the LSFL structure after the wear test. From this figure it can be observed that the surface morphologies on the CoCrMo surface remain intact during the given wear test. After 104 min of sliding the PE sphere over the LSFL textured surface with a 18 MPa load, 11 mm/s speed and BCS lubricant, the PE ball had a volume loss of 43 mm^3. This is nearly 9.5% of the total sphere volume. Hence, it can be concluded that LSFL cannot be used as a bearing surface of a hip joint, since in the end of high loading, it reduces the durability of the hip joint significantly. The other two textures lead to noticeably less wear on the PE ball, see Table 6. The CoF of LSFL was also higher than that of Grooves and TNP. However, the difference between the CoF LSFL and Grooves is much smaller than the wear PE experiences against LSFL and Grooves. The fact that LSFL show a higher hydrophobicity (see Table 5) may influence the wear rate as well. High friction in a joint will lead to more heat generation, which may cause performance degradation of the joint. However, no maximally defined CoF is stated for a hip joint. The wear recorded for TNP is actually very close to that found on the polished surface.

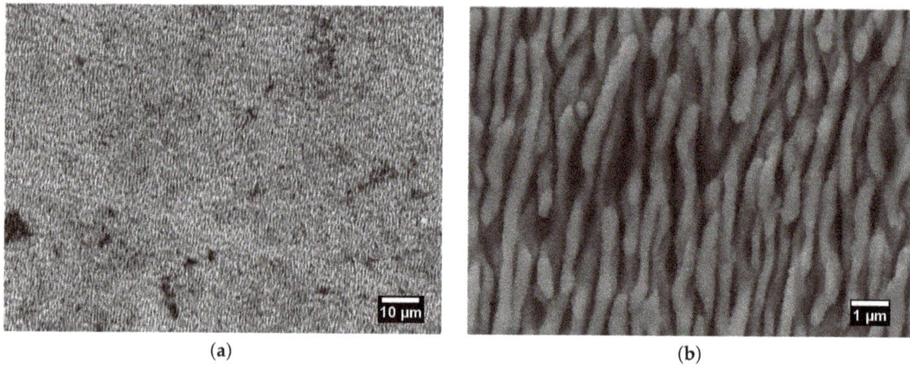

Figure 6. SEM micrographs of LSFL structure after wear test. (a) SEM image LSFL. (b) SEM image LSFL.

The wear conditions of the UMT, which are 18 MPa, 11 mm/s, reciprocal movement, are not comparable to the wear conditions in a natural hip joint, approximately 7.8 MPa and 21 mm/s during normal gait and rotational movement in all directions [21]. Since the surface structure TNP shows a periodicity in three directions (see Figure 4), instead of one in the cases of LSFL and Grooves, and also shows the lowest CoF and PE wear very close to the polished surface, TNP is the most promising candidate for a potential anti-bacterial surface structure on an artificial hip-joint.

3.6. Biocompatibility

Lutey et al. [15] showed that LSFL and TNP performed best on anti-bacterial properties regarding *E. coli* and *S. aureus* on stainless steel. A bacterial count reduction of 99.8% and 99.2% was found for *E. coli* and 84.7% and 79.9% was found for *S. aureus*, for the LSFL and TNP, respectively. Grooves (in [15] defined as Spikes) on the other hand, do not show improvement in anti-bacterial properties. However, to estimate the leaching of hazardous elements of the CoCrMo alloy into the human body, LSFL textured CoCrMo samples were used to perform a leaching test.

Release of Cobalt (Co) ions from the CoCrMo substrate may have an adverse affect on the patient's health. The Medicines and Healthcare products Regulatory Agency recommended a 7 µg/L threshold.

Concentrations above that threshold can be toxic for the patient [42]. Due to the increased surface area of the textured samples, when compared to the polished samples, textured samples may cause a higher ion release rate of Co and Ni ions. Chromium (Cr), Molybdenum (Mo) and Nickel (Ni) are also toxic in certain concentrations, but to the best of the authors knowledge no medical standardized regulations exist on acceptable concentration levels. The release of ions can be studied by means of a leaching experiment.

To that end, polished CoCrMo as well as LSFL textured CoCrMo samples were immersed in simulated body fluid (SBF, see Section 2.2.5) for nearly four weeks. Ion release was measured after 1, 7, 21 and 26 days. All samples were analyzed for the presence of Co, Cr, Mo and Ni elements by means of atomic emission spectroscopy analysis. No significant concentrations of Cr, Mo and Ni were found for any of the samples. Traces of Co were found in the SBF samples of the polished and the textured CoCrMo samples, see Figure 7. A gradual release of Co can be observed during the first two weeks for both polished and textured CoCrMo. After one day of immersion, the Co ion concentration is slightly higher for textured CoCrMo. Interestingly, there is no significant difference between the textured and untextured sample observed after one week and three weeks of immersion. After two weeks of immersion a larger Co concentration is found for polished CoCrMo and after four weeks of immersion the textured samples give a higher concentration, 27 ± 3 ppb vs. 17 ± 1 ppb. It was expected that the concentration of cobalt in the SBF would increase in time as more and more cobalt leaches from the surface into the fluid, until the equilibrium state is reached. The decrease in cobalt concentration of the untextured sample after 14 days could be explained by a change in pH due to a change in ion concentration in the SBF. The pH change could influence the equilibrium of Co ions. No precipitation of any element was observed at any point during and after the experiment. Unfortunately, the pH was not measured after the experiment. The difference in cobalt concentration after four weeks of immersion between polished and textured CoCrMo could be explained by the difference in surface area. According to Leyssens et al. [42], levels of Co lower than 300 µg/L will not cause health complications for individuals. The levels of Co in this study measured during 26 days of immersion, are well below this threshold. In the body the CoCrMo surface will be slightly larger. However, in the patients body, larger amounts of bodily fluids are present, and the human body does process low concentrations of Co [43]. However, it is questionable if this test can be compared with levels measured in patients. There are many factors which effect the leaching behavior of surfaces. To the best of our knowledge, no research on leaching of CoCrMo in SBF or similar circumstances has been conducted so far.

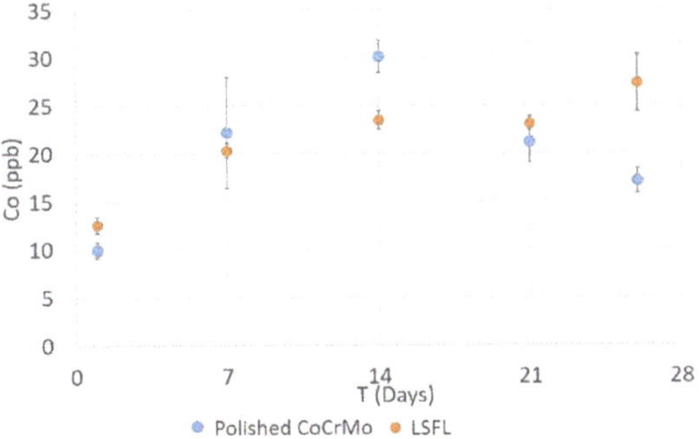

Figure 7. ICP-AES analysis of cobalt ion release of polished and LSFL textured CoCrMo samples in ppb as a function of time.

4. Conclusions

In this study, surface textures of nano and micrometer scale were produced on polished Cobalt–Chrome–Molybdenum alloy (CoCrMo) surfaces, using an infrared picosecond pulsed laser source. It was shown that the shape and size of the surface features can be controlled by adapting the laser fluence, the number of overscans of the laser spot over the surface and the type of polarization. To evaluate the wetting, tribological and leaching properties of laser-textured surfaces, three different types of textures were homogeneously produced on a large area (larger than the laser beam diameter), namely: low spatial frequency LIPSS, hierarchical grooves with superimposed low spatial frequency LIPSS, and triangular hexagonally packed nanopillars. The tribological behavior and the wettability of these three textures on CoCrMo were compared to a polished (i.e., untextured) CoCrMo surface. It was found that the textured surfaces caused higher friction in a CoCrMo-against-PE reciprocating contact compared to a polished reference. Moreover, only the LSFL textured surface showed a significantly higher wear of the PE counter surface. Furthermore, it was found that the hydrophobicity of the surface increases significantly due the micro-machined textures. Additionally, the biocompatibility of a LSFL textured surface on CoCrMo was compared to a polished CoCrMo surface. Both polished and textured surfaces release cobalt ions over a period of four weeks, but are still well below critical threshold levels reported in literature. Although, long term leaching experiments are recommended.

Based on the experimental conditions and results of this study, it is concluded that the laser textured surfaces on CoCrMo are not suitable for bearing surfaces in a metal-on-plastic contact. It is recommended to repeat the wear experiments at lower contact pressures, comparable to the conditions found in the hip joint, to study the friction and wear of PE under realistic conditions. The wear resistance, the antimicrobial activity and the effect on human cells of the processed surface textures would have to be investigated more thoroughly. It is recommended to look into other, possibly static, applications for antibacterial LIPSS surface textures on CoCrMo, e.g., dental implants.

Author Contributions: Conceptualization, S.H.v.d.P.; methodology, M.M. and S.H.v.d.P.; software, M.M.; validation, M.M., E.G.d.V. and S.H.v.d.P.; formal analysis, M.M., D.T.A.M. and S.H.v.d.P.; investigation, S.H.v.d.P.; resources, all authors; data curation, M.M. and S.H.v.d.P.; writing—original draft preparation, M.M. and S.H.v.d.P.; writing—review and editing, all authors; visualization, M.M. and S.H.v.d.P.; supervision, D.T.A.M. and G.-w.R.B.E.R.; project administration, S.H.v.d.P.; funding acquisition, D.T.A.M. and G.-w.R.B.E.R.

Funding: Parts of this study was funded by the European Union's Horizon 2020 research and innovation programme under the Marie Skłodowska-Curie grant agreement No. 675063 (Laser4Fun project, www.laser4fun.eu).

Acknowledgments: We thank Soheyla Ostvar Pour, University of Manchester School of Materials, for assistance with the execution of the leaching experiment, in particular the ICP-AES analysis.

Conflicts of Interest: The authors declare no conflict of interest.

Abbreviations

The following abbreviations are used in this manuscript:

CoCrMo	Cobalt Chrome Molybdenum
PE	Polyethylene
LIPSS	Laser-induced periodic surface structures
LSFL	low spatial frequency LIPSS
HSFL	high spatial frequency LIPSS
BCS	bovine calf serum
FFT	Fast Fourier transformation
SPP	surface plasmon polaritons
CoF	coefficient of friction
SBF	simulated body fluid
CA	contact angle
ICP-AES	Inductively Coupled Plasma Atomic Emission Spectroscopy

References

1. Löwik, C.A.; Wagenaar, F.C.; Van Der Weegen, W.; Poolman, R.W.; Nelissen, R.G.; Bulstra, S.K.; Pronk, Y.; Vermeulen, K.M.; Wouthuyzen-Bakker, M.; Van Den Akker-Scheek, I.; et al. LEAK study: Design of a nationwide randomised controlled trial to find the best way to treat wound leakage after primary hip and knee arthroplasty. *BMJ Open* **2017**, *7*, 1–6. [CrossRef] [PubMed]
2. Jaggessar, A.; Shahali, H.; Mathew, A.; Yarlagadda, P.K. Bio-mimicking nano and micro-structured surface fabrication for antibacterial properties in medical implants. *J. Nanobiotechnol.* **2017**, *15*, 1–20. [CrossRef] [PubMed]
3. Cunha, A.; Elie, A.M.; Plawinski, L.; Serro, A.P.; Botelho Do Rego, A.M.; Almeida, A.; Urdaci, M.C.; Durrieu, M.C.; Vilar, R. Femtosecond laser surface texturing of titanium as a method to reduce the adhesion of *Staphylococcus aureus* and biofilm formation. *Appl. Surf. Sci.* **2016**, *360*, 485–493. [CrossRef]
4. Getzlaf, M.A.; Lewallen, E.A.; Kremers, H.M.; Jones, D.L.; Bonin, C.A.; Dudakovic, A.; Thaler, R.; Cohen, R.C.; Lewallen, D.G.; van Wijnen, A.J. Multi-disciplinary antimicrobial strategies for improving orthopaedic implants to prevent prosthetic joint infections in hip and knee. *J. Orthop. Res.* **2016**, *34*, 177–186. [CrossRef] [PubMed]
5. Reshes, G.; Vanounou, S.; Fishov, I.; Feingold, M. Cell shape dynamics in *Escherichia coli. Biophys. J.* **2008**, *94*, 251–264. [CrossRef] [PubMed]
6. Harris, L.G.; Foster, S.J.; Richards, R.G.; Lambert, P.; Stickler, D.; Eley, A. An introduction to *Staphylococcus aureus*, and techniques for identifying and quantifying *S. aureus* adhesins in relation to adhesion to biomaterials: Review. *Eur. Cells Mater.* **2002**, *4*, 39–60. [CrossRef]
7. Puccio, F.D. Biotribology of artificial hip joints. *World J. Orthop.* **2015**, *6*, 77. [CrossRef] [PubMed]
8. Smet, K.D.; Campbell, P.; Straeten, C.V.D. *The Hip Resurfacing Handbook*; Woodhead Publishing: Sawston, UK, 2013; pp. 56–58.
9. Eichstädt, J.; Römer, G.R.B.E.; Huis in't Veld, A.J. Towards friction control using laser-induced periodic Surface Structures. *Phys. Procedia* **2011**, *12*, 7–15. [CrossRef]
10. Bonse, J.; Kirner, S.V.; Koter, R.; Pentzien, S.; Spaltmann, D.; Krüger, J. Femtosecond laser-induced periodic surface structures on titanium nitride coatings for tribological applications. *Appl. Surf. Sci.* **2017**, *418*, 572–579. [CrossRef]
11. Qin, L.; Wu, H.; Guo, J.; Feng, X.; Dong, G.; Shao, J.; Zeng, Q.; Zhang, Y.; Qin, Y. Fabricating hierarchical micro and nano structures on implantable Co–Cr–Mo alloy for tissue engineering by one-step laser ablation. *Colloids Surf. B Biointerfaces* **2018**, *161*, 628–635. [CrossRef]
12. Wu, B.; Zhou, M.; Li, J.; Ye, X.; Li, G.; Cai, L. Superhydrophobic surfaces fabricated by microstructuring of stainless steel using a femtosecond laser. *Appl. Surf. Sci.* **2009**, *256*, 61–66. [CrossRef]
13. Romano, J.M.; García-Girón, A.; Penchev, P.; Dimov, S. Triangular laser-induced submicron textures for functionalising stainless steel surfaces. *Appl. Surf. Sci.* **2018**, *440*, 162–169. [CrossRef]
14. Varlamova, O.; Reif, J.; Stolz, M.; Borcia, R.; Borcia, I.D. Wetting properties of LIPSS structured silicon surfaces. *Eur. Phys. J. B* **2019**, *92*, 91. [CrossRef]
15. Lutey, A.H.; Gemini, L.; Romoli, L.; Lazzini, G.; Fuso, F.; Faucon, M.; Kling, R. Towards laser-textured antibacterial surfaces. *Sci. Rep.* **2018**, *8*, 1–10. [CrossRef]
16. Rajab, F.H.; Liauw, C.M.; Benson, P.S.; Li, L.; Whitehead, K.A. Production of hybrid macro/micro/nano surface structures on Ti6Al4V surfaces by picosecond laser surface texturing and their antifouling characteristics. *Colloids Surf. B Biointerfaces* **2017**, *160*, 688–696. [CrossRef]
17. Kunz, C.; Müller, F.A.; Gräf, S. Multifunctional hierarchical surface structures by femtosecond laser processing. *Materials* **2018**, *11*, 789. [CrossRef]
18. Fadeeva, E.; Truong, V.K.; Stiesch, M.; Chichkov, B.N.; Crawford, R.J.; Wang, J.; Ivanova, E.P. Bacterial retention on superhydrophobic titanium surfaces fabricated by femtosecond laser ablation. *Langmuir* **2011**, *27*, 3012–3019. [CrossRef]
19. The MathWorks, Inc. *MATLAB® R2015b*; The MathWorks, Inc.: Natick, MA, USA, 2015.
20. Mezera, M.; Römer, G.R.B.E. Model based optimization of process parameters to produce large homogeneous areas of laser-induced periodic surface structures. *Opt. Express* **2019**, *27*, 6012–6029. [CrossRef]
21. International Organization for Standardization. *Implants for Surgery—Wear of Total Hip-Joint Prostheses—Part 2: Methods of Measurement*; ISO 14242-2:2016; ISO: Geneva, Switzerland, 2016; pp. 1–5.

22. Kokubo, T.; Takadama, H. How useful is SBF in predicting in vivo bone bioactivity? *Biomaterials* **2006**, *27*, 2907–2915. [CrossRef]

23. Bonse, J.; Höhm, S.; Kirner, S.; Rosenfeld, A.; Krüger, J. Laser-induced periodic surface structures (LIPSS)—A scientific evergreen. *IEEE J. Sel. Top. Quantum Electron.* **2016**, *23*, 9000615.

24. Bauer, F.; Michalowski, A.; Kiedrowski, T.; Nolte, S. Heat accumulation in ultra-short pulsed scanning laser ablation of metals. *Opt. Express* **2015**, *23*, 1035. [CrossRef]

25. Faas, S.; Bielke, U.; Weber, R.; Graf, T. Prediction of the surface structures resulting from heat accumulation during processing with picosecond laser pulses at the average power of 420 W. *Appl. Phys. A Mater. Sci. Process.* **2018**, *124*, 1–9. [CrossRef]

26. Faas, S.; Bielke, U.; Weber, R.; Graf, T. Scaling the productivity of laser structuring processes using picosecond laser pulses at average powers of up to 420 W to produce superhydrophobic surfaces on stainless steel AISI 316L. *Sci. Rep.* **2019**, *9*, 1933. [CrossRef]

27. Yasumaru, N.; Miyazaki, K.; Kiuchi, J. Fluence dependence of femtosecond-laser-induced nanostructure formed on TiN and CrN. *Appl. Phys. A Mater. Sci. Process.* **2005**, *81*, 933–937. [CrossRef]

28. Assion, A.; Husinsky, W.; Ganz, T.; Kudryashov, S.I.; Ajami, A.; Ionin, A.A.; Makarov, S.V.; Nathala, C.S. Experimental study of fs-laser induced sub-100-nm periodic surface structures on titanium. *Opt. Express* **2015**, *23*, 5915.

29. Liu, Q.; Zhang, N.; Yang, J.; Qiao, H.; Guo, C. Direct fabricating large-area nanotriangle structure arrays on tungsten surface by nonlinear lithography of two femtosecond laser beams. *Opt. Express* **2018**, *26*, 11718. [CrossRef]

30. Fraggelakis, F.; Mincuzzi, G.; Lopez, J.; Manek-Hönninger, I.; Kling, R. Controlling 2D laser nano structuring over large area with double femtosecond pulses. *Appl. Surf. Sci.* **2019**, *470*, 677–686. [CrossRef]

31. Ivanova, E.P.; Hasan, J.; Webb, H.K.; Truong, V.K.; Watson, G.S.; Watson, J.A.; Baulin, V.A.; Pogodin, S.; Wang, J.Y.; Tobin, M.J.; et al. Natural bactericidal surfaces: Mechanical rupture of *Pseudomonas aeruginosa* cells by cicada wings. *Small* **2012**, *8*, 2489–2494. [CrossRef]

32. Bixler, G.D.; Theiss, A.; Bhushan, B.; Lee, S.C. Anti-fouling properties of microstructured surfaces bio-inspired by rice leaves and butterfly wings. *J. Colloid Interface Sci.* **2014**, *419*, 114–133. [CrossRef]

33. Zimmermann, J.; Colombi Ciacchi, L. Origin of the selective Cr oxidation in CoCr alloy surfaces. *J. Phys. Chem. Lett.* **2010**, *1*, 2343–2348. [CrossRef]

34. Kofstad, P.K.; Hed, A.Z. High-temperature oxidation of Co-10 w/o Cr alloys. *J. Electrochem. Soc.* **1969**, *116*, 224–229. [CrossRef]

35. Jelovica Badovinac, I.; Kavre Piltaver, I.; Peter, R.; Saric, I.; Petravic, M. Formation of oxides on CoCrMo surfaces at room temperature: An XPS study. *Appl. Surf. Sci.* **2019**, *471*, 475–481. [CrossRef]

36. Liao, Y.; Hoffman, E.; Wimmer, M.; Fischer, A.; Jacobs, J.; Marks, L. CoCrMo metal-on-metal hip replacements. *Phys. Chem. Chem. Phys.* **2013**, *15*, 1–26. [CrossRef]

37. Huerta-Murillo, D.; García-Girón, A.; Romano, J.M.; Cardoso, J.T.; Cordovilla, F.; Walker, M.; Dimov, S.S.; Ocaña, J.L. Wettability modification of laser-fabricated hierarchical surface structures in Ti-6Al-4V titanium alloy. *Appl. Surf. Sci.* **2019**, *463*, 838–846. [CrossRef]

38. Wenzel, R.N. Surface roughness and contact angle. *J. Phys. Colloid Chem.* **1949**, *53*, 1466–1467. [CrossRef]

39. Cassie, A. Contact angles. *Discuss. Faraday Soc.* **1948**, *3*, 11–16. [CrossRef]

40. Sciancalepore, C.; Gemini, L.; Romoli, L.; Bondioli, F. Study of the wettability behavior of stainless steel surfaces after ultrafast laser texturing. *Surf. Coat. Technol.* **2018**, *352*, 370–377. [CrossRef]

41. Raimbault, O.; Benayoun, S.; Anselme, K.; Mauclair, C.; Bourgade, T.; Kietzig, A.M.; Girard-Lauriault, P.L.; Valette, S.; Donnet, C. The effects of femtosecond laser-textured Ti-6Al-4V on wettability and cell response. *Mater. Sci. Eng. C* **2016**, *69*, 311–320. [CrossRef]

42. Leyssens, L.; Vinck, B.; Van Der Straeten, C.; Wuyts, F.; Maes, L. Cobalt toxicity in humans—A review of the potential sources and systemic health effects. *Toxicology* **2017**, *387*, 43–56. [CrossRef]

43. Okazaki, Y.; Gotoh, E. Comparison of metal release from various metallic biomaterials in vitro. *Biomaterials* **2005**, *26*, 11–21. [CrossRef]

 © 2019 by the authors. Licensee MDPI, Basel, Switzerland. This article is an open access article distributed under the terms and conditions of the Creative Commons Attribution (CC BY) license (http://creativecommons.org/licenses/by/4.0/).

 *lubricants*

Article

On the Role of a ZDDP in the Tribological Performance of Femtosecond Laser-Induced Periodic Surface Structures on Titanium Alloy against Different Counterbody Materials

Jon Joseba Ayerdi [1,2,*], Nadine Slachciak [2], Iñigo Llavori [1], Alaitz Zabala [1], Andrea Aginagalde [1], Jörn Bonse [2] and Dirk Spaltmann [2]

[1] Surface Technologies, Mondragon University, Faculty of Engineering, Loramendi 4, 20500 Arrasate-Mondragon, Spain
[2] Bundesanstalt für Materialforschung und -prüfung (BAM), Unter den Eichen 87, 12205 Berlin, Germany
* Correspondence: jon.ayerdig@alumni.mondragon.edu

Received: 3 July 2019; Accepted: 30 August 2019; Published: 3 September 2019

Abstract: Laser-induced periodic surface structures (LIPSS, ripples) with ~500–700 nm period were produced on titanium alloy (Ti6Al4V) surfaces upon scan processing in air by a Ti:sapphire femtosecond laser. The tribological performance of the surfaces were qualified in linear reciprocating sliding tribological tests against balls made of different materials using different oil-based lubricants. The corresponding wear tracks were characterized by optical and scanning electron microscopy and confocal profilometry. Extending our previous work, we studied the admixture of the additive *2-ethylhexyl-zinc-dithiophosphate* to a base oil containing only anti-oxidants and temperature stabilizers. The presence of this additive along with the variation of the chemical composition of the counterbodies allows us to explore the synergy of the additive with the laser-oxidized nanostructures.

Keywords: lubricant additives; laser-induced periodic surface structures (LIPSS); wear; friction

---

1. Introduction

The material Ti6Al4V is one of the most widely used titanium alloys (e.g., in the aerospace industry and for implants in medicine) that are usually restricted in tribological applications due to the low surface hardness, the rather high coefficient of friction and the low abrasive wear. By means of surface modification technologies, the aforementioned shortcoming can be overcome. One of the most promising approaches is based on the use of lasers, which allow a contactless, fast, and reliable surface functionalization.

Ultrafast laser processing can be used to generate various surface morphologies, employing either direct contour-shaping or, for larger surface areas, functionalization via the generation of "self-organized" micro- and nanostructures [1]. This allows researchers to induce different surface functionalities in the fields of optics [2], fluidics [3], medicine [4,5], and tribology [6] and can be scaled up to the current demands for sizes and production rates in industrial applications.

The exposure to high-intensity pulsed laser radiation excites the treated materials into extreme conditions, which then return to equilibrium through various relaxation channels. Potentially, these involve phase transitions such as melting, evaporation (ablation), recrystallization, or accompanying chemical alterations in the ambient atmosphere (e.g., oxidation), etc., finally featuring different types of surface topographies [7]. One specific kind of these topographies currently gaining attention in various fields of applications is called laser-induced periodic surface structures (LIPSS, ripples). They are formed either perpendicular or parallel to the linear laser beam polarization and usually exhibit spatial periods in the sub-micrometer range [8]. These laser-induced nanostructures can be produced in a very reliable manner in a contactless single-step approach.

In a series of previous publications, we have explored the tribological performance of two different types of these surface nanostructures, high spatial frequency LIPSS (HSFL) [9] and low spatial frequency LIPSS (LSFL) [6,10,11], in the regime of mixed friction, reporting for the LSFL a remarkably positive effect on the coefficient of friction (CoF) and the associated wear of LIPSS-covered titanium or titanium alloy surfaces. They were tested in linear reciprocating sliding tribological tests (RSTT) in a fully formulated engine oil against a ball of hardened 100Cr6 steel and compared to polished reference surfaces. Via contrasting juxtaposition to identical RSTT conducted in non-additivated paraffin oil, it was speculated that this positive effect may be attributed to the presence of the additive (zinc dialkyl dithiophosphate, ZDDP [12]) contained in the engine oil (VP1) [10], able to efficiently bond to the laser-structured but not to the polished surfaces. These ZDDP molecules could then cover the laser-treated surfaces and, therefore, prevent a direct contact of the tested metals.

Apart from the topographic alterations, spatially and depth-resolved chemical analyses also revealed a fs-laser-induced oxidation process during the LIPSS formation on titanium [13]. In comparative oxidation experiments (thermal oxidation vs. anodic oxidation vs. LIPSS-induced oxidation) a beneficial tribological effect was found only for sufficiently large oxide layer thicknesses (>150 nm) and for high temperature surface oxidation [14]. The graded oxide layer present for the LSFL consisting mainly of amorphous TiO_2 and micro-crystalline Ti_2O_3 [13] may help to stabilize mechanically the oxygen containing near the surface region, preventing its delamination during the tribological demands/stresses. Moreover, with some hundreds of nanometers, the ball-sample deformation during the RSTT is of similar magnitude as the surface modulation depth of the LSFL [10], allowing a confinement of the lubricant in the tribological contact area during the RSTT.

In this work, we test our previous hypothesis that an additivated, fully formulated engine oil in combination with a laser-oxidized surface is beneficial for the tribological performance of Ti surfaces [10]. Specific emphasis is laid on the disclosure of the relevance of some specific anti-wear additives on the tribological performance of fs-laser-processed titanium alloy (Ti6Al4V) surfaces. By replacing the commercial engine oil (VP1) by an admixture of its base oil (VPX) with a single additive—here, the anti-wear additive (2-ethylhexyl zinc dithiophosphate)—its crucial role can be specifically addressed under our standard tribological testing conditions. Moreover, since the oils and additives are optimized for metallic (steel) surfaces, by varying the counterbody material, the role of the surface chemistry is further elucidated. Despite the use of engine oil, this work is an academic study and not one aimed at a specific engineering problem.

2. Experimental

Commercial grade-5 titanium alloy (Ti6Al4V) was purchased from Schumacher Titan GmbH (Solingen, Germany). The rods of 25 mm diameter were cut into circular slabs of 8 mm thickness. The top surfaces of the slabs were subsequently mechanically polished resulting in surface roughness parameters $R_a = 5$ nm and $R_{rms} = 6$ nm.

For large area surface processing, a linearly polarized commercial Ti:sapphire laser amplifier system was used (Femtolasers, Compact Pro, Vienna, Austria: $\tau = 30$ fs pulse duration, $\lambda = 790$ nm central wavelength, $\nu = 1$ kHz pulse repetition rate). The titanium alloy samples were mounted on a motorized x-y-z linear translation stage and placed perpendicular to the incident laser beam, realizing at the surface a Gaussian-like beam profile with a radius of w_0 $(1/e^2)$ ~140 μm. Square-shaped areas of 5×5 mm^2 were processed upon meandering movement of the sample ($v_x = 5$ mm/s scan velocity, $\Delta y = 0.1$ mm line-offset) under the focused laser beam at (35 ± 2) μJ/pulse. These laser-processing conditions correspond to a laser peak fluence of φ_0~0.11 J/cm^2 in front of the surface and an effective number of laser pulses per beam spot diameter of N_{eff} ~56 in the direction of the processed lines. Once processed, the samples were ultrasonically cleaned in acetone for 5 min. All samples were stored in a desiccator (up to a few months prior to the laser irradiation or the surface characterizations following the tribological tests). Given the very high temperatures of the titanium surface transiently exceeding 5000 K (estimated in [13]), repetitively reached during the fs-laser irradiation process, a significant storage-related post-oxidation of the laser treated samples at room temperature cannot be expected here.

The characteristics of the surface ripples were previously studied in detail on an identically processed sample revealing low-spatial frequency LIPSS (LSFL) with periodicities of (620 ± 80) nm along with height modulation depths of ±150 nm [10], typically manifested in a sine-like surface modulation. The modulation depths of the LSFL are similar to the ball-sample deformation during the RSTT (see Table 1). This may help to confine the lubricant in the tribological contact area of the nanostructured sample surface. For further details on the fs-laser processing and sample characterization, the reader is referred to [10].

RSTT measuring the coefficient of friction by the dissipated energy method were conducted with an in house-built tribometer [15] by sliding the LSFL-covered samples against either a hardened and polished steel ball (100Cr6, ø = 10 mm, R_a = 8 nm), a polished polycrystalline aluminium oxide ball (Al_2O_3, ø = 10 mm, R_a = 10 nm), or a polished polycrystalline silicon nitride ball (Si_3N_4, ø = 10 mm, R_a = 4 nm) as counterbodies. The normal load applied by dead weight is so small (1 N) that the stiffness of the holders is not compromised. The tribological tests were carried out in mixed lubrication conditions to have both, the effects of the laser-modified surface and the applied oil involved. The range of uncertainty in the friction coefficient measurements is ±0.02. Due to the limited availability of fs-laser-structured samples, each test was done once. The same RSTT conditions as in [10] were used in order to have directly comparable results. All relevant experimental parameters for the RSTT are summarized in Table 1.

Table 1. Linear reciprocating sliding tribological test conditions.

Test Parameter [Unit]	Value		
	Linear reciprocating sliding tribological test (RSTT)		
Tribosystem			
Samples	fs-LIPSS (LSFL) on Ti6Al4V		
Counterbodies	100Cr6/Al_2O_3 /Si_3N_4 balls of 10 mm diameter		
Young's Modulus [GPa]	100Cr6: ~200/Al_2O_3: ~365/Si_3N_4: ~300/Ti6Al4V: ~100		
Hardness [GPa]	100Cr6: ~7/Al_2O_3: ~17/Si_3N_4: ~16/Ti6Al4V: ~1		
Stroke [μm]	1000		
Frequency [Hz]	1		
Normal load [N]	1		
Number of cycles	1000		
Atmosphere	Laboratory air		
Temperature [°C]	~20–22		
Relative humidity [%]	~22–35		
Lubrication	VPX/VPX + 0.5% RC 3180		
Hertzian average/maximum contact pressure [MPa]	vs. 100Cr6 242/363	vs. Al_2O_3 272/407	vs. Si_3N_4 263/394
Hertzian radius of the sample-ball contact [μm]	vs. 100Cr6 36	vs. Al_2O_3 34	vs. Si_3N_4 35
Hertzian sample-ball deformation [μm]	vs. 100Cr6 0.26	vs. Al_2O_3 0.23	vs. Si_3N_4 0.24

The tribological tests were conducted using VPX oil as base lubricant, i.e., a variant of the polyalphaolefin (PAO) based factory fill Castrol engine oil SAE 0W-30 "VP1" (Castrol, Liverpool, UK) containing only anti-oxidation and anti-corrosion additives but no friction modifiers or wear protecting constituents. As a second variant, the commercial anti-wear additive RC 3180 purchased from LANXESS Deutschland GmbH (Business Unit Rhein Chemie, Cologne, Germany) was added to the base oil VPX. This additive consists of 2-ethylhexyl zinc dithiophosphate containing zinc, phosphorous, and sulphur by 9.5, 8, and 16 wt %, respectively. The optimum percentage of the additive content resulting in the lowest CoF and wear was identified as 0.5 wt % RC 3180 addition. That optimized mixture was then used for all RSTT experiments involving RC 3180-additivated VPX oil. The results for synthetic paraffin oil and for the fully additivated engine oil VP1 were already published in [10]. However, we will show them here along with the new results for direct comparison. After the RSTT, the samples were cleaned in benzene (petroleum ether) for 15 min using an ultrasonic bath in order to remove the residual lubricants.

The corresponding wear tracks were inspected by optical microscopy (OM, Carl Zeiss, Discovery V20/Keyence, VHX 5000, Oberkochen, Germany) and scanning electron microscopy (SEM, Carl Zeiss, Gemini Supra 40, Oberkochen, Germany). The volume of large wear scars was calculated by measuring the cross-sectional area and the width/length of the scars by means of a 3D confocal profilometer (Nanofocus, µ-surf Expert, Oberhausen, Germany) and subsequently applying analytical equations following the procedure described in the ASTM D7755-11 (2017) standard [16]. However, on the LSFL-covered and not completely worn surfaces the wear volume was estimated by assuming that roughly half of the wear track area (only at the peaks of the topography and not at the valleys) was removed over the depth measured by the profilometry (i.e., a few tens to hundreds of nanometers only).

In addition, selected parts of a wear track were analysed by scanning transmission electron microscopy (STEM). The sample was prepared by a focused ion beam (FIB) milling machine (FEI, Quanta 3D, Waltham, MA, USA) using an in situ lift-out technique. The preparation of the TEM lamella involved the deposition of a protective Pt cap layer at the region of interest, before thinning it by Ga-ions up to electron-transparency. The Pt layer protects the covered sample surface from being damaged during the FIB preparation of the lamella. This lamella of ~100 nm thickness was then characterized in a scanning transmission electron microscope (JEOL, JEM 2200FS, Akishima, Tokyo, Japan) operated at 200 kV with a point resolution better than 0.25 nm. The system is equipped with a field emission gun, an in-column energy filter and an energy-dispersive X-ray (EDX) spectroscopy system for elemental studies.

3. Results and Discussion

The tribological performance of the laser-processed surfaces was characterized by means of RSTT in two different lubricants (Section 3.1) and for three different counterbody materials (Section 3.2). Previous works under identical test conditions [10,11] disclosed no significant dependence on the CoF/wear results regarding the sliding direction relative to the orientation of the LIPSS for a metallic counterbody. Thus, the current tribological tests were always carried out in the direction perpendicular to the ablation lines and additionally on a non-irradiated or polished surface (without LIPSS) as a reference. Note that due to an update of the RSTT tribometer, the data acquisition rate was improved. Hence, the CoF results (for paraffin oil and VP1 oil) published in 2014 in Ref. [10] and displayed in the following section for comparison exhibit 10 times less data points.

3.1. Reciprocating Sliding Tribological Tests (RSTT) against 100Cr6 in Different Lubricants

In a first set of experiments the VPX base oil was used in RSTT under identical testing conditions previously used for paraffin oil [10]. This choice of lubricant allows us to take benefit of the anti-oxidation and temperature stabilizing components contained in VPX oil, which are not present in paraffin oil. Figure 1 compiles these new RSTT results (left column) and compares them to the previous results in paraffin oil (right column).

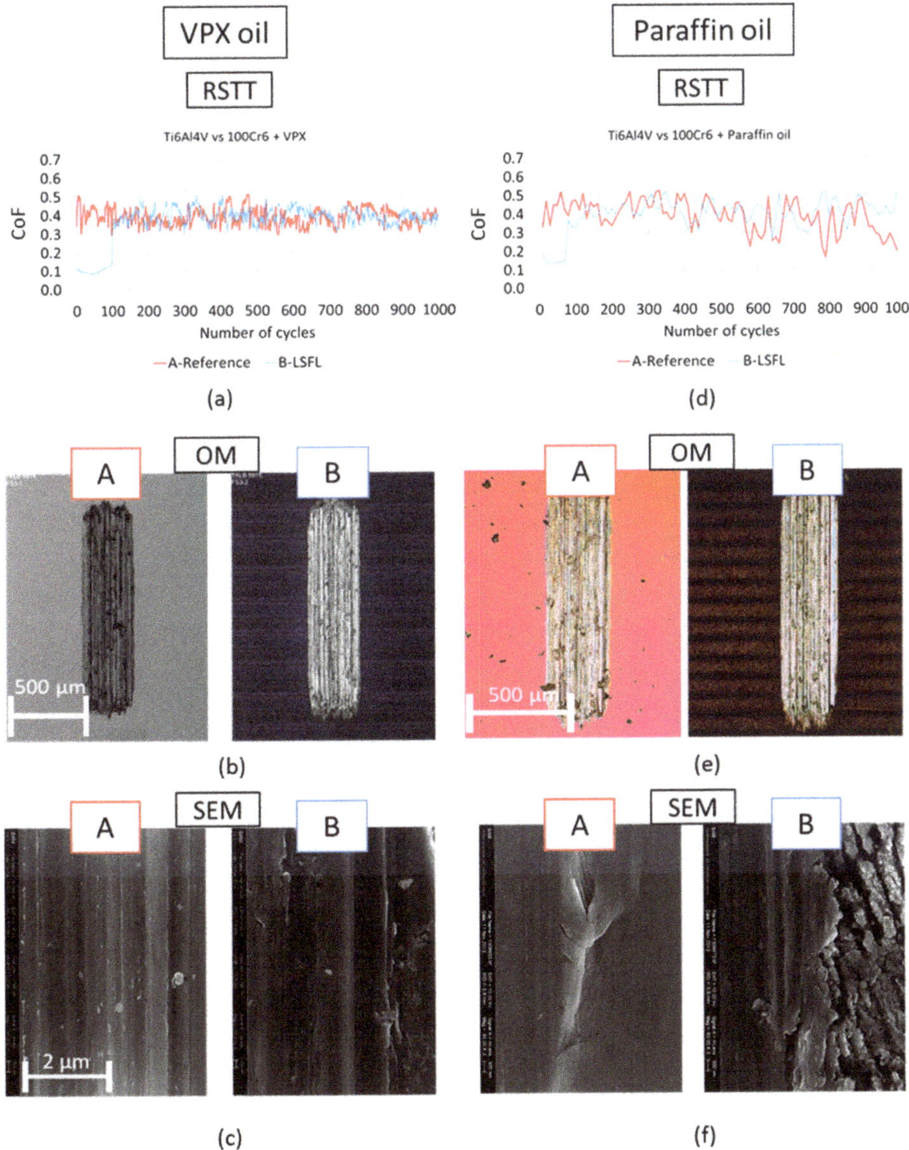

Figure 1. Tribological performance of fs-laser-processed Ti6Al4V titanium alloy after reciprocating sliding tribological tests (RSTT) against a 100Cr6 steel ball (normal force 1.0 N, stroke 1 mm, frequency 1 Hz, cycles 1000). The top row exhibits the coefficient of friction (CoF) as function of the number of sliding cycles in VPX oil (**a**) and paraffin oil (**d**). The middle row displays optical micrographs of the corresponding wear tracks after the tribological tests (**b**,**e**). The bottom row shows detailed scanning electron microscope (SEM) micrographs taken at the center (**c**) or at the edge (**f**) of the wear tracks. The data for the RSTT in paraffin oil are taken from [10]. The capitalized labels indicate tests performed on polished surfaces (**A**) or fs-laser-processed (low spatial frequency laser-induced periodic surface structures (LSFL)-covered) surfaces (**B**). Common scale bars are provided at the left.

Figure 1a presents the coefficient of friction as a function of the number of sliding cycles in VPX oil for the measurements made on the polished reference surface (red curve, A) and for tests performed in the fs-laser-processed area covered by LSFL (blue curve, B). The CoF acquired on the polished surface varies around 0.4 ± 0.1 through the entire test. In contrast, the CoF measured in the laser-processed region starts at a significantly lower value of $\sim$0.12 $\pm$ 0.02 before it sharply rises after $\sim$80 cycles for paraffin oil and $\sim$100 cycles for VPX oil to a similar level as seen for the reference curve. This sudden rise of the CoF is indicative of the starting damage of the surface when the protecting laser-induced oxide layer along with deeper lying material gets removed and the LSFL have no influence on the CoF behaviour anymore. This is in line with the visual inspections of the corresponding wear tracks shown in Figure 1b,c, where a severe surface damage is seen for both the polished and the laser-processed surfaces tested in VPX oil. The detailed SEM images further confirm that the LIPSS have not endured the tribological tests (compare Figure 1c, A, B). The mean values and corresponding standard deviations of the CoF and the wear volumes, quantified as 1070×10^{-6} mm^3 (reference) and $\sim$840 $\times$ 10^{-6} mm^3 (LSFL), are listed in Table 2. These new results obtained with VPX oil as lubricant are in line with our previous results reported for paraffin oil (compare with the right column of Figure 1).

Table 2. Mean values and standard deviations of the coefficient of friction (CoF) corresponding to the entire test (RSTT in Ti6Al4V against 100Cr6 ball, normal force 1.0 N, stroke 1 mm, frequency 1 Hz, cycles 1000) and resulting wear volumes for different lubricants and tested surfaces [reference (polished) and fs-laser-processed (LSFL-covered)]. The data for the RSTT in VP1 oil and in paraffin oil are taken from [10], wear volume n.a.—not available.

Lubricant	CoF		Wear Volume [10^{-6} mm^3]	
	Reference	LSFL	Reference	LSFL
Paraffin oil	0.388 ± 0.087	0.395 ± 0.087	n.a.	n.a.
VPX oil	0.391 ± 0.044	0.366 ± 0.096	1070	838
VP1 oil	0.362 ± 0.043	0.137 ± 0.004	n.a.	n.a.
VPX oil + 0.5% RC 3180	0.430 ± 0.054	0.088 ± 0.036	1039	0.35

In order to elucidate the role of the specific additive RC 3180, identical RSTT were performed in VPX oil + 0.5% RC 3180. These results are compared to the ones previously accomplished for the fully formulated commercial engine oil Castrol VP1, including the anti-wear additive zinc dialkyl dithiophosphate (ZDDP) [12]. Figure 2 collects the RSTT results obtained for the additivated VPX oil (left column) and our previous results for VP1 oil (right column).

Figure 2a displays the CoF vs. the number of sliding cycles in RC 3180-additivated VPX oil for the RSTT on the polished reference surface (red curve, A) along with the results obtained on the LSFL-covered surface (blue curve, B). Similar to the behaviour seen previously for VPX oil in Figure 1, the CoF recorded on the polished reference surface varies around 0.45 ± 0.15. However, the CoF recorded in the LSFL-covered region stays rather constant around $\sim$0.1 $\pm$ 0.05 through the entire test duration, with some additional spiked features between $\sim$200 and $\sim$600 cycles. The latter are probably caused by worn particles intermittently present in the tribological contact area. The remarkable difference in the CoF between the polished and the LSFL-covered Ti-alloy surfaces is also clearly visible in the wear tracks visualized after the RSTT by optical microscopy (OM) (Figure 2b) and SEM (Figure 2c). The SEM images confirm that the LIPSS widely endured the tribological tests (compare A, B in Figure 2c). Very different wear volumes were found, estimated as $\sim$1040 $\times$ 10^{-6} mm^3 (reference) and $\sim$0.35 $\times$ 10^{-6} mm^3 (LSFL), see Table 2. The results obtained with RC 3180-additivated VPX oil are again very similar to our previous results reported for the fully formulated VP1 oil (compare with the right column of Figure 2). Note that the use of RC 3180-additivated VPX oil results in a lower average CoF-level than that of VP1 oil, supposedly affected by improved data acquisition rate of the tribometer. Moreover, the periods of the partly worn LSFL in wear track B of Figure 2c are somewhat smaller than the ones for the corresponding track B presented in Figure 2f. This arises from somewhat differently laser fluences locally accumulated through the line processing with a Gaussian laser beam.

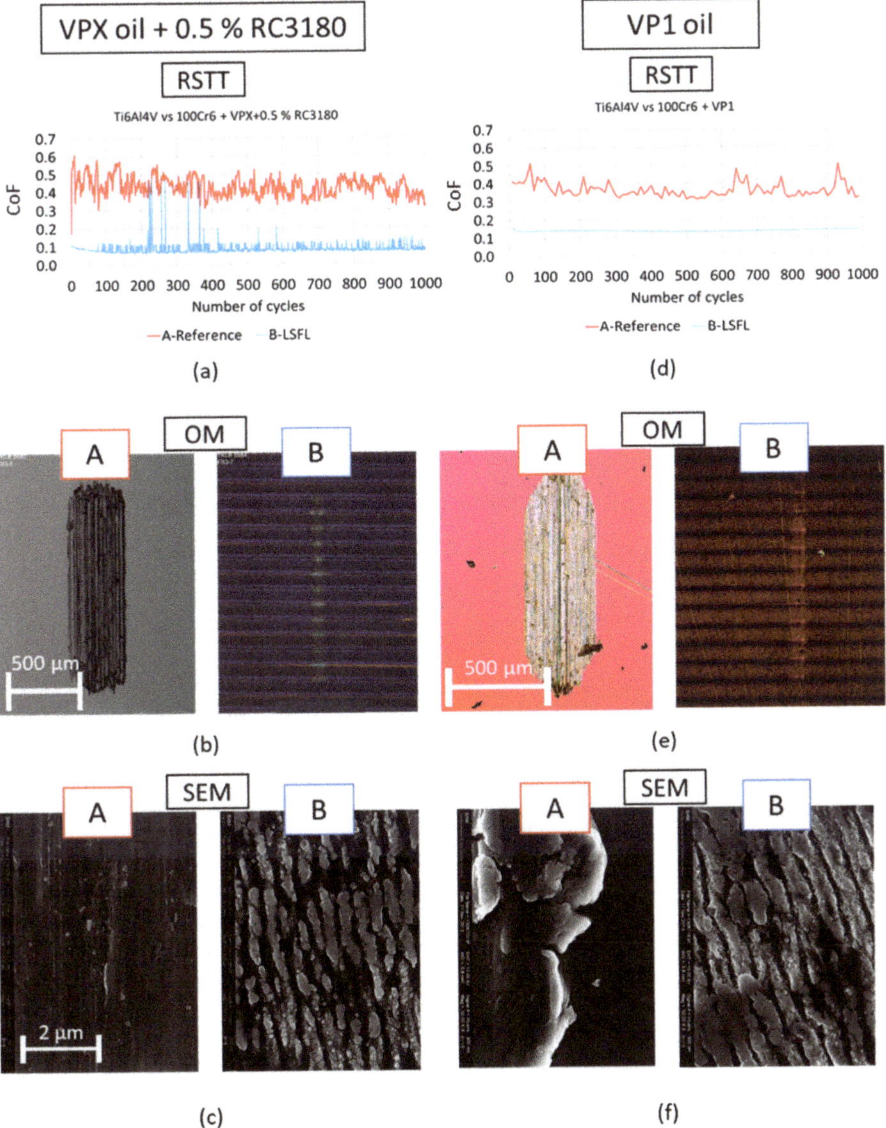

Figure 2. Tribological performance of fs-laser-processed Ti6Al4V titanium alloy after RSTT against a 100Cr6 steel ball (normal force 1.0 N, stroke 1 mm, frequency 1 Hz, cycles 1000). The top row exhibits the coefficient of friction (CoF) as function of the number of sliding cycles in RC 3180-additivated VPX oil (VPX oil + 0.5% RC 3180) (**a**) and VP1 oil (**d**). The middle row displays optical micrographs of the corresponding wear tracks after the tribological tests (**b,e**). The bottom row shows detailed SEM micrographs taken at the center of the wear tracks (**c,f**). The data for the RSTT in VP1 oil are taken from [10]. The capitalized labels indicate tests performed on polished surfaces (**A**) or fs-laser-processed (LSFL-covered) surfaces (**B**). Common scale bars are provided at the left.

Comparing the findings presented in Figures 1 and 2 for VPX oil and RC 3180-additivated VPX oil provides a direct evidence that the beneficial tribological performance featured at the fs-laser-processed surfaces covered with LSFL is caused by the anti-wear additive RC 3180 and not by their topographical characteristics. This is also seen in Figure 3, which visualizes the values compiled in Table 2. However, this improved tribological performance is likely promoted by the laser-induced graded oxide layer as it is not present for the polished Ti-alloy surface. While the roughened fs-laser-induced oxide layer on the LSFL consists mainly of amorphous TiO_2 and extends ~200 nm into depth, the native oxide layer (<10 nm thick) [13] on the smooth polished surface is too thin to interact efficiently with the 2-ethylhexyl zinc dithiophosphate molecules. It is also interesting to note that compared to the fully formulated/additivated VP1 oil, this beneficial reduction in CoF and wear can already be achieved by solely using RC 3180.

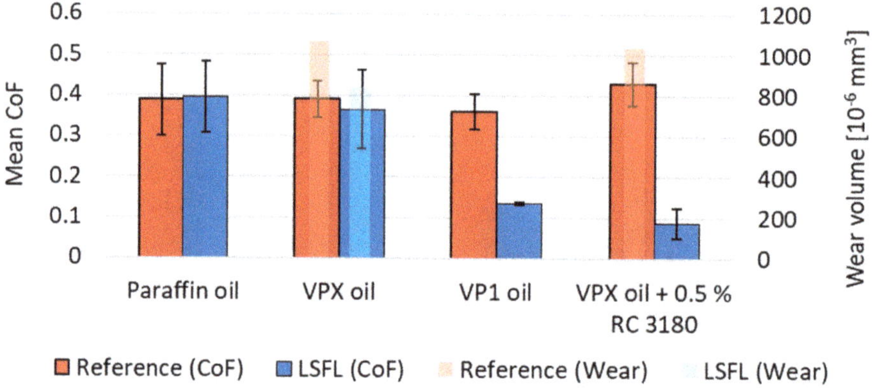

Figure 3. Visualization of the data compiled in Table 2.

It is worth noting that the initial values of the CoF in the LSFL-covered regions are all comparable at a level of 0.10–0.15 indicating the presence of a protecting oxide layer during the first hundred sliding cycles (see Figure 1a,d and Figure 2a,d). For the lubricants without anti-wear additives (paraffin oil, VPX oil), the oxide layer is worn, while with such additives (VP1 oil, VPX oil + 0.5% RC 3180) the laser-induced oxide layer endures the entire tribological test.

For a deeper characterization of a wear track obtained with an anti-wear additive, we extended our previous STEM analysis [14] by adding spatially information on the elemental contribution underneath the laser-irradiated and worn surface. The corresponding wear track was obtained in the same processed LSFL area as previously shown in Figure 2d,e(B),f(B), but it was tribologically tested parallel to the ablation lines, i.e., perpendicular to the LSFL, with otherwise unchanged RSTT conditions. Figure 4a shows an optical micrograph of the worn surface after an RSTT on LSFL-covered titanium alloy surfaces in VP1 oil. Note that the dark vertical lines arise from the line wise laser processing, while the bright lines in the centre correspond to the wear track. From this tribologically tested surface, a FIB lamella was prepared. The location it was taken from is marked in green in Figure 4b. A cross-sectional STEM image was taken in bright field mode from this lamella, which is presented in Figure 4c. The wear track is marked with a black double arrow. The depth of the zone affected by the laser structuring process and tribological testing can be estimated to be less than 500 nm. This zone is due to the local chemistry in the tribological contact area that creates a protective surface layer in the LSFL. It prevents a direct contact between the two sliding bodies, resulting in reduced friction and wear (Figure 2). The STEM image clearly shows the grain structure of the polycrystalline Ti6Al4V alloy. In all images, the protective Pt layer is marked, which consists of two sublayers. Figure 4d is a magnified detail of Figure 4c, of which elemental maps (EDX) were established for Fe (Figure 4e), Ti (Figure 4f), and Zn (Figure 4g). Signals in the top part of these images are reflections of the Pt protective layer. Close to the Ti surface, material transfer (Fe) of the steel counterbody is visible (Figure 4e), as well as Zn (Figure 4g) of the ZDDP additive. Figure 4h shows a superposition of the STEM image (Figure 4d) with the elemental maps of Fe, Ti, and Zn. It reveals that within the wear track, small pits have formed, which contain more or less loose debris. In these locations, Zn could be found, which is an indicator for the presence of ZDDP.

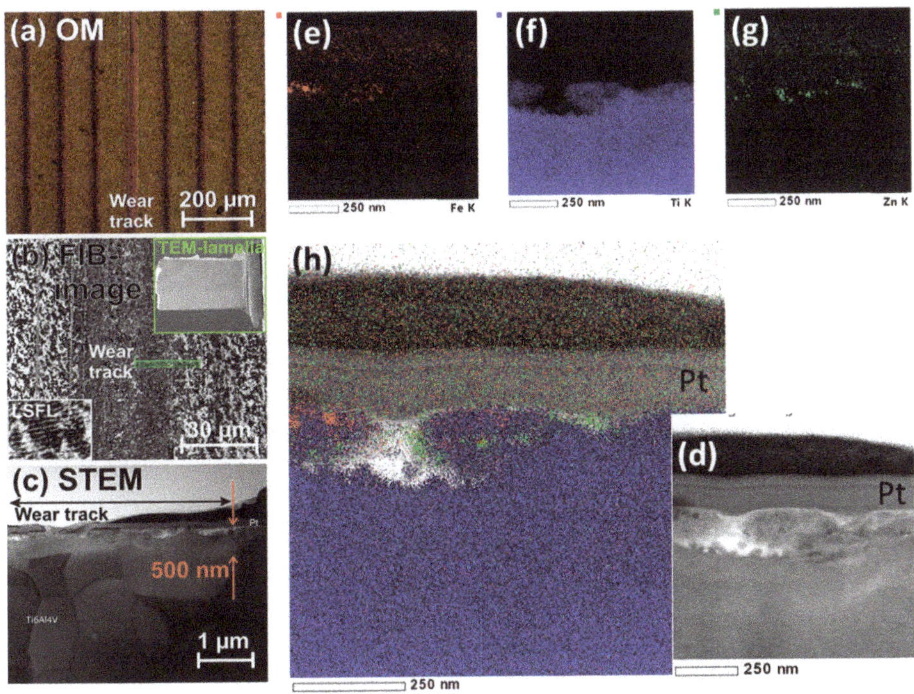

Figure 4. Characterization of a wear track on LSFL-covered Ti4Al6V after RSTT [identical processed LSFL area as previously shown in Figure 2e(B), but tribologically tested parallel to the ablation lines (perpendicular to the LSFL) with otherwise unchanged RSTT conditions]. (**a**) optical micrograph. (**b**) Focused ion beam (FIB) image with respective lamella and magnified detail. (**c**) STEM cross-sectional image of this lamella. (**d**) magnified detail of (**c**), of which energy-dispersive X-ray (EDX) spectroscopy elemental maps were taken for Fe (**e**), Ti (**f**) and Zn (**g**), (**h**) is an overlay of (**d–g**). Images (**a–c**) are taken from [14].

3.2. RSTT against Different Counterbody Materials

The additives ZDDP are highly optimized to work efficiently on metallic (mostly steel) surfaces as anti-wear agents [12,17–19]. Since in the previous experiments (shown in Section 3.1) balls made of 100Cr6 steel were used as counterbodies, the question remains open, whether the beneficial tribological effect originates from the binding of the additive molecules to the oxidized titanium alloy surface or to the surface of the steel ball. Hence, in a second set of experiments the material of the counterbody was systematically varied in order to use also non-metallic, i.e., oxide- (Al_2O_3) or nitride-based (Si_3N_4) ceramics, which feature different chemical compositions and were available in a similar surface quality as the 100Cr6 steel balls.

Figure 5 gathers the results by showing the graphs of the CoF and OM and SEM micrographs of the corresponding wear tracks on polished and fs-laser-processed Ti6Al4V surface after the RSTT in VPX oil + 0.5% RC 3180 against balls made of 100Cr6 steel [left column, (a)], Al_2O_3 ceramic [middle column, (b)] and Si_3N_4 ceramic [right column, (c)].

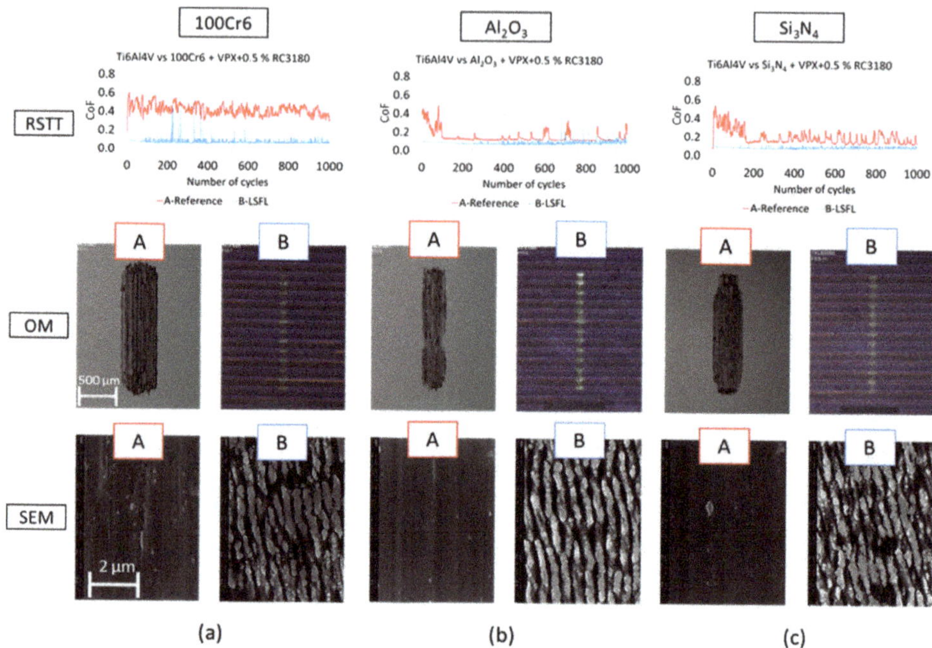

Figure 5. Tribological performance of fs-laser-processed Ti6Al4V surfaces after RSTT (normal force 1.0 N, stroke 1 mm, frequency 1 Hz, cycles 1000) in RC 3180-additivated VPX oil (VPX oil + 0.5% RC 3180) against 10 mm balls made of different materials. RSTT against 100Cr6 steel (metallic) counterbody (**a**), Al$_2$O$_3$ (oxide ceramic) counterbody (**b**) and Si$_3$N$_4$ (non-oxide ceramic) counterbody (**c**). The top row provides the coefficient of friction (CoF) as function of the number of sliding cycles. The middle row shows optical micrographs of the corresponding wear tracks after the tribological tests and the bottom row depicts detailed scanning electron microscope (SEM) micrographs taken at the centre of the wear tracks. The capitalized labels indicate tests performed on polished surfaces (**A**) or fs-laser-processed (LSFL-covered) surfaces (**B**). Common scale bars are provided at the left.

For a direct comparison, Figure 5a displays again the results obtained for the 100Cr6 steel balls previously shown in Figure 2 (left column). The CoF's measured in the fs-laser-processed regions (B, blue curves) with both ceramic counterbodies (Al$_2$O$_3$, Si$_3$N$_4$) are as small as those obtained for the metallic steel balls, see Table 3. In all cases, the LSFL endured the RSTT, see the high-resolution SEM images for the associated wear tracks (B) in Figure 5, bottom row. The CoF's recorded in the polished reference surfaces (A, red curves) always start between ~0.4–0.6. However, after less than 200 cycles the CoF's are remarkably lower for the ceramic counterbodies when compared to the metallic steel ball. Nevertheless, the averaged CoF always shows a clear reduction between the polished surfaces and the fs-laser-processed regions, although the reduction ratio is smaller for both ceramic counterbodies. The reduced CoF's on the polished surfaces when using ceramic balls compared to the steel balls are also reflected in the size of the corresponding wear tracks (compare the tracks A in the OM, middle row, Figure 5).

Table 3. Mean values and standard deviations of the coefficient of friction (CoF) corresponding to the entire test (RSTT in Ti6Al4V, normal force 1.0 N, stroke 1 mm, frequency 1 Hz, cycles 1000) and resulting wear volumes for different lubricants and tested surfaces [reference (polished) and fs-laser-processed (LSFL-covered)] using different counterbody materials.

Lubricant	Counterbody	CoF		Wear Volume [10^{-6} mm^3]	
		Reference	LSFL	Reference	LSFL
VPX oil	100Cr6	0.391 ± 0.044	0.366 ± 0.096	1070	838
	Al$_2$O$_3$	0.167 ± 0.053	0.148 ± 0.078	175	295
VPX oil + 0.5% RC 3180	100Cr6	0.430 ± 0.054	0.088 ± 0.036	1039	0.35
	Al$_2$O$_3$	0.158 ± 0.063	0.094 ± 0.022	169	7
	Si$_3$N$_4$	0.206 ± 0.077	0.094 ± 0.012	452	4

The quantitative values of the averaged CoF's along with the estimated wear volumes provided in Table 3 prove that the difference in the counterbody material (chemical composition as well as hardness) and the tribological contact characteristics estimated from an elastic (Hertzian) deformation model [20] (contact pressure, contact radius, sample-ball deformation, see Table 1) are not the dominating effect in our RSTT featuring the beneficial tribological effect. Hence, it is likely that the laser-induced oxidation and the interplay with surface topography are the crucial aspects here. Inferring that a minimum oxygen-containing layer thickness and an enlarged surface area are required together with the additives ZDDP [14] to form a sufficiently thick anti-wear layer during RSTT, which prevents a direct contact of the two sliding tribological bodies. These findings are supported by Figure 6, which visualizes the values gathered in Table 3.

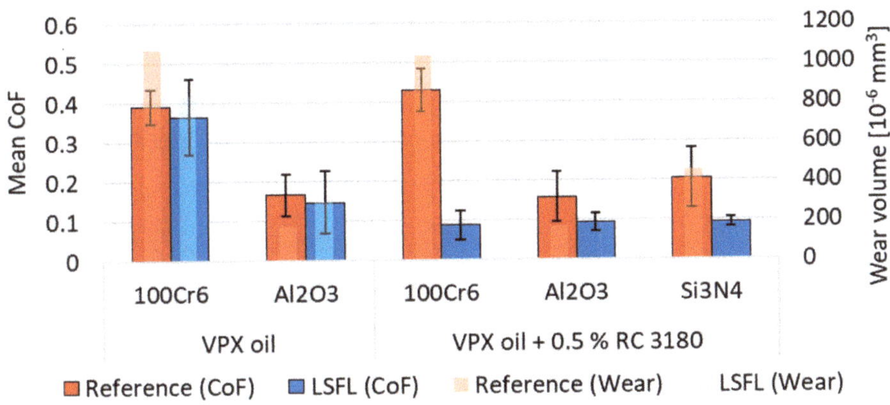

Figure 6. Visualization of the data compiled in Table 3.

4. Conclusions

Low spatial frequency LIPSS (LSFL) were uniformly processed by Ti:sapphire fs-laser pulses on Ti6Al4V titanium alloy samples. The tribological performance of the surfaces [reference (polished) vs. fs-laser-processed (LSFL-covered)] was determined in linear reciprocating sliding tribological tests against 100Cr6, Al$_2$O$_3$ and Si$_3$N$_4$ balls as counterbodies in two different lubricants [VPX oil and VPX oil + 0.5% of an anti-wear (ZDDP) additive, RC 3180]. Subsequently, the wear tracks were characterized by OM, SEM, STEM, and confocal profilometry. For the specific testing conditions here, a reduction by a factor of 4–5 for the coefficient of friction and by a factor >2500 for the wear volume was observed in RC 3180-additivated VPX oil, while the tribological tests against 100Cr6 balls showed no beneficial influence of LSFL-covered surfaces with VPX oil lubrication. These new results proved the similar behaviour between VPX oil and paraffin oil (free of any additives) and between RC

3180-additivated VPX oil and commercial VP1 oil containing ZDDP as anti-wear agent. This clearly evidences our previous speculation that only the admixture of the specific ZDDP molecules account for the reproduction of the beneficial tribological behaviour on the laser-processed Ti6Al4V surfaces. Additionally, no significant influence of the counterbody material (metal vs. ceramics) was observed, thereby implying that the positive effect is mainly caused by the presence of the 2-ethylhexyl zinc dithiophosphate molecules in the lubricant along with a nanostructured and oxidized layer on the laser-processed surfaces. The interplay between the sample topography (featuring an enlarged surface area and a confinement of the lubricant) and the local chemistry in the tribological contact area—via formation of a protective surface layer on the LSFL—prevents a direct contact of the two sliding bodies, finally resulting in reduced friction and wear.

Author Contributions: Conceptualization, D.S. and J.B.; Methodology, J.B. and D.S.; Validation, all authors; Formal analysis, J.J.A. and N.S.; Investigation, all authors; Resources, J.B., N.S., and D.S.; Data curation, J.J.A. and N.S.; Writing—original draft preparation, J.J.A. and J.B.; Writing—review and editing, all authors; Visualization, J.J.A., N.S., and J.B.; Supervision, D.S., J.B., A.Z., I.L., and A.A.; Project administration, D.S., J.B., A.Z., and I.L.; Funding acquisition, J.J.A., A.Z., J.B. and D.S.

Funding: This work was supported by the ERASMUS+ program of the European Union (KA103). This project has received funding from the European Union's Horizon 2020 research and innovation programme under grant agreement No. 814494 (project "i-TRIBOMAT") and grant agreement No. 665337 (project "LiNaBioFluid"). The Surface Technologies research group (Mondragon University, Faculty of Engineering) gratefully acknowledges the financial support given by the Red Guipuzcoana de Ciencia Tecnología e Innovación 2018 program through the project ASEFI (Orden Foral Número 218/2018).

Acknowledgments: The authors would like to thank C. Neumann (BAM 6.3) for help with profilometric sample characterizations, S. Benemann (BAM 6.1) for the SEM characterizations, R. Koter (BAM 6.4) for the laser surface processing of a sample copy, and S. Binkowski (BAM 6.3) for polishing the titanium samples. The STEM analysis was performed by I. Dörfel (BAM 5.1). The femtosecond laser processing was performed within the frame of German Science Foundation (Deutsche Forschungsgemeinschaft, DFG) project under Grant KR 3638/1-2.

Conflicts of Interest: The authors declare no conflict of interest.

References

1. Abere, M.; Zhong, M.; Krüger, J.; Bonse, J. Ultrafast laser-induced morphological transformations. *MRS Bull.* **2016**, *41*, 969–974. [CrossRef]
2. Liu, H.; Lin, W.; Hong, M. Surface coloring by laser irradiation of solid substrates. *APL Photonics* **2019**, *4*, 051101. [CrossRef]
3. Kirner, S.V.; Hermens, U.; Mimidis, A.; Skoulas, E.; Florian, C.; Hischen, F.; Plamadeala, C.; Baumgartner, W.; Winands, K.; Mescheder, H.; et al. Mimicking bug-like surface structures and their fluid transport produced by ultrashort laser pulse irradiation of steel. *Appl. Phys. A* **2017**, *123*, 754. [CrossRef]
4. Epperlein, N.; Menzel, F.; Schwibbert, K.; Koter, R.; Bonse, J.; Sameith, J.; Krüger, J.; Toepel, J. Influence of femtosecond laser produced nanostructures on biofilm growth on steel. *Appl. Surf. Sci.* **2017**, *418*, 420–424. [CrossRef]
5. Heitz, J.; Plamadeala, C.; Muck, M.; Armbruster, O.; Baumgartner, W.; Weth, A.; Steinwender, C.; Blessberger, H.; Kellermair, J.; Kirner, S.V.; et al. Femtosecond laser-induced microstructures on Ti substrates for reduced cell adhesion. *Appl. Phys. A* **2017**, *123*, 734. [CrossRef]
6. Bonse, J.; Kirner, S.V.; Griepentrog, M.; Spaltmann, D.; Krüger, J. Femtosecond laser texturing of surfaces for tribological applications. *Materials* **2018**, *11*, 801. [CrossRef] [PubMed]
7. Bäuerle, D. *Laser Processing and Chemistry*, 4th ed.; Springer: Berlin/Heidelberg, Germany, 2011.
8. Bonse, J.; Höhm, S.; Kirner, S.V.; Rosenfeld, A.; Krüger, J. Laser-induced periodic surface structures—A scientific evergreen. *IEEE J. Sel. Top. Quantum Electron.* **2017**, *23*, 9000615. [CrossRef]
9. Bonse, J.; Höhm, S.; Koter, R.; Hartelt, M.; Spaltmann, D.; Pentzien, S.; Rosenfeld, A.; Krüger, J. Tribological performance of sub-100-nm femtosecond laser-induced periodic surface structures on titanium. *Appl. Surf. Sci.* **2016**, *374*, 190–196. [CrossRef]
10. Bonse, J.; Koter, R.; Hartelt, M.; Spaltmann, D.; Pentzien, S.; Höhm, S.; Rosenfeld, A.; Krüger, J. Femtosecond laser-induced periodic surface structures on steel and titanium alloy for tribological applications. *Appl. Phys. A* **2014**, *117*, 103–110. [CrossRef]

11. Bonse, J.; Koter, R.; Hartelt, M.; Spaltmann, D.; Pentzien, S.; Höhm, S.; Rosenfeld, A.; Krüger, J. Tribological performance of femtosecond laser-induced periodic surface structures on titanium and a high toughness bearing steel. *Appl. Surf. Sci.* **2015**, *336*, 21–27. [CrossRef]
12. Nicholls, M.A.; Do, T.; Norton, P.R.; Kasrai, M.; Bancroft, G.M. Review of the lubrication of metallic surfaces by zinc dialkyl-dithiophosphates. *Tribol. Int.* **2005**, *38*, 15–39. [CrossRef]
13. Kirner, S.V.; Wirth, T.; Sturm, H.; Krüger, J.; Bonse, J. Nanometer-resolved chemical analyses of femtosecond laser-induced periodic surface structures on titanium. *J. Appl. Phys.* **2017**, *122*, 104901. [CrossRef]
14. Kirner, S.V.; Slachciak, N.; Elert, A.M.; Griepentrog, M.; Fischer, D.; Hertwig, A.; Sahre, M.; Dörfel, I.; Sturm, H.; Pentzien, S.; et al. Tribological performance of titanium samples oxidized by fs-laser radiation, thermal heating, or electrochemical anodization. *Appl. Phys. A* **2018**, *124*, 326. [CrossRef]
15. Klaffke, D.; Hartelt, M. Tribological characterization of thin hard coatings by reciprocating sliding tests. *Tribol. Lett.* **1995**, *1*, 265–276. [CrossRef]
16. ASTM D7755-11 2017. *Standard Practice for Determining the Wear Volume on Standard Test Pieces Used by High-Frequency, Linear-Oscillation (SRV) Test Machine*; ASTM: West Conshohocken, PA, USA, 2017.
17. Itoa, K.; Martin, J.M.; Minfray, C.; Katoa, K. Low-friction tribofilm formed by the reaction of ZDDP on iron oxide. *Tribol. Int.* **2006**, *39*, 1538–1544. [CrossRef]
18. Qu, J.; Blau, P.J.; Howe, J.Y.; Meyer, H.M. Oxygen diffusion enables anti-wear boundary film formation on titanium surfaces in zinc-dialkyl-dithiophosphate (ZDDP)-containing lubricants. *Scipta Mater.* **2009**, *60*, 886–889. [CrossRef]
19. Hsu, C.; Stratmann, A.; Rosenkranz, A.; Gachot, C. Enhanced growth of ZDDP-based tribofilms on laser-interference patterned cylinder roller bearings. *Lubricants* **2017**, *5*, 39. [CrossRef]
20. Stachowiak, G.W.; Batchelor, A.W. *Engineering Tribology*; Tribology Series 24; Elsevier: Amsterdam, The Netherlands, 1993; p. 349.

© 2019 by the authors. Licensee MDPI, Basel, Switzerland. This article is an open access article distributed under the terms and conditions of the Creative Commons Attribution (CC BY) license (http://creativecommons.org/licenses/by/4.0/).

 *lubricants*

Article

Tribological Properties of High-Speed Uniform Femtosecond Laser Patterning on Stainless Steel

Iaroslav Gnilitskyi [1,2,3,*], Alberto Rota [4,5,6], Enrico Gualtieri [4,5], Sergio Valeri [4,5,6] and Leonardo Orazi [1,5]

1   University of Modena and Reggio Emilia (UNIMORE), Amendola 2, 42122 Reggio Emilia, Italy; leonardo.orazi@unimore.it
2   NoviNano Lab, Pasternaka 5, 79015 Lviv, Ukraine
3   Department of Photonics, Lviv Polytechnic National University, Stepana Bandery 14, 79000, Lviv, Ukraine
4   Department of Physics, Computer Science and Mathematics, University of Modena and Reggio Emilia, Giuseppe Campi 213/a, 41125 Modena, Italy; alberto.rota@unimore.it (A.R.); enrico.gualtieri@unimore.it (E.G.); valeri@unimore.it (S.V.)
5   Centro Interdipartimentale per la Ricerca Applicata e i Servizi nella Meccanica Avanzata enella Motoristica Intermech-Mo.Re., Università di Modena e Reggio Emilia, Vignolese 905/b, 41125, Modena, Italy
6   Istituto CNR-NANO S3, Giuseppe Campi 213, 41125 Modena, Italy
*   Correspondence: iaroslav.gnilitskyi@novinano.com; Tel.: +380955764330

Received: 18 July 2019; Accepted: 17 September 2019; Published: 24 September 2019

Abstract: In this work, an analysis of the tribological performance of laser-induced periodic surface structures (LIPSS) treated X5CrNi1810 stainless steel was conducted. The approach followed by authors was to generate LIPSS-patterned circular tracks, composed of radial straight grooves with uniform angular periodicity. This permitted to measure the tribological properties in a pin-on-flat configuration, keeping fixed the orientation between the grooves and the sliding direction. A Stribeck curve was measured, as well as the consequent wear. A deep analysis of the sub-surface conditions after LIPSS generation was moreover performed using Focused Ion Beam (FIB) cross-section.

Keywords: LIPSS; laser surface texturing; tribological properties; wear; Stribeck curve

---

1. Introduction

Nature shows examples of how periodic and self-organized surface structures decrease the coefficient of friction (CoF), for example, the skin of a snake, pangolin, and other animals. Such examples inspired surface engineering to generate surfaces with low CoF. Bearing in mind the processing time, which is requested by the different kinds of lithography, LIPSS-based process can be a potentially strong candidate for improving tribological performances. The mutual configurations of micro-and-nanostructures can improve tribological properties by guiding wear particles along the direction of the grooves or by inducing the particles rolling between them [1].

Material processing by lasers with a pulse duration in the femtosecond range (10^{-15} s) has recently drawn a lot of attention due to nonthermal ablation mechanisms [2]. This characteristic makes this class of laser particularly appealing in material treatments, related to the absence of heat-induced material damage on dielectric, semiconductors, and metallic material [3]. Femtosecond pulses have also been exploited to induce periodic surface structures known as laser-induced periodic surface structures (LIPSS), firstly observed in 1965 [4]. As it is commonly accepted, LIPSS are divided into 2 groups: Low Frequency LIPSS, where period of structures is around laser wavelength, and High Frequency LIPSS, where period of structures by order of magnitude less than laser wavelength. Further, in paper Low Frequency LIPSS will be noted like LIPSS, while High Frequence LIPSS will be indicated as it is. LIPSS are already used and have a potential to be applied in numerous fields, including improving adhesion [5], wettability [6], surface colorization [7], better proliferation and adhesion of tissue cells [8], and others.

The shortcomings to apply LIPSS for tribology were recognized by Yu and Lu in 1999 [9]. They obtained microstructures with a nanosecond excimer laser and demonstrated the improvement of tribological characteristics in comparison to a non-treated surface. Furthermore, Honda at al. [10] used femtosecond laser-generated LIPSS on diamond-like carbon (DLC) and tested friction properties with an atomic-force microscope. This test showed a reduction of friction on the film with LIPSS compared with the untreated ones. Yasumaru et al. [11] also tested DLC, additionally coated with a thin layer of MoS_2, showing how to decrease and increase CoF. Using different beam scanning strategies, they generated LIPSS and demonstrated a CoF reduction compared to a flat surface. The study in [12] showed a CoF decrease even without lubricant in a pin-on-disc configuration on TiC-coated steel surface. The first tribological tests on LIPSS-textured semiconductors were made by Eichstadt et al. in [13], where an increase of the CoF was observed in both lubricated and non-lubricated conditions.

The tribological performances were investigated on large LIPSS-treated area of 100Cr6 stainless steel, high toughness bearing steel X30CrMoN15, pure titanium and titanium alloy (Ti6Al4V) by using a ball-on-disc configuration under paraffin and engine oil as lubricants in [14] and [15]. No effect was measured on CoF for LIPSS-treated samples tested with both the mentioned lubricants. While a significant reduction of CoF and wear for Ti and Ti6Al4V was shown with engine oil in the case of ripples oriented perpendicularly to the sliding motion, no beneficial influence appeared with paraffin oil. It should be outlined that these were the first experiments demonstrating the dependence of the CoF by LIPSS orientation, while the mechanisms relating to the influence of nanometric structures on the CoF have not been correctly explained.

The first research on wear and tribological performances for High Spatial LIPSS (HSFL) was made by Bonse et al. [16], where the large Titanium samples were covered by LIPSS and tested with paraffin and engine oil as lubricants. The related tribological analysis did not demonstrate the effects of HSFL on CoF, concluding that the depth of LIPSS played a crucial role in friction behavior.

In [17], the authors investigated the effects of LIPSS on AISI 316, obtaining a sensible CoF reduction on both dry and lubricated conditions. Another important tribological quantity is the Stribeck curve. In fact, in a lubricated contact, the CoF is not a univocal quantity independent on the load and velocity, as in the case of dry contacts. In fact, in the presence of lubricants, the CoF is a function of the Stribeck number, given by the relation:

$$\text{Stribeck number} = \eta\ v/L$$

where η is the viscosity of the oil, v is the sliding velocity, and L is the applied load.

The typical Stribeck curve is divided into three regimes, the boundary lubrication (BL), the mixed lubrication (ML), and the elastohydrodynamic (EH) regimes. At low Stribeck number, the two counterparts are completely in contact, and a small amount of lubricant is present between them; this situation represents the BL regime. This regime is characterized by high CoF and high wear. Increasing the Stribeck number increases the oil thickness, leading to a decrease in the CoF down to its minimum; this situation represents the ML regime. Increasing the Stribeck number further leads to the transition into the EH regime, where the lubricant thickness is continuous, and the friction is ruled by the viscosity characteristics of the oil. In this last regime, the CoF slightly increases with the Stribeck parameter due to viscous effects. The building of the Stribeck curve enables to define the three different Stribeck regimes as a function of the load and the sliding velocity.

One of the main drawbacks of the cited works is that, in case of linear textures, only reciprocating friction can be investigated, with the known issues about control of speed and, in the case of lubricant, the fluid dynamic conditions. In the present paper, the tribological properties of LIPSS-textured stainless steel samples were investigated, using advanced techniques to generate highly regular LIPSS (HR-LIPSS) [18]. The textures were obtained with radial symmetry, permitting to investigate the Stribeck curve in a pin-on-disc configuration, with a fixed orientation of the LIPSS texture with respect to the sliding direction. In particular, we were interested in surface texturing as a tool to improve the tribological performance of tribo-pairs in BL and ML regimes, the most critical regimes in a mechanical device. To the author's knowledge, this approach was never applied in the aforementioned literature.

2. Materials and Methods

2.1. Laser Set-Up

Periodic radial straight grooves patterned with LIPSS were generated on 30 mm diameter discs made of X5CrNi1810 steel. The discs were previously lapped to have a 30 nm average roughness (Root Mean Square).

In the experiment, Yb:KGW a chirped-pulse application laser system (model PHAROS 20W from Light Conversions Ltd. was used. Optical pulses (central wavelength 1030 nm, pulse width 213 fs, repetition rate 600 kHz) were forwarded to a galvanometric scanning head (Cambridge Technology). Alignment of linearly polarized laser light was controlled by a half-wave plate. The surface treatment was done in air at room temperature by scanning a laser beam across the sample surface. The laser beam was focused by an F-theta lens with a focal length of 56 mm that produced an approximate diameter of the irradiation spot of 10.6 microns ($1/e^2$ of peak intensity) on the sample. The transmittance of the focusing system was measured independently prior to the experiments and was found to be 80% at the laser wavelength. With this data, a peak fluence of 0.51 J/cm^2 was estimated on the surface.

Figure 1 shows the strategy of laser treatment. For the scanning approach, 5 mm long scanlines were formed in the radial direction by moving the laser spot with the galvanometric scanner at a scanning speed v_s equal to 3 m/s, while rotational motorized stage with circumferential speed ω of 1.45–2.5 rad/s generated circular patterned tracks with 4 microns of interline spacing. The polarization plane was set-up to uniformly orient LIPSS in the circumferential directions.

Figure 1. (**a**) Schematic of the laser treatment as a laser beam is scanned over a sample. (**b**) Details of laser patterning strategy: the laser beam is moving forth and back, while the sample is rotating by angular stage.

2.2. Focused Ion Beam (FIB) Analysis

Each disc was patterned to have three concentric textured areas with different radius. In each circle, two tribo-tests were performed at most. Similarly, the tribo-tests on the flat surface were made in the textured-free area of the discs.

FIB was employed to explore local properties, allowing to remove and observe trenches of the surface without introducing relevant additional damage, and thus to detect sub-surface micro-structure (buried cracking, defects, contaminations, etc.) [19–21]. Such characterization was performed by using a dual-beam system (FEI StrataTM DB235), combining a high-resolution FIB column equipped with a Ga Liquid Metal Ion Source (LMIS) and a Secondary Electron Microscope (SEM) column equipped with Schottky Field Emission Gun (SFEG) electron source. We obtained perpendicular "micro-cross sections" of samples using a FIB (e-beam = 30 KeV) as a micro-machining miller, setting a 1 nA ion beam current for the first rough trench, and a 300 pA ion beam current for final polishing. Thus, a step-by-step milling procedure was carried out to extract tranches of surface: preliminarily, to prevent the topmost material from mechanical intermixing upon contact with energetic ions [22], the surface sample was capped by a thin platinum layer (1 μm thickness). A Pt-shield had been grown, starting from a gas precursor and using a 300 pA ion beam current that assisted the local deposition. Finally, by tilting the sample holder, we obtained images of discovered walls collecting secondary electrons generated by electron column as a primary beam; in particular, we used electron column as a primary beam for non-destructive secondary electron analysis (e-beam = 15 KeV).

2.3. Tribology Test

The tribological tests were performed in a dedicated chamber filled with 20 mL commercial motor oil Shell Elix Synthetic 5w-40. The tests consisted of building the Stribeck curves of the textured and untextured surfaces, the last one used as a reference.

To check the load range that has to be applied to fall in BL and ML regimes, some preliminary tests were performed at different loads and velocity, not reported in the manuscript. We realized that, when falling in the BL regime, wear was very severe, and the texture was completely destroyed. To avoid this, we decided to focus on ML regime, assuming that the effect in the BL regime was similar.

In the calculation of the Stribeck parameter, the value of the oil viscosity was set to 1 Pa·s for simplicity, considering that all the tests were performed with the same oil. To change the Stribeck number, two strategies can be applied, namely changing the load or changing the sliding velocity. In the present study, the load was varied from 0.5 to 7.5 N in steps of 0.5 N, keeping the sliding velocity constant at 7 mm/s. The sliding configuration was pin-on-disc, where a flat pin of 2 mm diameter was loaded on a rotating disc (Figure 2). The range of the track radius was between 8.5 and 14 mm. The pins are made of 100Cr6 steel. Each test was repeated at least 3 times, and the results were averaged, leading to the Stribeck curves shown in the Results and Discussions section. The oil was replaced every 3 tests with the new one.

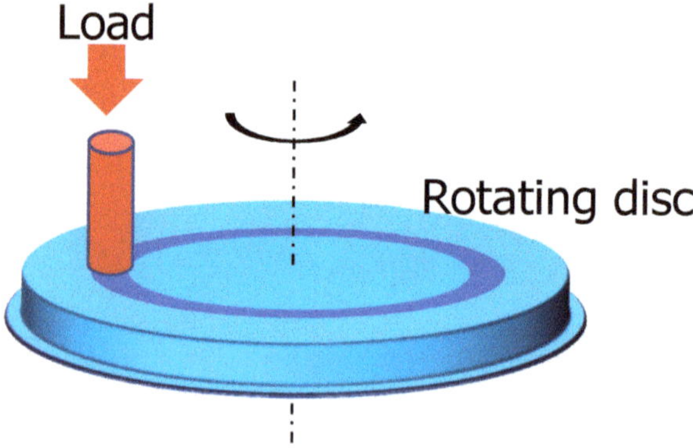

Figure 2. Scheme of the pin-on-flat tribological configuration. The flat pin is loaded on the rotating flat disc.

After the Stribeck tests, the line cross-sections of the related wear tracks were measured by a stylus profilometer, to quantify the wear. In Figure 3, a wear-track profile is reported, by way of example. It is possible to identify the worn material (red area) and the pile-up (green area). For each wear track,

4 profiles were acquired, corresponding to the 4 cardinal points of the circular track. The corresponding wear areas were averaged, enabling the calculation of the average wear volume.

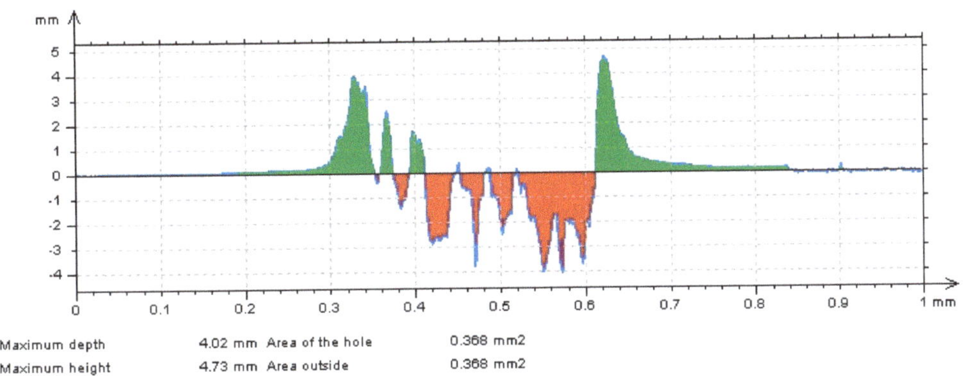

Maximum depth	4.02 mm	Area of the hole	0.368 mm2
Maximum height	4.73 mm	Area outside	0.368 mm2

Figure 3. Line profile of one of the wear track after a tribo-test, by way of example. The red area represents the worn materials, while the green one is the pile-up.

3. Results and Discussion

Figure 4a,b show the top view and the cross-section, respectively, of LIPSS-treated surface of the steel, imaged by SEM. Figure 4a shows the typical morphology of the related texture, characterized by the periodic alternation of parallel grooves and valleys, separated by about 600 nm. This morphology was uniformly distributed all over the processed surface, proving the accuracy and reproducibility of the method. The observation of the texture at a 52° tilt angle with respect to the normal revealed a second-order pattern, composed of parallel straight nanostructures perpendicular to the primary-order grooves (Figure 4b).

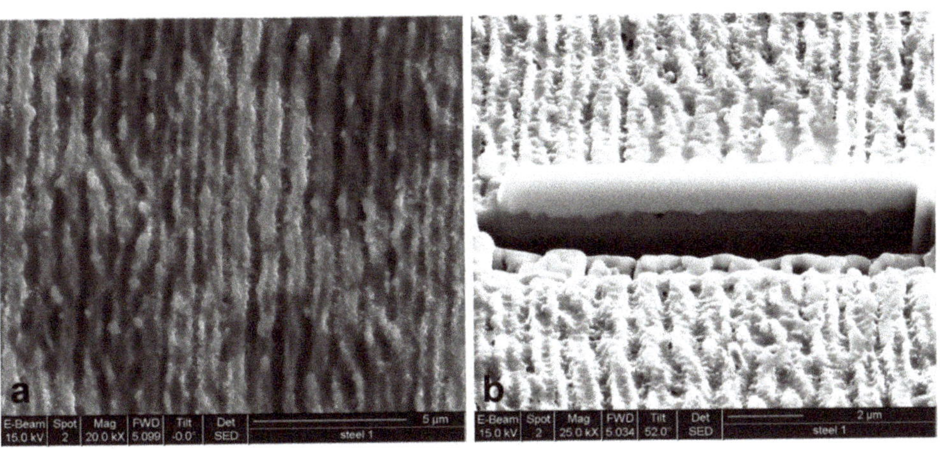

Figure 4. Top (**a**) and cross-section (**b**) view of the LIPSS (laser-induced periodic surface structures) textured surface. E-beam was 15 kV, and the magnification is 20 k.

In the FIB cross-section, the bright Pt protection layer was visible on top of the section. Just under the Pt layer, the height variation of the texture was evident. The peak-to-valley distance was measured for a large number of structures, resulting in between 120 and 300 nm. The sub-surface portion of the texture appeared uniform and compact, with no presence of void or crack, indicating that the laser effect was localized in the very outermost part of the material.

In Figure 5, the results of the tribological investigation were reported, for the patterned surface and the unpatterned flat one, used as a reference. The test was referred to one pin, but similar results were found for other similar pins (not shown). The curves represented the CoF variation in ML regime. As evident from the curves, at the lowest Stribeck number, the system was in the minimum of the ML regime and progressively increased towards the EH region (not evident from the graph). All over the investigated range, the CoF on the patterned surface was sensibly lower than on the unpatterned one; the CoF reduction went from a minimum of 10% for 4–8×10^{-7} m^{-1}, to a maximum of 25% for 5–9×10^{-7} m^{-1}. This was a very important result because the most effective reduction appeared in the most severe condition of the exploited range (namely at low Stribeck number). As already reported in many other studies on surface texturing, this could be ascribed to the reservoir effect of the grooves, which release lubricant even when the two counterparts are in contact, maintaining effective lubrication [23–25].

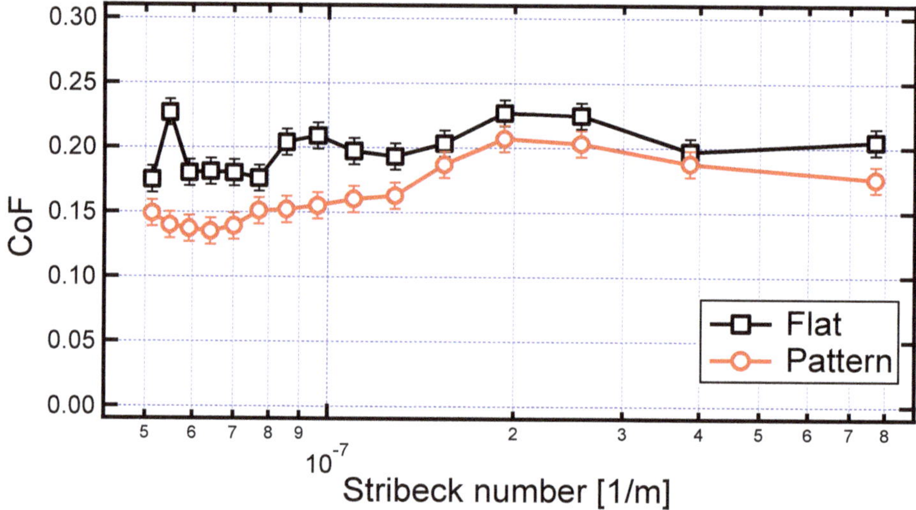

Figure 5. Stribeck curves for flat surface (open black squares) and the textured one (open red circles). Tests were performed in a box filled of oil.

The effect of LIPSS has been studied in terms of wear reduction as well. As evident from Figure 6, the wear of the disc during the Stribeck tests was less pronounced in the case of the textured surface, leading to a wear reduction of about 65%. This was probably related to the role played by the grooves as debris pocket, which limits the amount of particulates all over the interface, and so their effect as the abrasive body.

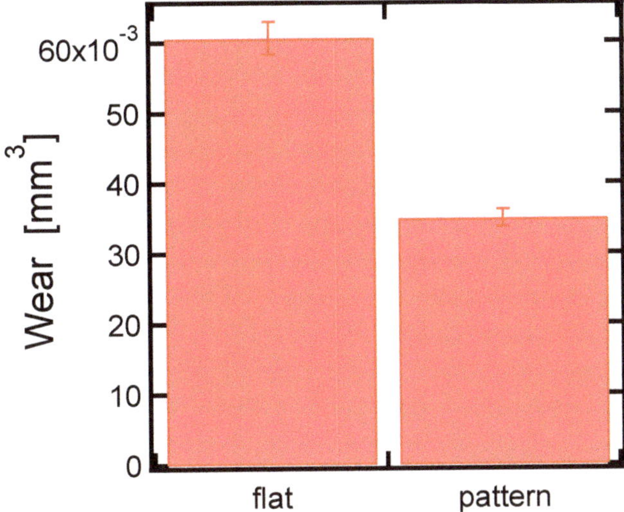

Figure 6. Wear of the disc after the Stribeck tests, for the textured and untextured surfaces.

4. Conclusion

Laser-induced periodic surface structures with good uniformity were imprinted on X5CrNi18010 stainless steel by using a unique scanning strategy in the radial direction. Such a strategy has been allowed to investigate the mixed lubrication regime of the Stribeck curve. All over the investigated range, the CoF on the LIPSS-patterned surface was substantially lower than on the unpatterned one. The CoF reduction went from a minimum of 10% for 4–8×10^{-7} m^{-1}, to a maximum of 25% for 5–9×10^{-7} m^{-1}. Also, the wear of the disc was less pronounced in the case of the textured surface, leading to a wear reduction of about 65%. These results are in perfect agreement with most of the studies on the tribological effect of surface texturing [23–25]; but in the present work, the texture was generated in a faster and efficient way, enabling the processing of a large number of pieces and/or of large areas.

Author Contributions: Conceptualization, I.G., A.R. and L.O.; Methodology, I.G., A.R. and E.G.; Resources, S.V. and L.O.; Supervision, S.V. and L.O.; Validation, A.R.; Writing—original draft, I.G. and A.R.; Writing—review & editing, E.G., S.V. and L.O.

Conflicts of Interest: The authors declare no conflict of interest.

References

1. Müller, F.; Kunz, C.; Gräf, S. Bio-inspired functional surfaces based on laser-induced periodic surface structures. *Materials* **2016**, *9*, 476. [CrossRef] [PubMed]
2. Gattass, R.R.; Mazur, E. Femtosecond laser micromachining in transparent materials. *Nat. Photonics* **2008**, *2*, 219. [CrossRef]
3. Sundaram, S.K.; Mazur, E. Inducing and probing non-thermal transitions in semiconductors using femtosecond laser pulses. *Nat. Mater.* **2002**, *1*, 217–224. [CrossRef] [PubMed]
4. Birnbaum, M. Semiconductor surface damage produced by ruby lasers. *J. Appl. Phys.* **1965**, *36*, 3688–3689. [CrossRef]
5. Rotella, G.; Orazi, L.; Alfano, M.; Candamano, S.; Gnilitskyi, I. Innovative high-speed femtosecond laser nano-patterning for improved adhesive bonding of Ti6Al4V titanium alloy. *Cirp. J. Manuf. Sci. Tec.* **2017**, *18*, 101–106. [CrossRef]
6. Vorobyev, A.Y.; Guo, C. Multifunctional surfaces produced by femtosecond laser pulses. *J. Appl. Phys.* **2015**, *117*, 033103. [CrossRef]

7. Dusser, B.; Sagan, Z.; Soder, H.; Faure, N.; Colombier, J.P.; Jourlin, M.; Audouard, E. Controlled nanostructrures formation by ultra fast laser pulses for color marking. *Opt. Express* **2010**, *18*, 2913–2924. [CrossRef] [PubMed]

8. Gnilitskyi, I.; Pogorielov, M.; Viter, R.; Ferraria, A.M.; Carapeto, A.P.; Oleshko, O.; Orazi, L.; Mishchenko, O. Cell and tissue response to nanotextured Ti6Al4V and Zr implants using high-speed femtosecond laser-induced periodic surface structures. *Nanomed-Nanotechnol.* **2019**, 102036. [CrossRef] [PubMed]

9. Yu, C.; Yu, H.; Liu, G.; Chen, W.; He, B.; Wang, Q.J. Understanding topographic dependence of friction with micro-and nano-grooved surfaces. *Tribol. Lett.* **2014**, *53*, 145–156. [CrossRef]

10. Mizuno, A.; Honda, T.; Kikuchi, J.; Iwai, Y.; Yasumaru, N.; Miyazaki, K. Friction properties of the DLC film with periodic structures in nano-scale. *Tribol. Int.* **2006**, *1*, 44–48. [CrossRef]

11. Yasumaru, N.; Miyazaki, K.; Kiuchi, J. Control of tribological properties of diamond-like carbon films with femtosecond-laser-induced nanostructuring. *Appl. Surf. Sci.* **2008**, *254*, 2364–2368. [CrossRef]

12. Pfeiffer, M.; Engel, A.; Gruettner, H.; Guenther, K.; Marquardt, F.; Reisse, G.; Weissmantel, S. Ripple formation in various metals and super-hard tetrahedral amorphous carbon films in consequence of femtosecond laser irradiation. *Appl. Phys. A.* **2013**, *110*, 655–659. [CrossRef]

13. Eichstädt, J.; Römer, G.R.; Huis, A.J. Towards friction control using laser-induced periodic surface structures. *Phys. Procedia.* **2011**, *1*, 7–15. [CrossRef]

14. Bonse, J.; Koter, R.; Hartelt, M.; Spaltmann, D.; Pentzien, S.; Höhm, S.; Rosenfeld, A.; Krüger, J. Femtosecond laser-induced periodic surface structures on steel and titanium alloy for tribological applications. *Appl. Phys. A.* **2014**, *117*, 103–110. [CrossRef]

15. Bonse, J.; Koter, R.; Hartelt, M.; Spaltmann, D.; Pentzien, S.; Höhm, S.; Rosenfeld, A.; Krüger, J. Tribological performance of femtosecond laser-induced periodic surface structures on titanium and a high toughness bearing steel. *Appl. Surf. Sci.* **2015**, *336*, 21–27. [CrossRef]

16. Bonse, J.; Höhm, S.; Koter, R.; Hartelt, M.; Spaltmann, D.; Pentzien, S.; Rosenfeld, A.; Krüger, J. Tribological performance of sub-100-nm femtosecond laser-induced periodic surface structures on titanium. *Appl. Surf. Sci.* **2016**, *374*, 190–196. [CrossRef]

17. Gnilitskyi, I.; Rotundo, F.; Martini, C.; Pavlov, I.; Ilday, S.; Vovk, E.; Ilday, F.Ö.; Orazi, L. Nano patterning of AISI 316L stainless steel with Nonlinear Laser Lithography: Sliding under dry and oil-lubricated conditions. *Tribol. Int.* **2016**, *99*, 67–76. [CrossRef]

18. Gnilitskyi, I.; Derrien, T.J.; Levy, Y.; Bugakova, N.M.; Mocek, T.; Orazi, L. High-speed manufacturing of highly regular femtosecond laser-induced periodic surface structures: Physical origin of regularity. *Sci. Rep.* **2017**, 8485. [CrossRef]

19. Shakhvorostov, D.; Pöhlmann, K.; Scherge, M. Structure and mechanical properties of tribologically induced nanolayers. *Wear.* **2006**, *260*, 433–437. [CrossRef]

20. Xie, Z.H.; Munroe, P.R.; Moon, R.J.; Hoffman, M. Characterization of surface contact-induced fracture in ceramics using a focused ion beam miller. *Wear.* **2003**, *255*, 651–656. [CrossRef]

21. Xie, Z.H.; Hoffman, M.; Moon, R.J.; Munroe, P.; Cheng, Y.B. Scratch damage in ceramics: Role of microstructure. *Am. Ceram. Soc. Bull.* **2003**, *86*, 141–148. [CrossRef]

22. Wong, V.; Tian, T.; Smedley, G.; Moughon, L.; Takata, R.; Jocsak, J. Low-engine-friction technology for advanced natural-gas reciprocating engines. *M.I.T.* **2006**.

23. Borghi, A.; Gualtieri, E.; Marchetto, D.; Moretti, L.; Valeri, S. Tribological effects of surface texturing on nitriding steel for high-performance engine applications. *Wear.* **2008**, *265*, 1046–1051. [CrossRef]

24. Gualtieri, E.; Borghi, A.; Calabri, L.; Pugno, N.; Valeri, S. Increasing nanohardness and reducing friction of nitride steel by laser surface texturing. *Tribol. Int.* **2009**, *42*, 699–705. [CrossRef]

25. Vilhena, L.M.; Sedláček, M.; Podgornik, B.; Vižintin, J.; Babnik, A.; Možina, J. Surface texturing by pulsed Nd: YAG laser. *Tribol. Int.* **2009**, *42*, 1496–1504. [CrossRef]

© 2019 by the authors. Licensee MDPI, Basel, Switzerland. This article is an open access article distributed under the terms and conditions of the Creative Commons Attribution (CC BY) license (http://creativecommons.org/licenses/by/4.0/).

MDPI
St. Alban-Anlage 66
4052 Basel
Switzerland
Tel. +41 61 683 77 34
Fax +41 61 302 89 18
www.mdpi.com

Lubricants Editorial Office
E-mail: lubricants@mdpi.com
www.mdpi.com/journal/lubricants

www.ingramcontent.com/pod-product-compliance
Lightning Source LLC
LaVergne TN
LVHW070926170726
843515LV00005B/876